Extremal Optimization

Fundamentals, Algorithms, and Applications

Extremal Optimization

Fundamentals, Algorithms, and Applications

Yong-Zai Lu • Yu-Wang Chen
Min-Rong Chen • Peng Chen
Guo-Qiang Zeng

Chemical Industry Press

CRC Press
Taylor & Francis Group
Boca Raton   London   New York

CRC Press is an imprint of the
Taylor & Francis Group, an **Informa** business

CRC Press
Taylor & Francis Group
6000 Broken Sound Parkway NW, Suite 300
Boca Raton, FL 33487-2742

International Standard Book Number-13: 978-1-4987-0565-3 (Hardback)

Library of Congress Cataloging-in-Publication Data

Names: Lèu, Yongzai. | Chen, Yu-Wang. | Chen, Min-Rong. | Chen, Peng (Optimizaton specialist) | Zeng, Guo-Qiang.
Title: Extremal optimization : fundamentals, algorithms, and applications / authors, Yong-Zai Lu, Yu-Wang Chen, Min-Rong Chen, Peng Chen, and Guo-Qiang Zeng.
Description: Boca Raton : Auerbach Publications, 2015. | Includes bibliographical references and index.
Identifiers: LCCN 2015039223 | ISBN 9781498705653 (alk. paper)
Subjects: LCSH: System analysis--Data processing. | Mathematical optimization. | Multidisciplinary design optimization.
Classification: LCC T57 .L82 2015 | DDC 620/.0042--dc23
LC record available at http://lccn.loc.gov/2015039223

Visit the Taylor & Francis Web site at
http://www.taylorandfrancis.com

and the CRC Press Web site at
http://www.crcpress.com

Contents

Preface

With the high demand and the critical situation of solving hard optimization problems we are facing in social, environment, bioinformatics, traffic, and industrial systems, the development of more efficient novel optimization solutions has been a serious challenge to academic and practical societies in an information-rich era. In addition to the traditional math-programming-inspired optimization solutions, computational intelligence has been playing an important role in developing novel optimization solutions for practical applications. On the basis of the features of system complexity, a new general-purpose heuristic for finding high-quality solutions to NP-hard (nondeterministic polynomial-time) optimization problems, the so-called "extremal optimization (EO)," was proposed by Boettcher and Percus. In principle, this method is inspired by the Bak–Sneppen model of self-organized criticality describing "far-from-equilibrium phenomena," from statistical physics, a key concept describing the complexity in physical systems. In comparison with other modern heuristics, such as simulated annealing, genetic algorithm (GA), through testing on some popular benchmarks (TSP [traveling salesman problem], coloring, K-SAT, spin glass, etc.) of large-scale combinatory-constrained optimization problems, EO shows superior performance in the convergence and capability of dealing with computational complexity, for example, the phase transition in search dynamics and having much fewer tuning parameters.

The aim of this book is to introduce the state-of-the-art EO solutions from fundamentals, methodologies, and algorithms to applications based on numerous classic publications and the authors' recent original research results, and to make EO more popular with multidisciplinary aspects, such as operations research, software, systems control, and manufacturing. Hopefully, this book will promote the movement of EO from academic study to practical applications. It should be noted that EO has a strong basic science foundation in statistical physics and bioevolution, but from the application point of view, compared with many other metaheuristics, the application of EO is much simpler, easier, and straightforward. With more studies in EO search dynamics, the hybrid solutions with the marriage of EO and other metaheuristics, and the real-world application, EO will be an additional weapon to deal with hard optimization problems. The contents of this book cover the following four aspects:

1. General review for real-world optimization problems and popular solutions with a focus on computational complexity, such as "NP-hard" and the "phase transitions" occurring on the search landscape.
2. General introduction to computational extremal dynamics and its applications in EO from principles, mechanisms, and algorithms to the experiments on some benchmark problems such as TSP, spin glass, Max-SAT (maximum satisfiability), and graph partition. In addition, the comparisons of EO with some popular heuristics, for example, simulated annealing and GA, are given through analytical and simulation studies.
3. The studies on the fundamental features of search dynamics and mechanisms in EO with a focus on self-organized optimization, evolutionary probability distribution, and structure features (e.g., backbones) are based on the authors' recent research results. Moreover, modified extremal optimization (MEO) solutions and memetic algorithms are presented.
4. On the basis of the authors' research results, the applications of EO and MEO in multiobjective optimization, systems modeling, intelligent control, and production scheduling are presented.

The authors have made great efforts to focus on the development of MEO and its applications, and also present the advanced features of EO in solving NP-hard problems through problem formulation, algorithms, and simulation studies on popular benchmarks and industrial applications. This book can be used as a reference for graduate students, research developers, and practical engineers when they work on developing optimization solutions for those complex systems with hardness that cannot be solved with mathematical optimization or other computational intelligence, such as evolutionary computations. This book is divided into the following three parts.

Section I: Chapter 1 provides the general introduction to optimization with a focus on computational complexity, computational intelligence, the highlights of EO, and the organization of the book; Chapter 2 introduces the fundamental and numerical examples of extremal dynamics-inspired EO; and Chapter 3 presents the extremal dynamics–inspired self-organizing optimization.

Section II: Chapter 4 covers the development of modified EO, such as population-based EO, multistage EO, and modified EO with an extended evolutionary probability distribution. Chapter 5 presents the development of memetic algorithms, the integration of EO with other computational intelligence, such as GA, particle swarm optimization (PSO), and artificial bee colony (ABC). Chapter 6 presents the development of multiobjective optimization with extremal dynamics.

Section III includes the applications of EO in nonlinear modeling and predictive control, and production planning and scheduling, covered in Chapters 7 and 8, respectively.

MATLAB® is a registered trademark of The MathWorks, Inc. For product information, please contact:

The MathWorks, Inc.
3 Apple Hill Drive
Natick, MA 01760-2098 USA
Tel: 508 647 7000
Fax: 508-647-7001
E-mail: info@mathworks.com
Web: www.mathworks.com

Acknowledgments

This book was written based on the published pioneering research results on extremal dynamics-inspired optimization, and the authors' recent research and development results on the fundamentals, algorithms, and applications of EO during the last decade at both Shanghai Jiao Tong University (SJTU) and Zhejiang University, China. We wish to thank the Department of Automation, SJTU, the Research Institute of Cyber Systems and Control, Zhejiang University, and the Research Institute of Supcon Co. for their funding of PhD programs and research projects. We are most grateful to Professor J. Chu at Zhejiang University, Professors Y. G. Xi and G. K. Yang at SJTU, and Directors Y. M. Shi and Z. S. Pan of Supcon Co. for their strong support and encouragement.

In particular, we are deeply indebted to the members of Chinese Academy of Engineering: Professor C. Wu and Professor T. Y. Chai; and SJTU Professor G. K. Yang who have freely given of their time to review the book proposal and write suggestions and recommendations.

We are grateful to Chemical Industrial Press for providing the funding to publish this book, and also to Ms. H. Song (commissioning editor, Chemical Industrial Press) and the staff of CRC Press for their patience, understanding, and effort in publishing this book.

This book was also supported by the National Natural Science Foundation of China (Nos. 61005049, 51207112, 61373158, 61472165, and 61272413), Zhejiang Province Science and Technology Planning Project (No. 2014C31074), National High Technology Research and Development Program of China (No. 2012AA041700), National Major Scientific and Technological Project (No. 2011ZX02601-005), National Science and Technology Enterprises Technological Innovation Fund (No. 11C26213304701), Zhejiang Province Major Scientific and Technological Project (No. 2013C01043), and the State Scholarship Fund of China.

Finally, we thank the relevant organizations for their permission to reproduce some figures, tables, and math formulas in this book. See specific figures for applicable source details.

FUNDAMENTALS, METHODOLOGY, AND ALGORITHMS

Chapter 1

General Introduction

1.1 Introduction

With the revolutionary advances in science and technology during the last few decades, *optimization* has been playing a more and more important role in solving a variety of real-world systems for modeling, optimization, and decision problems. The major functions of optimization are to provide one or multiple solutions that are able to optimize (e.g., minimize or maximize) the desired objectives subject to the given constraints in the relevant search space. The optimization techniques have been popularly applied in business, social, environmental, biological, medical, man-made physical and engineering systems, etc. Due to the increase in computational complexity, traditional mathematics-inspired optimization solutions, such as mathematical programming (e.g., linear programming [LP], nonlinear programming [NLP], mix-integer programming [MIP]), can hardly be applied to solving some real-world complex optimization problems, such as the NP-hard (nondeterministic polynomial-time hard) problems (Korte and Vygen, 2012) defined in computational complexity theory.

To make optimization solutions applicable, workable, and realistic for those complex systems with roughing search landscape, and limited or no mathematical understanding (i.e., knowledge) between decision variables, desired criteria, and constraints, a number of alternative multidisciplinary optimization approaches have been developed. Instead of traditional optimization methodologies and algorithms, computer science and computational intelligence (CI)-inspired *meta-heuristics search* optimization solutions (Patrick and Michalewicz, 2008) have been developed. The CI (Engelbrecht, 2007) applied in optimization is to stimulate a set of natural ways to deal with complex computational problems in optimization solutions. The major features of CI are to model and stimulate the behaviors and features of natural evolution, the human body, artificial life, biological and physical systems with computer algorithms, such as evolutionary computations, for example, genetic algorithms (GAs) (Holland, 1992), genetic programming (GP) (Koza, 1998), artificial neural network (ANN) (Jain and Mao, 1996), fuzzy logic (FL) (Zadel, 1965; Xu and Lu, 1987), artificial life (Langton, 1998), artificial immune system (De Castro and Timmis, 2002), DNA computing (Daley and Kari, 2002), and statistical physics (Hartmann and Weigt, 2005). The major advantages of CI in solving complex optimization problems are (1) the requirement of limited or no knowledge of mathematical first principles to describe the quantitative relations of system behaviors between decision variables, desired criteria, and constraints of a system under study; (2) the "point-to-point" gradient-based search is replaced by "generation" and "population"-based search, and consequently, it is not necessary to calculate the gradient during the search process and the search space can be significantly enlarged; (3) the introduction of probability-inspired "mutation," "crossover," and "selections" operations significantly improves the search capability and efficiency; (4) the uses of natural features of "self-learning" and "self-organizing" in modeling, data mining, clustering, classification, and decisions to enhance the search power, robustness, and adaptation, particularly for those systems with variable environments. As a result, the solutions of complex problems might not be "optimal" under the sense of mathematical definition, but may provide a satisfactory result with much less search costs in *memory, communications*, and *time*. In addition to the highlighting of mathematical optimization, Section 1.2 will focus on presenting the concepts of optimization from practical aspects.

1.2 Understanding Optimization: From Practical Aspects

1.2.1 Mathematical Optimization

A well-known standard mathematical (continuous) optimization problem or mathematical programming (Luenberger, 1984) can be represented as

$$\text{Given a function } f: A \to R \text{ from some set to the real numbers} \tag{1.1}$$

Sought an element X_0 in A such that

$f(X_0) \leq f(X)$ for all X in A ("minimization") or
$f(X_0) \geq f(X)$ for all X in A ("maximization")

Many real-world, benchmark, or theoretical optimization problems can be modeled and then solved by relevant computer programming. However, A is a subset of the Euclidean space R^n, usually specified by a set of equality or inequality constraints; consequently, the unconstrained optimization problem as shown in Equation 1.1 then turns into a constrained optimization problem as presented in the following:

$$\text{minimize } f(X), \quad X = [x_1, \ldots, x_n]^T \in R^n \tag{1.2}$$

$$\text{subject to } g_i(X) \le 0, \quad i = 1, \ldots, m$$

$$h_i(X) = 0, \quad i = 1, \ldots, p$$

where
   $f(X): R^n \to R$ is an objective function to be minimized over the variable X
   $g_i(X) \le 0$, inequality constraints
   $h_i(X) = 0$, equality constraints

   In addition to continuous or numerical optimization problems, there are many decision problems that can be represented as a *combinatorial optimization problem* (*COP*) (Korte and Vygen, 2012). The COP is a subset of mathematical optimization in applied mathematics and computer science, for example, artificial intelligence and machine learning. The set of feasible solutions of COP is either discrete or sequential, and the objectives or fitness become the function of the relevant discrete solutions. The most popular example of combinatorial optimization problems (COPs) is the *traveling salesman problem* (*TSP*) in which the solution X is a sequence (i.e., order) of the cities to be visited, and the desired criterion is defined as the total traveling distance. The constraints for feasible solutions may be defined as the feasible order of the cities to be visited. When the size of the problem is big enough, the "exhaustive search" is not feasible, and then the problem could turn to an NP-hard problem that is hardly possible to solve within polynomial time by a deterministic *Turing machine*. The relevant computational complexity will be discussed in the coming sections.
   On the other hand, when there are multiple criteria desired, then the scalar objective function f turns to a vector of multiple functions, that is, $[f_1, f_2, \ldots, f_m]^T$, the problem becomes a *multiobjective optimization* that will be discussed in detail in Chapter 6.

1.2.2 Optimization: From Practical Aspects

To better understand and extend the fundamental concept and frame of mathematical optimization from a wider vision and from practical aspects, the following perspectives are vital particularly in applications and development of novel optimization solutions:

1. In real-world systems, the decisions or optimized solution, X, might not be a continuous or discrete valued vector as shown in Equation 1.1; instead, there can be a much more complicated solution, such as a 3D-designed-architecture, bio-gene-array, molecular design, enterprise business plan, or colored medical imaging. Consequently, how to encode those solutions to a proper type of decisions, X, becomes a key factor in building an optimization model for a complex system. In fact, the choice of this encoding heavily relies on the forms of $f(X)$ and the search mechanism to be used.

2. The mathematical function $f(X)$ defined in Equations 1.1 and 1.2 may be a valued function, for example, the profit of an enterprise under a certain time window, or could be replaced by a certain correspondence between two or more parameters, plans, or patterns, for instance, the correspondence between profit and carbon emission level in a manufacturing enterprise, or the design of architecture for a bridge. Consequently, the complexity in computation of $f(X)$ will increase in computation, communication, and data storage costs plus human preference, as a result, the design of $f(X)$ becomes much more difficult in building a realistic optimization model.

3. It should also be noted that building a proper optimization model and finding a feasible solution is a difficult and sophisticated job. The key to success is to define or understand *what to do*? or so-called *requirement for optimization*, and *knowledge acquisition* with collecting enough domain knowledge and data, then determining *what optimization model, solution and platform* will be used to effectively solve the relevant optimization problems.

1.2.3 Example Applications of Optimization

The applications of optimization have been creating great economic and social benefits and becoming a driving force in human civilization and industrial modernization. To better understand the concepts and critical positions of optimization in human life, let us discuss the following three example application cases to show the critical position of optimization in industries.

Case 1: Natural gas pipeline network (NGPN): An example of network optimization
The schematic layout of NGPN both interstate and intrastate in the United States is shown in Figure 1.1 (available at http://www.eia.gov/pub/oil_gas/natural_gas/analysis_publications/ngpipeline/compressorMap.html). To design and operate an NGPN more efficiently with regard to capital investment and operational costs subject to changes in supply and demand, the relevant optimization problems can be divided into two:

1. *The NGPN design optimization*: The solutions of NGPN design optimization are to provide an optimal NGPN topology (i.e., the distribution of its nodes

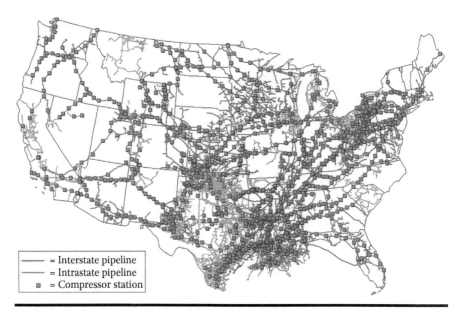

= Interstate pipeline
= Intrastate pipeline
■ = Compressor station

Figure 1.1   Natural gas pipeline network in the United States.

and connections) and the engineering design (e.g., the physical and engineering parameters) that minimizes the capital investment and energy efficiency under the given supply and demand figures. Since the NGPN covers both interstate and intrastate pipeline network architecture, and the detailed engineering design, the problem formulation and optimization solutions become very complicated.

2. *The NGPN operational optimization*: The solutions of NGPN operational optimization are to generate an optimal control strategy (e.g., select the control nodes with the relevant control parameters) that may optimize the desired objectives, such as energy efficiency under the changes in supply and demand.

In fact, these two problems can be viewed as "pipeline network design" with layout and parameter optimization and the "control over gas pipeline network," respectively. Consequently, the optimization solutions should be multifunctional and multiscale with offline and online modes.

Case 2: Optimization for manufacturing enterprises: A multiscaled optimization problem
In the era of *information economy*, any manufacturing enterprise being a business entity lives in a *global business ecosystem*. To be a winner and not a survivor or loser, the enterprise must collect both external global information and internal business-manufacturing information, and then make timely and right decisions to adapt to the changes in a variable business environment. A global manufacturing enterprise can be viewed as a complex system. Here, we only list a few examples in decisions

that are able to make an enterprise robust and profitable under an unpredictable business environment.

1. The decisions on the *product-mix*, the type and amount of core products to be produced with *order management* and/or *inventory control* for both *make-to-order* and *make-to-inventory* business models, respectively. This is to answer *what to produce* under a certain business environment and given time window.
2. The decisions on *procurement plan* are to define the suppliers for purchasing the *bill of materials* (*BOM*) and the amount of BOM to be purchased. This also involves a hybrid optimization with discrete and continuous decisions in selecting suppliers and determining the size of purchasing.
3. The decisions on *advanced production planning and scheduling* (*APPS*) are to determine where and when to prefill the work (manufacturing) orders, namely the *order-to-site* assignment and *"production scheduling"* under a desired time window. These problems could be formulated as multilevel COPs.
4. *Real-time optimization* and *advanced control* make real-time decisions and control under a variable business and production environment, namely, answer the question: *"How to produce?"* to reach the goals.

In addition, many other optimization subsystems, such as production logistics, inventory control, energy management, and profit optimization, will also be integrated. The applications of extremal optimization (EO) in nonlinear systems modeling and control, and APPS will be presented in details in Chapters 7 and 8, respectively.

Case 3: Protein folding and microarray gene optimization: An example of bioinformatics
The research and development on the applications of optimization techniques in protein folding (Dobson, 2003) and microarray gene ordering (Moscato et al., 2007) have been an attractive challenging problem in computational biology and theoretical molecular biology. Understanding the mechanism of missing folding plays an important role in finding the solutions for diseases, for example, Parkinson and Alzheimer, caused by missing folding, since there are a huge number of different types of proteins with complicated sequential structures in the human body. Consequently, the relevant problems become an NP-hard one, the modeling and prediction of protein structures has been an attractive area in bioinformatics using optimization, data mining, and machine learning techniques.

1.2.4 Problem Solving for Optimization

This section will address the general issues for the optimization problem solving from knowledge acquisition, data gathering, problem formulation or modeling, the development of problem solving, simulation or offline tests to execution, etc. The general working paths could follow the following steps:

1. *Problem definition*: The problem definition can be considered as a multi-scaled problem proposal that includes (i) the *functional scale*, that is, "what to do?" or "for what?," namely, the tasks of the system to be built, for example, building a production planning and scheduling system that may create proper decisions in terms of external and internal information; (ii) the *environment or geographical scale*, for example, the system will cover production line, multiple lines, entire enterprises, or multienterprises in different industrial sectors, for example, a manufacturing enterprise associated with third-party logistics; and (iii) the *time scale*, for example, the decisions could be under annual, monthly, or daily time windows, or in a real-time mode.

2. *Modeling or problem formulation*: To build an optimization model or complete the problem formulation, the three elements (objectives, decisions, constraints) have to be clearly defined. The objective could be a single scalar value, vector (either qualitative or quantitative), two-dimensional or three-dimensional images encoded from the system under studies, etc. The decisions could be continuous, discrete, sequence numbers, an image, an architecture, etc. The constraints for feasible regions could be modeled mathematically or on some rule-base. In the problem formulation, the elegantly using "encoding" and "decoding" for information and data is the key to success. To make it easier in problem solving, we may try to map the real system to an optimization benchmark, such as TSP.

3. *Problem-solving architecture and algorithm*: Based on the task of the problem under study and the form of the system model, a proper optimization solution (algorithm) should be selected to reach the desired goals under the limits in problem size, efficiency, precision, etc.

4. *Simulation test and implementation*: The simulation tests may provide developers with valuable knowledge and decisions to make the comparison of different solutions with system tuning to reach the design targets.

1.3 Phase Transition and Computational Complexity

1.3.1 Computational Complexity in General

With the rapid development in science and technology, the optimization community faces serious challenges mainly in two ways: (1) the computational complexity in real-world optimization problems and (2) how to apply the latest IT technology and its development environments, such as clouds, big data, and internet of things in solving complex optimization problems.

Computational complexity theory is an attractive subject in computer science and mathematics. As is well known, any optimization problem finally must be solved by a proper computer algorithm. A problem is considered as a difficult one if its solution requires significant computational resources, whatever the algorithm used. Usually, an NP-hard problem in combinatorial optimization has such an instance. The computational complexity for optimization problems could be measured by the required computation time, data storage, and communications. In other words, whether the applications of an algorithm are able to create an optimal, satisfactory, or even a feasible solution within a reasonably short time and using limited storage. Consequently, developing a more efficient optimization solution has been an attractive multidisciplinary subject in computer science and mathematics. The effectiveness of an algorithm can be measured with the rate of growth of time or space as the size of the problem increases, for example, the number of cities in TSP. The relationship between problem size n and required computation time can be represented by $O(n)$, $O(\log n)$, $O(n \log n)$, $O(n^2)$, $O(2^n)$, $O(n^{\log n})$, $O(n^n)$, or $O(n!)$, they can be divided into *polynomial time algorithm* and *exponential time algorithm* categories, the details are given in https://en.wikipedia.org/wiki/Time_complexity. From the three example optimization problems as noted in Section 1.2, we may find the computational complexity in optimization does not only involve the problem size and the required huge computational costs, but also depends on the *complexity or hardness in the search landscape*. When using gradient-inspired search, the search process might stop when dropping to a *local minimum*.

1.3.2 Phase Transition in Computation

On the other hand, from a statistical mechanics point of view, the *phase transition* is nothing but the onset of nontrivial macroscopic (collective) behavior in a system composed of a large number of "elements" that follow simple microscopic laws (Martin et al., 2001). For example, at atmospheric pressure, water boils at a "critical" temperature $Tc = 100°C$. The water turns from its liquid phase to its gas phase. This process is called *phase transition*. In complex system theory, the *phase transition* is related to the *self-organized process* that the system transforms from disorder to order. It is interesting to note that some notions and methods from statistical physics can also foster a deeper understanding of computational phenomena (Hartmann and Weigt, 2005; Selman, 2008). In fact, *phase transitions* (Hartmann and Weigt, 2005) and *simulated annealing* (SA) (Kirkpatrick et al., 1983) are two typical examples related to physics-in-optimization. Many experimental results (Gent and Walsh, 1996; Bauke et al., 2003; Xu and Li, 2006) have shown that for a class of NP-hard problems one or more *order parameters* can be defined, and hard instances occur around the critical values of these order parameters. In general, such critical values form a boundary that separates the search space of problems into two regions. One is called *underconstrained* with many feasible solutions in the region; the other is called *overconstrained* with fewer feasible solutions in the region.

In reality, the real hard problems just occur on the boundary between these two regions, where the probability of a solution is low but nonnegligible (Cheesman et al., 1991). In fact, the phenomena of *easy–hard–easy* and *easy–hard* transitions in a hard search process usually come from the property of *phase transitions* existing in many NP-hard problems. The probability of solutions and computational complexity governed by phase transitions has been studied in statistical physics for a 3-SAT problem being a typical NP-complete problem that shows the connection between NP-complete problems and phase transitions. Many relevant research results on this topic have been published in special issues of *Artificial Intelligence* (Hogg et al., 1996), *Theoretical Computer Science* (Dubolis and Dequen, 2001), *Discrete Applied Mathematics* (Kirousis and Kranakisdeg, 2005), and *Science* and *Nature* (Monasson et al., 1999; Selman, 2008).

Typical examples to study on phase transitions in hard optimization problems include TSP (Gent and Walsh, 1996), spin glasses (Boettcher, 2005a,b), graph partitioning (Boettcher and Percus, 1999) problems, etc. In addition to these examples, phase transition also exists in the job shop (Beck and Jackson, 1997) and project scheduling (Herroelen and Reyck, 1999) problems, where the resource parameters exhibit a sharp *easy–hard–easy* transition behavior. It should be noted that *the capability of dealing with the computational complexity with phase transition is one of the key advanced features of EO.*

1.4 CI-Inspired Optimization

The CI may be defined as a set of nature-inspired computational methodologies and approaches to address complex real-world problems to which traditional approaches, that is, first-principles modeling or explicit statistical modeling, are ineffective or infeasible. Namely, many such real-world problems can hardly be formulated and solved in terms of traditional mathematical approaches. However, with the recent multidisciplinary research and development in information technology, bioinformatics, and computer sciences, some complex optimization problems may be formulated and solved with CI. The CI-inspired optimization solutions may be divided into the following major categories.

1.4.1 Evolutionary Computations

Evolutionary computations (Engelbrecht, 2007) being one of the most important subfields in CI mimics the population-based sexual evolution through reproduction of generations performed with relevant computer programs. It also mimics *genetic* so-called *evolutionary algorithms* (*EAs*) that mainly consist of *genetic algorithms* (*GAs*), *evolutionary strategies* (*ES*), *genetic programming* (*GP*), etc. They may be applied in solving different types of optimization problems. In addition to being applied in solving functional and COPs, GP can also be used to generate explicit mathematical models with *symbolic regressions* and further perform "computer programming" automatically through evolutionary procedures: *mutation, crossover,* and *selection.*

The theoretical foundation of evolutionary computation relies on Darwinian evolutionary principles for the biological mechanism of evolution. To emulate the bio-evolution process, the search processes performed in EAs are population-based with the probability-driven genetic operators in chromosomes: *mutation, crossover,* and *selection.* In fact, the individuals with higher fitness (i.e., the winners) in the population pool may have higher probability to survive as parents for the next generation. During the evolution process, there are multiple individuals (chromosomes or solutions) in the population pool of each generation being qualified as feasible solutions. Consequently, the user may select a preferred solution through viewing the entire path of the evolution. The evolution computations have been widely applied in solving many real-world optimization problems.

1.4.2 Swarm Intelligence

Swarm intelligence (SI) is a subset of artificial intelligence with collective behaviors of decentralized, self-organized natural or artificial systems. The concept of SI is employed in work on optimization search. On the other hand, in computer sciences, the SI belongs to a kind of *multiagent systems* (MASs). Both *particle systems optimization* and *ant colony optimization* (ACO) are the most popular algorithms in SI. Particle swarm optimization is a search method which utilizes a set of agents that move through the search space to find the global minimum or maximum of an objective function (Eberhart and Kennedy, 1995). The trajectory of each particle is determined by a simple rule incorporating the current particle velocity and exploration histories of the particle and its neighbors. Since its introduction by Kennedy and Eberhart (1995), PSO has been an attractive research field in both operations research (OR) and computer science communities due to its superior search efficiency even for those complex optimization problems with a complicated search landscape, such as having high dimensions and/or multiple local optima. Many research contributions to PSO search dynamics and its applications to large-scale optimization problems have been published (Coello et al., 2004; Fan and Zahara, 2007; Liu et al., 2007; Senthil Arumugam et al., 2009; Chen et al., 2014b).

In the SI family, ACO (Dorigo and Gambardella, 1997) is another popular intelligent optimization search solution that emulates the natural behavior of real ants in finding food through *indirect* communications between ants via the released *pheromone* representing the food found by ants. As a result, when multiple ants work together cooperatively, they may find maximum food within a shorter time, that is, the shortest road-traveling path. In fact, the ACO is a man-made computer program for performing an optimization search similar to the natural behavior and capabilities of ants in a food search. From computer science aspects, the ACO can be viewed as another typical multiagent system (MAS).

1.4.3 Data Mining and Machine Learning

It has been a well-known concept that in addition to first principles, "data" play an important role in information and knowledge discovery. Traditionally, the chain of *data–information–knowledge* has been widely recognized and applied in sciences and engineering. Eventually, statistical analysis, system identification, data mining, information fusion, and machine learning have been used in system modeling, knowledge discovery, and decision making. However, when we are entering the era of *big data* and *internet of things*, unprecedented data in size, type, speed, and global interconnections have become ubiquitous. The concept of data has enlarged from *structured* data to *unstructured* data. Instead of numerical (i.e., analog, digital, or discrete) numbers, the unstructured data could be image, voice, audio, etc. This brings computer and optimization scientists new challenges and opportunities: *the development of novel techniques of data mining, information fusion, and machine learning for system modeling, optimization, and decisions.*

The most popular data mining and machine learning solutions are "artificial neural network (ANN)," "fuzzy logic (FL)," and "support vector machine (SVM)." In addition to their applications in system modeling, forecast, and signal processing, they are also capable of dealing with some optimization problems. Instead of the traditional meta-heuristic search methods, the solutions of data mining and machine learning-inspired optimization may be generated through supervised or unsupervised learning-oriented data pattern clustering and classification. However, it should be noted that enough (historical and/or model-based forecasted) valuable data that describe the quantitative and/or qualitative correspondences between decisions, constraints, and objectives are vital to find satisfactory results within a feasible region. For the fundamentals and applications of data mining and machine learning in model-free decisions, readers may refer to Lu (1996).

1.4.4 Statistical Physics

Recently, the investigation of the relationship between physics (particularly statistical mechanics) and optimization has been an attractive subject in computer science, statistical physics, and optimization communities (Hartmann and Rieger, 2004). The critical positions and applications of optimization in physics have been popularly realized. On the other hand, the applications of statistical physics in leading novel optimization solutions and the probabilistic analysis of combinatorial optimization have also been recognized. The statistical physics or thermal dynamics-inspired *simulated annealing* (SA) (Kirkpatrick et al., 1983) is a generic probabilistic meta-heuristic method for a global optimization problem in a large search space. One of the advanced features of SA is to force the escape from the local minima during the search process. SA has been applied in solving combinatorial optimization benchmarks, for example, the random graph, the satisfiability, and the traveling salesman problems.

As we discussed in Section 1.4, the *phase transition* being a general physical phenomenon plays an important role in finding the real hardness in optimization search. The computation becomes "hard" when the search space exits phase transitions, for example, the transitions between "weak (or under)" constraints and "strong (or over)" constraints. As a result, the path of general search dynamics including SA could be hardly possible to jump over the constraint valley in the search space. Consequently, the optimal solutions found are usually only local, not global. Many efforts have been made to deal with the hardness in optimization search, such as *genetic algorithms* that change the search from "point-to-point" or "individual-to-individual" to "population-to-population" in order to significantly enlarge the search space during the evolution generation after generation.

Moreover, to find high-quality solutions, particularly dealing with such kind of hardness, a general-purpose method, so-called *extremal optimization* (EO), has been proposed by Boettcher and Percus (1999). The proposed optimization solutions are inspired by statistical physics, coevolution, and the self-organizing process often found in nature. The EO search process successively eliminates extremely undesirable components of suboptimal solutions to perform the search with *far-from-equilibrium* dynamics. Compared to SA being a search with "equilibrium" dynamics, the moves in an EO search are likely to jump over the transition valley in the search space, and a global optimal solution might be found within a reasonably short time. The solutions and applications of EO will be introduced in detail throughout this book.

1.5 Highlights of EO

1.5.1 Self-Organized Criticality and EO

To build the bridge between statistical physics and computational complexity and find high-quality solutions for hard optimization problems, an extremal dynamics-oriented local-search heuristic, the so-called *extremal optimization* (EO) has been proposed (Boettcher and Percus, 1999). EO was originally developed from the fundamental of statistical physics. More specifically, EO is inspired by *self-organized criticality* (SOC) (Bak et al., 1987), which is a statistical physics concept to describe a class of systems that have a critical point as an attractor. Moreover, as also indicated by Bak et al. (1987), the concept of SOC may be further described and experimented on by a sandpile model as shown in Figure 1.2. If a sandpile is formed on a horizontal circular base with any arbitrary initial distribution of sand grains, if a steady state of the sandpile is formed by slowly adding sand grains gradually, the surface of the sandpile makes on average a constant angle with the horizontal plane. The addition of each sand

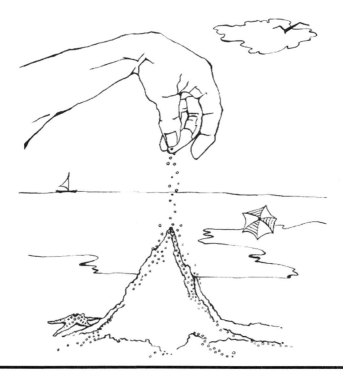

Figure 1.2   Experiment of sandpile model for SOC.

grain results in some activity on the surface of the pile: an avalanche of sand mass follows, which propagates on the surface of the sandpile. In the stationary regime, avalanches are of many different sizes that they would have a power law distribution. If one starts with an initial uncritical state, initially most of the avalanches are small, but the range of sizes of avalanches grows with time. After a long time, the system arrives at a critical state, in which the avalanches extend overall length and time scales.

Their macroscopic behavior exhibits the spatial and temporal scale-invariance characteristics of the critical point of a phase transition. SOC is typically observed in slowly driven nonequilibrium systems with extended degrees of freedom and a high level of nonlinearity. It is interesting to note that in SOC, there is no need to tune control parameters with precise values. Just inspired by this principle, EO drives the system far from equilibrium: aside from ranking, there exists no adjustable parameter, and new solutions are accepted indiscriminately. Consequently, the nature of SOC performed in EO may result in better solutions for those hard optimization problems consisting of phase transitions. The mechanism of EO can be characterized from the perspectives of statistical physics, biological evolution, and ecosystems (Lu et al., 2007).

1.5.2 Coevolution, Ecosystems, and Bak–Sneppen Model

Eventually, the fundamental of EO strategy was motivated by the Bak–Sneppen (BS) model (Bak and Sneppen, 1993). In this model, the high degree of adaptation of most species is obtained by the elimination of badly adapted ones. The basic idea behind this model is that of "coevolutionary avalanches." It is well known that "competitive" and "coevolutionary" activities are regarded as two important factors which help organisms evolve generation by generation in nature. Although coevolution does not have optimization as its exclusive goal, it serves as a powerful paradigm for EO. Following the spirit of the BS model, EO merely updates those components with worst fitness in the current solution, replacing them by random values without ever explicitly improving them. At the same time, EO changes the fitness of all the species connected to the weakest one randomly. Thus, the fitness of the worst species and its neighbors will always change together, which can be considered a coevolutionary activity. This coevolutionary activity gives rise to chain reactions or "avalanches": the large (nonequilibrium) fluctuations that rearrange major parts of the system, potentially making any configuration accessible. Large fluctuations allow the solution to escape from local minima and explore the configuration space efficiently, while the extremal selection process enforces frequent returns to a near-optimal solution.

In addition, the mechanism of EO can also be studied from the perspective of ecosystems. An ecosystem is defined as a biological community of interacting organisms and their surrounding environment. In other words, the fitness values of any species living in the same ecosystem are coupled with each other. Consequently, the changes in the fitness of any species will affect the fitness landscape (i.e., environment) of the whole ecosystem. The interactive relationship between species in an ecosystem can be regarded as the inherent fundamental mechanism which drives all the species to coevolve. According to the natural selection or "survival of the fittest" proposed by Darwin, those species with higher fitness will survive while those with lower fitness will die out. Similarly, EO considers that those species with lower fitness die out more easily than others. Hence, EO always selects those "weakest" to be updated or mutated. The change in the fitness of the worst species will impact the fitness landscape of the whole system. At the same time, the fitness of those species connected to the weakest species will also be affected by the altered environment and be changed simultaneously. The superior features of SOC and the search from the far-from-equilibrium region provide EO with the capability to deal with those hard optimization problems with phase transition.

Note that there is no single tuning parameter to adjust for the selection of a better solution. The algorithm operates on a single configuration S with only a mutation operation. In fact, the EO algorithm was developed based on the fundamental of the BS model.

1.5.3 Comparing EO with SA and GA

In this section, we will make comparisons between EO and some other popular meta-heuristic optimization solutions, such as GA and SA. All these meta-heuristic search methods share a common goal: finding an optimal or satisfactory solution with reasonable computation costs and accuracy subject to a set of given constraints, but each of them has its own features. It is well known that GA was proposed based on the fundamentals of natural selection. The population (i.e., a pool of chromosomes)-based evolutionary global search and the operation cycle of "generation, selection, reproduction (with genetic operators: crossover and mutation) and evaluation" are the key features for GA while EO stands on individual-based local search and mutation alone. On the other hand, to make GA work properly, there are a number of tuning parameters, such as population size, mutation and crossover probability rates, and elitist breeding policies, while basic EO does not have a single tuning parameter or the τ-EO (a modified version of EO) only needs a single tuning parameter. From the performance point of view, EO has more power to solve those hard optimization problems with phase transitions. However, when applying EO in practice, namely, encoding a real problem to a formulation required by EO, each component or gene in evolution sense has to be assigned a fitness value according to its contribution to the whole objective criterion. In other words, the global fitness has to be decomposed into multiple subterms being contributed by relevant components or genes in a solution or a chromosome as defined.

Physically, annealing is the process of cooling a material very slowly, allowing it to settle to its most stable (minimum energy) state. The SA (Kirkpatrick et al., 1983) uses the Monte Carlo Markov chain process to tackle optimization problems with local minima. The SA search algorithms are based on an analogy taken from thermodynamics and the behavior of equilibrium with a scheduled decreasing annealing temperature. If carefully setting a decrease function for the annealing temperature, it might be possible for the SA search to reach a global optimum (ground state) slowly via annealing-driven escape from local minima. In contrast, EO may drive the system from a local minimum to a far-from-equilibrium area across or near the phase transition boundary, while SA works in an equilibrium manner. For large graphs of low connectivity, EO has shown much better performance in convergence rate and accuracy than GA and SA. A numerical study (Boettcher and Percus, 2000) has shown that EO's performance relative to SA is particularly strong near phase transitions. In order to improve EO performance and extend its application areas, some modified versions of EO have been developed. The details will be addressed in the Section II of this book.

1.5.4 Challenging Open Problems

As a novel optimization approach introduced from statistical mechanism, EO has been moving from the physics community to AI and optimization communities to tackle computational complexity with phase transitions existing in many NP-hard problems. The natures of SOC and coevolution play an important role in making EO work well, particularly for those COPs. The theoretical analysis has shown that at the phase boundary (e.g., easy–hard–easy), EO descends sufficiently fast to the ground state with enough fluctuations to escape from jams. However, since EO is still under its early stage for both theoretical and practical perspectives, there are the following challenge open research topics:

1. Why does EO provide such superior performance near phase transitions of some NP-hard problems? More fundamental research is needed to explore the detailed functionality of SOC in the EO search process and fundamental comparisons between EO and many other approaches to intelligent computations.
2. To our best knowledge, there are limited research results about EO convergence analysis, for example, the analysis of the search dynamics of EO with random walk and/or Markov chain. In mathematics, the theory about the convergence of Markov chains has been developed and applied to analyze the convergence for stochastic, metropolis, and genetic algorithms. However, the studies on EO search dynamics are still an open problem in order to understand EO fundamentals deeply.
3. The EO has tested for many classical NP-hard benchmarks, but has had quite limited applications in solving real-world hard problems. Moving EO from academic society to real-world applications is one of the critical challenges. Even many real-world problems may be converted to those classical NP-hard benchmarks through simplification; however, the elegant maps is not an easy job and the real-world problems are usually much more complicated than benchmarks.
4. Finally, the study on the theoretical discovery for the fusion of self-organizing, probability-inspired mutation, and coevolutions is an open challenge to make EO more applicable in solving real complex optimization problems.

1.6 Organization of the Book

The objective of this book is to provide readers in the communities of optimization developers and practitioners with the fundamental, methodology, algorithms, and applications of EO solutions. Moreover, some critical subjects and contents in problem formulation and problem solving are also addressed in association with many popular benchmarks and real-world application cases. In addition to introducing the concepts, fundamentals, and solutions of EO, this book also presents the authors' recent research results on extremal dynamics–oriented self-organizing optimization, multiobjective optimization, and mimic algorithms from fundamentals, algorithms, and applications. The main topics are as follows:

1. To challenge the computational complexity usually occurring on the boundary of phase transition, an analytical solution for combinatorial optimization with microscopic distribution and fitness network to characterize the structure of configuration space is one of the main topics in computational intelligence. Moreover, based on the natural connections between the multiparticle systems in statistical physics and multivariate or combinatorial optimization, finding the minimum cost solution in a combinatorial optimization problem (COP) is analogous to finding the lowest energy state in a physical system, a general-purpose self-organizing optimization method with extremal dynamics is proposed.

2. Based on the concept of "selfish-gene" proposed by Dawkins (1976), a novel self-organizing evolutionary algorithm (EA), so-called *gene optimization* is developed to find high-quality solutions for those hard computational systems. The proposed algorithm successively eliminates extremely undesirable components of suboptimal solutions based on the local fitness of the genes. A near-optimal solution can be quickly obtained by the self-organizing coevolution processes of computational systems. Since the evolutionary computation methodology under study takes aim at the natures and micromechanisms of those hard computational systems, and guides the evolutionary optimization process at the level of genes, it may create much better EAs and affect the conceptual foundations of evolutionary computation.

3. Novel multiobjective extremal optimization (MOEO) algorithms are proposed to solve unconstrained or constrained multiobjective optimization problems in terms of introducing the fitness assignment based on Pareto optimality to EO. MOEO is highly competitive with three state-of-the-art multiobjective evolutionary algorithms (MOEAs), that is, Nondominated Sorting Genetic Algorithm-II (NSGA-II), Strength Pareto Evolutionary Algorithm2 (SPEA2), and Pareto Archived Evolution Strategy (PAES). MOEO is also successfully extended to solve multiobjective 0/1 knapsack problems, mechanical components design problems, and portfolio optimization problems.

4. A number of hybrid memetic algorithms with the marriage of "extremal optimization" and other optimization solutions, such as PSO-EO (particle swarm optimization-extremal optimization), ABC-EO (artificial bee colony-extremal optimization), "EO-Levenberg–Marquardt," and EO-SQP, are developed from fundamentals, algorithms to applications.

The book starts with a general introduction to optimization that covers understanding optimization with selected application fields, challenges faced in an information-rich era, and state-of-the-art problem-solving methods with CI. The book is structurally divided into three sections. Section I covering Chapters 1, 2, and 3 focuses on the introduction to fundamentals, methodologies, and algorithms of extremal dynamics and SOC-inspired EO. Section II consisting of Chapters 4, 5, and 6 presents the modified EO and the integration of EO with other computational intelligent solutions, such as population-based EO (PEO), PSO-EO, ABC-EO, and multiobjective EO. Section III covering Chapters 7 and 8 focuses on the applications of EO in nonlinear systems modeling and controls, and production planning and scheduling.

Chapter 2

Introduction to Extremal Optimization

2.1 Optimization with Extremal Dynamics

In statistical physics, the behavior of a multientity system can be described by its dynamics on the whole state space S. Generally, the total number of the possible microscopic configurations $N = |S|$, where $|S|$ represents the cardinality of the state space S, is extremely large. It is hardly possible to find the ground state $s^* \in S$ with minimum energy. However, in thermodynamics, many physical systems with numerous interacting entities can spontaneously organize toward a critical state, that is, a state that presents long-range correlations in space and time. That is the so-called "self-organized criticality (SOC)" (Bak et al., 1987), which is a paradigm for the description of a wide range of dynamical processes. Usually, those systems that exhibit SOC often consist of a large number of highly interrelated entities, and the collective behaviors of the overall system can be statistically analyzed at the macroscopic level. Correspondingly, the system configurations can also be regulated by the move class for updating the microscopic states of its entities, and the optimal configuration may naturally emerge from extremal dynamics (Gabrielli et al., 1997; Boettcher and Percus, 2003), simply through a selection against the "worse" elements, just like descending water always breaks through the weakest barriers, and biological systems eliminate the least-fit species.

Various models relying on extremal dynamics have been proposed to explain the self-organized critical phenomenon. The BS model (Bak and Sneppen, 1993; Paczuski et al., 1996) is perhaps the simplest prototype of SOC under extremal dynamics, and it is defined on a d-dimensional lattice with L^d sites and periodic boundaries. In the evolutionary interpretation of the BS model, each site represents a species, and has an associated "fitness" value between 0 and 1 (randomly sampling from a uniform distribution). At each updating step, the extremal species, that is, the one with the smallest value is selected. Then, that species and its interrelated species are replaced with new random numbers. After a sufficient number of update steps, the system reaches a highly correlated SOC, and the fitness of almost all species has transcended a certain fitness threshold. However, the dynamical systems maintain punctuated equilibrium: the species with the lowest fitness can undermine the fitness of those interrelated neighbors while updating its own state. This coevolutionary activity gives rise to chain reactions called "avalanches," large fluctuations that rearrange major parts of the system, potentially making any configuration accessible (Boettcher and Frank, 2006).

In close analogy to the self-organizing process of the BS model, the basic EO algorithm proceeds as follows (Boettcher and Percus, 2000):

1. Randomly generate an initial configuration S and set the optimal solution $S_{best} \leftarrow S$
2. For the current configuration S,
   a. Evaluate λ_i for each variable x_i
   b. Find j with $\lambda_j \geq \lambda_i$ for all i, that is, x_j has the "worst" fitness
   c. Choose at random a $S' \in N(s)$ such that the "worst" x_j changes its state
   d. If the global objective $F(S') < F(S_{best})$, then set $S_{best} \leftarrow S'$
   e. Accept $S \leftarrow S'$ unconditionally, independent of $F(S') - (S)$
3. Repeat step 2 until some termination criteria (e.g., running time) are satisfied
4. Return S_{best} and $F(S_{best})$

From the above EO algorithm, it can be seen that unlike GAs working with a population of candidate solutions, EO evolves a single solution S and makes local modification to the worst components. This requires that a suitable representation should be selected that permits individual solution components (or the so-called decision variables) to be assigned a quality measure (i.e., fitness). This differs from holistic approaches such as EAs that assign equal fitness to all components of a solution based on their collective evaluation against an objective function. In EO, each decision variable in the current solution S is considered as a "species" (Lu et al., 2007).

To avoid getting stuck into a local optimum (Boettcher and Percus, 2000), a single parameter is introduced into EO and the improved algorithm is called τ-EO. In τ-EO, according to fitness λ_i, all x_i are ranked as a permutation Π of the variable labels i with $\lambda_{\Pi(1)} \leq \lambda_{\Pi(2)} \leq \cdots \leq \lambda_{\Pi(n)}$. The worst variable x_j is of rank 1, $j = \Pi(1)$, and the best variable is of rank n. Then the variable is selected stochastically, in terms of a probability distribution over the rank order, rather than always selecting the "worst" variable at step 2. The variable with the kth highest fitness is selected with the probability $P_k \propto k^{-\tau}$ ($1 \leq k \leq N$), given that there are N entities in a computational system.

2.2 Multidisciplinary Analysis of EO

EO is a multidisciplinary technique that is based on the fundamental knowledge of statistical physics, biological evolution, and ecosystems, as shown in Figure 2.1 (Lu et al., 2007).

From the point of view of statistical physics, EO is motivated by the BS model of SOC, which is able to describe evolution via punctuated equilibrium, thus modeling evolution as a self-organized critical process. SOC is a statistical physics concept to describe a class of dynamical systems that have a critical point as an attractor. Their macroscopic behavior exhibits the spatial and temporal scale-invariance characteristics of the critical point of a phase transition (Bak et al., 1987). It is interesting to note that in SOC, there is no need to tune control parameters to precise values. SOC is typically observed in slowly driven nonequilibrium systems with extended degrees of freedom and a high level of nonlinearity (Bak et al., 1987). Inspired by SOC, EO drives the system far from equilibrium: aside from ranking, there exists no adjustable parameter, and new configurations are accepted indiscriminately. EO quickly approaches near-optimal solutions by persistently performing mutation on the worst species.

From the perspective of biological evolution, the EO heuristic was motivated by the BS model that shows the emergence of SOC in ecosystems. The fundamental idea behind this model is that of coevolutionary avalanches. It is well known that competitive and coevolutionary activities are regarded as two important factors that help organisms evolve generation by generation in nature. Although coevolution does not have optimization as its exclusive goal, it serves as a powerful paradigm for EO (Boettcher and Percus, 2001a). EO follows the spirit of the BS model in that it merely updates those components with worst fitness in the current solution, replacing them by random values without ever explicitly improving them. At the same time, EO changes the fitness of all the species connected to the weakest

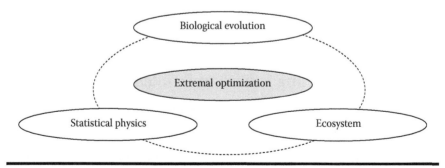

Figure 2.1 EO: a multidisciplinary technique. (From Lu, Y. Z. et al. Studies on extremal optimization and its applications in solving real world optimization problems. *Proceedings of the 2007 IEEE Symposium on Foundations of Computational Intelligence (FOCI 2007),* **Hawaii, pp. 162–168. © 2007, IEEE.)**

one randomly. Thus, the fitness of the worst species and its neighbors will always change together, which can be considered a coevolutionary activity. This coevolutionary activity gives rise to chain reactions or "avalanches": large (nonequilibrium) fluctuations that rearrange major parts of the system, potentially making any configuration accessible. Large fluctuations allow the method to escape from local minima and explore the configuration space efficiently, while the extremal selection process enforces frequent returns to near-optimal solutions.

Furthermore, EO can be analyzed from the ecosystem point of view. An ecosystem is defined as a biological community of interacting organisms and their surrounding environment. That is to say, the fitness of any species living in an ecosystem will be affected by the fitness of any other species in the same ecosystem, whereas the change in the fitness of any species will affect the fitness landscape (i.e., environment) of the whole ecosystem. The interaction relationship between any two species in the ecosystem can be regarded as the inherent fundamental mechanism that drives all the species to coevolving. The food chain may be one of the ways in which the interaction between any two species takes place. The food chain provides energy that all living things in the ecosystem must have to survive. In the food chain, there exist direct or intermediate connections between any species. According to natural selection or "survival of the fittest" proposed by Darwin, those species with higher fitness will have a higher probability to survive while those with a lower fitness will die out. In other words, the species with the lower fitness will die out with a higher probability than other species. When one species with a lower fitness dies out, those species above the extinct species in the food chain will also be under the threat of extinction, no matter how high their fitness value is. Similarly, EO considers those species with a lower fitness to die out more easily than others. Hence, EO always selects those "weakest" to update, or mutate. The change in the fitness of the worst species will impact the fitness landscape of the whole system. At the same time, the fitness of those species connected to the weakest species will also be affected by the altered environment and be changed simultaneously.

2.3 Experimental and Comparative Analysis on the Traveling Salesman Problems

In this section, the optimization method with extremal dynamics is presented to solve the traveling salesman problem (TSP) for experimental and comparative analysis. The TSP requires finding the shortest tour between n cities, with the salesman visiting each only once and returning to the starting point, that is, a directed Hamiltonian cycle containing all cities, and it is a classic combinatorial optimization conundrum in OR and computational physics due to the combinatorial complexity of the tours and the strong nonconvexity of the objective function in the hyperdimensional solution space. The TSP has often served as a test bed for a variety of combinatorial optimization algorithms with conceptual simplicity and wide applicability to many fields.

2.3.1 EO for the Symmetric TSP

2.3.1.1 Problem Formulation and Algorithm Design

The TSP is to find a closed tour of minimum length through a set of n cities randomly distributed in Euclidean space. In the simplest case, the cities lie on a two-dimensional map. In principle, the TSP can be solved exactly for any finite n, but in practice, it will turn to be a computational hard problem for a large n because the number of possible paths grows as $(n-1)!/2$ (Méndez et al., 1996). Thus, the objective is translated into finding near-optimal solutions with reasonable computational costs.

Let us consider a TSP instance with n cities. Thus, each city will have $n-1$ possible successors, and it can be ideally connected to its first nearest neighbor (Chen and Zhang, 2006). However, it is often "frustrated" by the competition of other cities, causing it to be connected to its kth ($1 \leq k \leq n-1$) nearest neighbor instead (Boettcher and Percus, 2000). As scaling the microscopic states of entities, for example, positions, energies, magnetizations, etc., in a physical system, each possible connection starting from a city is considered as a state (degree of freedom) of the city. More specifically, the state variable of city i is defined as $s_i = k$ if it is connected to the kth nearest neighbor. In the utopian society, each city resides in its ground state, that is, $s_i = 1$ for all i ($1 \leq i \leq n$).

However, in a feasible TSP tour, all cities possess their deterministic states, and may not always stay on their ground state. Consequently, each city has an underlying impetus for driving it to its ground state, and the impetus is relevant to the length of its forward-directed edge. In this section, the concept of potential energy from physical mechanics is introduced to evaluate the kinetic characteristics of all cities.

Let $D = \{d_{ij}\}$ be the distance matrix, where d_{ij} is the intercity distance between the ith and jth cities. Assuming p_i to be the length of the edge starting from city i in a TSP tour, the potential energy of city i is defined as

$$e_i = p_i - \min_{j \neq i}(d_{ij}) \tag{2.1}$$

Finally, the energy function for any feasible TSP tour s can be expressed as

$$F(s) = \sum_{i=1}^{n} \min_{j \neq i}(d_{ij}) + \sum_{i=1}^{n} e_i \tag{2.2}$$

Obviously, the first part of Equation 2.2 is a constant for a specific TSP instance, and it can be viewed as the internal energy for an isolated physical system. Thus, the optimal solution for combinatorial optimization is equivalent to the system with minimal free energy, that is, the whole computational system reaches its ground state.

On the basis of the definition of potential energy, the optimization method with extremal dynamics is proposed as follows (Chen et al., 2007):

1. The cities are numbered $i = 1, 2, ..., n$. A possible configuration is naturally represented as the order in which the cities are visited. In this path representation, the cities to be visited are put into a permutation of n elements (Larrañaga et al., 1999), so that if city i is the jth element of the permutation, it means that city i is the jth city to be visited.

   The optimization process starts from a random solution s. Let s_{best} represent the best solution found so far, and set $s_{best} \leftarrow s$ initially.

2. To the "current" solution s, the potential energy e_i for each city i ($1 \leq i \leq n$) and the overall energy $E(s)$ of the computational system can be calculated, respectively, by Equations 2.1 and 2.2. For self-organization to occur, the computational system must be driven by sequentially updating those cities with a high potential energy. Since a great variety of physical systems have generic statistical properties characterized by power-law distributions as discussed in Tsallis and Stariolo (1996) and Mosetti et al. (2007), in the optimization dynamics, a city having kth "high" potential energy is selected from the set of cities according to a scale-invariance power-law distribution:

$$P(k) \propto k^{-\alpha} \qquad (2.3)$$

where the rank k is going from $k = 1$ for the city with the highest potential energy to $k = n$ for the city with the lowest potential energy, and the power-law exponent α is an adjustable parameter. For $\alpha = 0$, randomly selected cities get forced to update, resulting in merely a random walk through the configuration space. For $\alpha \to \infty$, only the extremal cities with the highest potential energy get updated that may trap the computational system to a metastable state (i.e., near-optimal solution).

For updating the state of the selected cities, now, let us focus on the move class, also called neighborhood relation, that is, to each solution s, a set of neighbors $N(s)$ is defined (Franz and Hoffmann, 2002). If the move class is a reversible process, that is, if $s' \in N(s)$, then also, $s \in N(s')$, and the state space will be defined as an undirected network structure. On this complex network of configuration space, the degrees of freedom of each solution $s \in S$ is the cardinality of its neighborhood $|N(s)|$, and a random walk takes places for finding the ground state. In the TSP, the neighborhood $N(s)$ can be easily constructed by the 2-opt move (Fredman et al., 1995; Helsgaun, 2000): construct new Hamiltonian cycles by deleting two edges and reconnecting the two resulting paths in a different way. Simply, the 2-opt move keeps the tour feasible and corresponds to a reversal of a subsequence of the cities. Since the potential energy of the selected city is expected to be updated, the 2-opt move must replace its forward-directed edge, and so, there will be $n - 1$ possible neighbors, that is, the cardinality of the neighborhood $|N(s)| = n - 1$. Being in the current state s, the random walker in our optimization dynamics chooses a new state s' having μth "low" energy out of its neighbors $N(s)$ by another power-law distribution $P_\mu \propto \mu^{-\beta}$ ($1 \leq \mu \leq |N(S)|$), and accepts $s \leftarrow s'$ unconditionally, if $E(s') < E(s_{best})$, sets $s_{best} \leftarrow s'$. The external control parameter β can be used to regulate the extent of avalanche-like fluctuations in those highly susceptible states (near the ground state), that is, uphill moves may be accepted, and it can help the random walker pass through barriers of the energy landscape. In principle, the dynamical system can jump from one metastable state (corresponding to a near-optimal solution in combinatorial optimization) to another by avalanche dynamics.

3. Repeat the update step 2 until a given termination criterion is satisfied, which is either a certain number of iterations or a predefined amount of *CPU* time or the convergence of the performance improvement. Return s_{best} and $E(s_{best})$ as the resulting solution.

It is worth noting that the 2-opt move of updating the states of the selected city with a high potential energy also has an impact on the potential energy of its interacting cities. Consequently, the extremal dynamics and coevolutionary processes can drive the computational system (TSP tour) to its ground state gradually.

To get an explicit conception of the parameters α and β, it is useful to conduct some numerical simulations. First, in Figure 2.2, the average energy and the errors for 10 runs with the same initial configurations are presented for the $n = 64$ random Euclidean TSP instance. It can be seen that the energy isn't very sensitive to the value of α, but numerous experiments demonstrate that a value $\alpha \sim 1 + 1/\ln(n)$ seems to work best as discussed in Boettcher and Percus (2003).

Second, the same initial configurations and $\alpha = 1 + 1/\ln(n)$ are used, and Figure 2.3 shows typical runs for different values of the parameter β. In the simulation, the parameters are set as $\alpha = 1 + 1/\ln(n)$ and (a) $\beta = 1$, (b) $\beta = 2$, (c) $\beta = 3$, and (d) $\beta \to \infty$. Starting with the case $\beta = 1$, it is apparent that the convergence is slightly slow and fluctuations are large. As the value of β increases, the fluctuations are considerably reduced, and drastically damped while $\beta \to \infty$. The experiments show that the ground state with minimal energy can be reached when $\beta = 3$.

2.3.2 SA versus Extremal Dynamics

SA updates the system configuration according to the Metropolis algorithm, which is the classical Monte Carlo method for simulating the behavior of a physical system at temperature T. The solid theoretical foundation of SA techniques in optimization is the connection between statistical mechanics (the behavior of systems with many degrees of freedom in thermal equilibrium at a finite temperature) and multivariate or

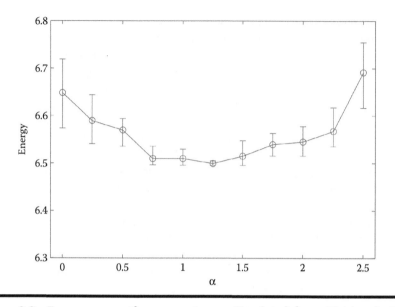

Figure 2.2 Energy versus the parameter α. (Reprinted from *Physica A*, 385, Chen, Y. W. et al. Optimization with extremal dynamics for the traveling salesman problem. 115–123. Copyright 2007, with permission from Elsevier.)

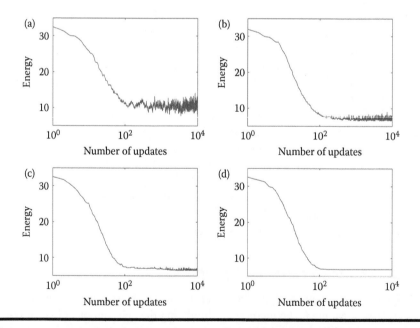

Figure 2.3 Evolution of the energy function. (Reprinted from *Physica A*, 385, Chen, Y. W. Lu et al. Optimization with extremal dynamics for the traveling salesman problem. 115–123. Copyright 2007, with permission from Elsevier.)

combinatorial optimization (finding the minimum of a given function depending on many parameters). In the SA computational method, a neighboring state is selected at random, and then, the candidate state is accepted with the Boltzmann probability $\exp(-\Delta E/T)$, where ΔE is the energy change associated with the move (Franz and Hoffmann, 2002). When the simulated temperature T being a tuning parameter decreases, the computational system moves consistently toward the ground state, yet still jumps out of the local optima due to the occasional acceptance of upward moves. When the temperature is sufficiently close to zero through an exponential cooling regime, the uphill moves will no longer be accepted and then an approximation solution can be obtained. Obviously, the successful application of SA requires a proper decreasing temperature schedule with careful tuning case by case. For improving the optimization efficiency and maximizing the ground-state probability, the modifications of the acceptance probability distribution including Tsallis statistics had been extensively studied as in De Menezes and Lima (2003). SA and its variations as powerful optimization tools have been applied to a variety of disciplines. In its implementation to the TSP, a typical SA optimization dynamics is as shown in Figure 2.4.

The above simulation is also for the $n = 64$ random Euclidean TSP instance. As illustrated in Figure 2.4, the system moves consistently toward a quasi-equilibrium

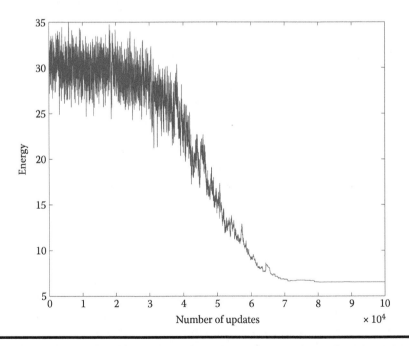

Figure 2.4 Evolution of the energy function in a typical run of SA. (Reprinted from *Physica A*, 385, Chen, Y. W. et al. Optimization with extremal dynamics for the traveling salesman problem. 115–123. Copyright 2007, with permission from Elsevier.)

state (a near-optimal TSP tour), while the tuned temperature parameter T decreases. However, SA samples numerous states far from the ground state with the lowest energy, and its searching capacity is inefficient in the last portion of the optimization process.

In contrast to SA, which only dwells on the macroscopic behavior of computational systems (i.e., global energy function), and does not investigate the micromechanism of solution configurations, the extremal dynamics simulates a complex multientity system in statistical physics. Both the collective behavior and the individual states of particles are considered simultaneously during the optimization dynamics. As shown previously in Figure 2.3, a near-optimal solution can be quickly obtained by the greedy-searching process at first, and enough fluctuations (ergodic walk) can help to escape from the local optima and explore new regions of the configuration space. Combining the greedy search and the fluctuated explorations near the backbone of optimization problems, this optimization process can be viewed as an ideal search dynamics for those computational systems with rugged-energy landscapes.

2.3.3 Optimizing Near the Phase Transition

Phase transition of complex systems is a dramatic change of some system properties when an order or control parameter crosses a critical value, for example, water changing from ice (solid phase) to water (liquid phase) to steam (gas phase) when temperature increases. Phase transition has been extensively studied in physics and computer science (Gent and Walsh, 1996; Biroli et al., 2002). In combinatorial optimization, NP-complete problems can be summarized by at least one-order parameter and the hard cases occur at a critical value of such a parameter. This critical value separates two regions of characteristically different properties. Computational complexity occurs on the boundary between these two regions, where the probability of a solution is low but nonnegligible.

Random instances of the TSP can be easily generated by placing n cities on a square of area A uniformly, and the expected optimal tour length l_{opt} approaches the limit

$$\lim_{n \to \infty} \frac{l_{opt}}{\sqrt{n}} = k\sqrt{A} \qquad (2.4)$$

where k is a constant, and the best current estimate for k is 0.7124 ± 0.0002 (Gent and Walsh, 1996). This asymptotic result suggests that a natural control parameter is the dimensionless ratio $l/\sqrt{n \cdot A}$. At large values of this parameter, that is, the tour length is large compared to the number of cities to be visited, the solutions can be easily found. More accurately speaking, the system configuration can be constructed arbitrarily when the free energy $\sum_{i=1}^{n} e_i$ of the computation system is high. From this aspect, it can be inferred that the numerous sampling in the preceding

Table 2.1 Comparisons of Simulated Annealing and Extremal Dynamics Applied to the TSP

	Simulated Annealing			Extremal Dynamics		
n	Mean $(l/\sqrt{n \cdot A})$	Standard Deviation $(l/\sqrt{n \cdot A})$	Number of Updates	Mean $(l/\sqrt{n \cdot A})$	Standard Deviation $(l/\sqrt{n \cdot A})$	Number of Updates
16	0.9023	0.0217	32228	0.9023	0.0217	5000
32	0.8394	0.0237	70901	0.8336	0.0211	6000
64	0.8295	0.0208	147360	0.8075	0.0202	7500
128	0.8134	0.0095	245600	0.7774	0.0163	10000
256	0.8134	0.0045	368400	0.7724	0.0140	15000
512	0.8988	0.0076	526300	0.7709	0.0065	20000
1024	1.181	0.0105	801190	0.7699	0.0052	30000

Source: Reprinted from *Physica A*, 385, Chen, Y. W. et al. Optimization with extremal dynamics for the traveling salesman problem. 115–123. Copyright 2007, with permission from Elsevier.

phase of SA is a fruitless effort. Contrarily, the optimization dynamics studied in this chapter, which descends sufficiently fast to the ground state with enough fluctuations to escape the local optima, is more reasonable and effective.

In Table 2.1, the computational results on a set of random TSP instances are compared analytically. The computational results for each value of n are averaged over 10 instances. The computational results indicate that the proposed method with extremal dynamics significantly outperforms the state-of-the-art SA, particularly when the number of cities is larger. Especially, the ergodic fluctuation near the phase transition can help find the inapproachable ground state. Moreover, the potential energy defined for each particle is completely consistent with the global energy function, and the extent of fluctuations can be adjusted by the exponent β; as discussed in Section 3.2, the proposed optimization dynamics can effectively organize the system to the optimal configuration. It can dispel the misconception that EO is not competitive for the TSP (Boettcher and Percus, 2000), and open the door for the utilization of extremal dynamics to solve COPs in fields ranging from computational physics to scientific and engineering computation.

2.3.4 EO for the Asymmetric TSP

In contrast to the symmetric TSP, the distance from one city to another in asymmetric TSP may not necessarily be the same as the distance on the reverse direction.

The asymmetric TSP appears to be more difficult than its symmetric counterparts, both with respect to optimization and approximation (Gutin and Punnen, 2002; Laporte, 2010). Generally, the asymmetric TSP heuristics can be divided into the following three categories (Cirasella et al., 2001): (1) tour construction methods, such as nearest-neighbor search and greedy algorithm, (2) algorithms based on patching subcycles together in a minimum cycle cover, such as branch and bound algorithms (Miller and Pekny, 1989), and (3) tour improvement methods, mainly referring to local search methods based on rearranging segments of the tour, such as 3-opt search (Lawler et al., 1985), hyperopt search (Burke et al., 2001), and Lin–Kernighan algorithm (Lin and Kernighan, 1973; Kanellakis and Papadimitriou, 1980; Helsgaun, 2000).

2.3.4.1 Cooperative Optimization

In the EO algorithm for the asymmetric TSP, the potential energy of each city is defined in the same way for the symmetric TSP. Then, a city having kth "high" energy is selected from the set of cities according to the following scale-free distribution: $P_k \propto k^{-\tau}$ $(1 \leq k \leq N)$.

Then, 3-opt move is used to update the states of selected cities. In 3-opt, the rearrangement process breaks the tour into three segments C_1, C_2, and C_3 by deleting three edges, and then reorders the segments as C_2, C_1, and C_3 and relinks them to form a new tour (Cirasella et al., 2001), as illustrated in Figure 2.5. Each 3-opt move simultaneously updates the states of three cities: p, q, and r.

In 3-opt move, there are $(N-1) \times (N-2)$ possible updating options that can be chosen to update the states of the "bad" city. Each of these states leads to a neighbor solution, and then, a cooperative optimization process can be implemented. After selecting a "bad" city with high energy (occasionally a "good" city with low energy), the updating strategy should be implemented. For the asymmetric TSP, a city cannot update its state without affecting the states of others. The selected "bad" city should cooperate with other cities to achieve a collective optimal state

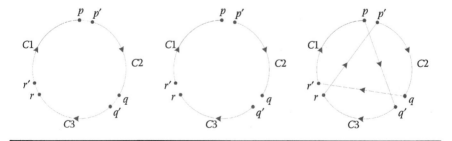

Figure 2.5 **3-Opt move for the asymmetric TSP. (Reprinted from *Physica A, 390,* Chen, Y. W. et al. Improved extremal optimization for the asymmetric traveling salesman problem. 4459–4465. Copyright 2011, with permission from Elsevier.)**

considering all the updating cities. In the 3-opt move, there are $(N-1) \times (N-2)$ possible options to update the state of the selected city. Each of these states leads to a neighbor solution. Furthermore, a cooperative optimization strategy is implemented as follows:

1. If the local optimum solution S_{local} in the 3-opt neighbor space satisfies $F(S_{local}) < F(S_{best})$, then set $S_{best} \leftarrow S_{local}$ and accept $S \leftarrow S_{local}$.
2. Else, randomly select a new solution S' from the 3-opt neighbor space with a probability $p(0 \leq p \leq 1)$, and accept $S \leftarrow S'$. That is to say, a random 3-opt move is applied to update the state of the selected city with the probability p. With the probability $1 - p$, a greedy strategy is used to select the local optimum solution S_{local}, and set $S \leftarrow S_{local}$.

The choice of a scale-free distribution for P_k ensures that no rank of fitness gets excluded from further evolution while maintaining a bias against entities with bad fitness (Boettcher, 2005a). The probability parameter p is used to introduce some "noise" or "random walk" to the algorithm so that the optimization dynamics can escape from metastable states (i.e., local–optimal solutions) more easily (Selman et al., 1994).

2.3.4.2 Parameter Analysis

In this section, a benchmark asymmetric TSP instance, ftv170 in TSPLIB (Reinelt, 1991), is taken as an example to analyze the parameters of the above cooperative EO algorithm. In the simulation, a couple of simple measures are used to evaluate the algorithm's abilities of exploring and exploiting the solution space. All the computational results are simulated with the probability of $p = 0.2$. Figure 2.6 illustrates the EO algorithm's exploring capability with respect to the parameter τ.

In Figure 2.6, the distance between two asymmetric TSP solutions is defined as the number of cities that reside in different states. The time lag is calculated as step lag. Suppose that a cooperative EO optimization process has T iterative steps. To calculate the average distance with time lag t, for example, it can be calculated as the distance between the two asymmetric TSP solutions at all the two steps of having the lag t, for example, $(1, 1 + t)$, $(2, 2 + t)$, ..., and $(T - t, T)$. The average distance can be used to measure the capability of exploring the solution space. Each curve in Figure 2.6 corresponds to a specific value of the parameter τ. It shows that the exploring capability or the so-called "mobility" of the algorithm decreases with the increase of the parameter value τ.

Furthermore, Table 2.2 compares the simulation results of SA, τ-EO, and the proposed cooperative EO algorithm on a set of benchmark asymmetric TSP instances with difference sizes. Ten runs are carried out for each instance. $F(S_{opt})$ denotes the known optimal solution. The mean error $\sigma(\%) = 100 \times (\overline{F}(S) - F(S_{opt}))/F(S_{opt})$ is used to measure the optimization performance and effectiveness, where $\overline{F}(S)$

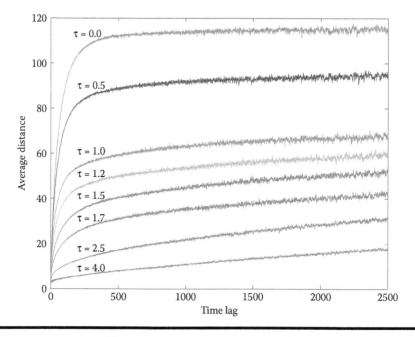

Figure 2.6 Average distance between solutions at different time lags (ftv170). (Reprinted from *Physica A*, 390, Chen, Y. W. et al. Improved extremal optimization for the asymmetric traveling salesman problem. 4459–4465. Copyright 2011, with permission from Elsevier.)

Table 2.2 Simulation Results on Benchmark Asymmetric TSP Instances

		SA		τ-EO		Cooperative EO		
Problem	$F(S_{opt})$	σ(%)	t_{CPU}	σ(%)	t_{CPU}	Success Rate	σ(%)	t_{CPU}
ftv70	1950	0.56	15	0.10	7	10/10	0.00	2.14
kro124p	36230	1.17	23	0.21	13	10/10	0.00	6.04
ftv170	2755	1.60	39	0.87	35	5/10	0.07	41.17
rbg358	1163	1.55	57	0.34	53	10/10	0.00	2.62
rbg403	2465	2.19	60	0.51	60	9/10	0.01	17.87
rbg443	2720	1.77	60	0.74	60	10/10	0.00	24.35

Source: Reprinted from *Physica, A,* 390, Chen, Y. W. et al. Improved extremal optimization for the asymmetric traveling salesman problem. 4459–4465. Copyright 2011, with permission from Elsevier.

denotes the average of the best solutions found by a specific algorithm over 10 runs. The average computation time (t_{CPU}) on an Intel Pentium PC is used to evaluate the efficiency of optimization algorithms.

It can be concluded from Table 2.2 that the proposed EO optimization method provides superior optimization performances in both computational effectiveness and efficiency.

2.4 Summary

Physics has provided systematic viewpoints and powerful methods for optimization problems. By mapping the optimization problems into physical systems, the EO method with self-organizing dynamics can be used effectively to solve COPs. There are many advantages of EO such as extremal dynamics mechanism, coevolution, only-mutation operator, and long-term memory. Thus, EO can be considered as a good heuristic method that is competitive with or outperforms many state-of-the-art heuristics. The experimental results on both the symmetric and asymmetric TSPs demonstrate that the EO algorithm performs very well and provides much better performance than the existing stochastic search methods developed from statistical physics, such as SA. For the algorithmic equivalence of NP-complete problems, this interdisciplinary optimization method can be extended to solve a wide variety of combinatorial and physical optimization problems particularly with phase transitions on search space.

Chapter 3

Extremal Dynamics–Inspired Self-Organizing Optimization

3.1 Introduction

Combinatorial optimization is pervasive in most fields of science and engineering. Its aim is to optimize an objective function on a finite set of feasible solutions. For example, the TSP (Gutin and Punnen, 2002), one of the classical COPs, can be described as a problem of search for the shortest tour among a set of cities. Generally speaking, most of COPs in practice are deemed as computationally intractable, and have been proven to belong to the class of NP-complete problems (Garey and Johnson, 1979), where NP stands for "nondeterministic polynomial time." For NP-complete problems, although the optimality of a possible solution can be verified in polynomial time, the computational time for finding the optimal solution grows exponentially with the dimension of input variables in the worst case. Furthermore, if a polynomial time algorithm can be found to solve one NP-complete problem, then, all the other NP-complete problems will become solvable in polynomial time (Papadimitriou, 1994).

However, in modern computer science, it has been commonly conjectured that there are no such polynomial time algorithms for any NP-complete problems (Cormen et al., 2001). Alternatively, a variety of nature-inspired optimization techniques have been developed for finding near-optimal solutions of NP-complete problems within a reasonable computational time. Examples of such algorithms include SA (Kirkpatrick et al., 1983), GA (Forrest, 1993), ACO (Bonabeau et al.,

2000), and particle swarm optimization (Kennedy and Eberhart, 1995), among others.

It is interesting to note that most of the existing optimization methods employ a centralized control model, and rely on a global objective function for evaluating intermediate and final solutions. In dealing with hard COPs, however, it would be difficult and time consuming to collect global information due to the interaction in, and the dimensionality of, computation. In order to overcome the limitation of the existing centralized optimization, several decentralized, self-organized computing methods have been studied in recent years with the help of complex systems and complexity science.

Boettcher and Percus (2000) presented a stochastic search method called EO. The method is motivated by the BS evolution model (Bak and Sneppen, 1993; Sneppen, 1995), in which the least-adapted species are repeatedly mutated following some local rules. To further improve the adaptability and performance of the EO algorithm, a variation of EO called τ-EO (Boettcher and Percus, 2000) was subsequently presented by introducing a tunable parameter τ. So far, EO and its variants have successfully addressed several physical systems with two-degree-of-freedom entities, that is, the COPs with binary-state variables, such as graph bipartitioning problems (Boettcher and Percus, 2000), Ising spin glasses (Middleton, 2004), community detection in complex networks (Duch and Arenas, 2005), etc. Recently, Liu and Tsui (2006) presented a general *autonomy-oriented computing* (AOC) framework for the optimization of self-organized distributed autonomous agents. AOC is also a bottom-up computing paradigm for solving hard computational problems and for characterizing complex systems behavior. Computational approaches based on AOC systems have been applied to distributed constraint satisfaction problems, image feature extraction (Liu et al., 1997), network community-mining problems (Yang and Liu, 2007), etc. In addition, Han (2005) introduced the concept of the local fitness function for evaluating the state of autonomous agents in computational systems.

In order to design a reasonable self-organized computing method, there are generally several important issues that should be considered:

1. How can we construct a mapping from COPs to multientity complex systems?
2. What types of local fitness function should be used to guide the computational systems to evolve toward the global optimum?
3. What are the microscopic characteristics of optimal solutions?
4. How can we characterize the computational complexity of such a method in searching a solution space?

Much literature has been focused on one or some of the above questions over the past few years (Mézard et al., 2002; Han and Cai, 2003; Goles et al., 2004; Achlioptas et al., 2005). The aim of this chapter is to answer each of those questions, and furthermore, provide a solid theoretical foundation for the self-organized computing method under study. First of all, COPs are modeled into multientity systems, in which a large number of self-organizing, interacting agents are involved. Then, the microscopic characteristics of optimal solutions are examined with respect to the notions of the discrete-state variable and local fitness function. Moreover, the complexity of search in a solution space is analyzed based on the representation of the fitness network and the observation of phase transition. Finally, based on the analysis, a self-organized computing algorithm is described for solving hard COPs.

3.2 Analytic Characterization of COPs

In this section, an analytic characterization of combinatorial optimization is presented from the point of view of a self-organizing system. The characterization will provide us with the insights into the design of a decentralized, self-organized computing approach to tackling COPs. In particular, it points out ways of improving the efficiency of solution searching based on such an approach.

3.2.1 Modeling COPs into Multientity Systems

Combinatorial optimization is concerned with finding the optimal combination of a set of discrete constrained variables (Papadimitriou and Steiglitz, 1998). As summarized by Blum and Roli (2003), a COP, $P = (S, F)$, can be defined in terms of

- A set of discrete variables, $X = \{x_1, \ldots, x_n\}$, with the relevant variable domains, $D_1, \ldots, D_n$
- Multiple constraints among variables
- An objective function F to be optimized, where $F: D_1 \times \cdots \times D_n \to R$

Correspondingly, the solution space can be represented as

$$S = \{s = (x_1, \ldots, x_n) \mid x_i \in D_i, s \text{ satisfies all the constraints}\}$$

where each element s is a candidate solution. Given a COP, the aim is to find the optimal solution, $s^* \in S$, such that the global objective, $F(s^*) \leq F(s)$ ($\forall s \in S$).

Since combinatorial solutions often depend on a nontrivial combination of *multiple elements* with specific states, a COP can be straightforwardly translated into a *multientity* computational system, if it is formulated from the viewpoint of a complex system. In a complex computational system, each entity can be viewed as an *autonomous agent*, and the agent can collect local (limited) information from the environment, and acts with other interacting agents to achieve the system objective

(Liu et al., 2005). Specifically, an autonomous computational agent, a, can be characterized based on the following properties (Liu et al., 2002):

- A finite number of possible states and some behavioral rules
- Local information about the computational system and its conditions
- Certain local objectives

Thus, a multientity computational system can be further modeled based on

- A group of agents, $A = \{a_1, \ldots, a_n\}$
- Interactions between agents
- A set of reactive rules, governing the interactions among agents

Drawing on the above intuitive idea of mapping between the representation of multientity computational systems and the definition of combinatorial optimization, various methodologies for complex systems modeling and analysis can be explored and applied to handle combinatorial computing problems. Generally speaking, the modeling and analysis techniques for complex systems can be classified into top-down and bottom-up approaches (Liu et al., 2005). Top-down approaches follow the idea of centralized control, and mainly study the macroscopic system-level characteristics of a complex system. On the contrary, bottom-up approaches, which were developed more recently, employ the concept of decentralized control, and apply the entity-oriented models to characterize the microscopic-emergent properties. In order to address the complexity of practical computational systems, this chapter focuses on the study of a novel decentralized, self-organized computing method, as opposed to the earlier work on centralized combinatorial optimization.

3.2.2 Local Fitness Function

In a multientity computational system, the process of self-organization in individual entities is crucial for searching the global optimal solution of the whole system. At any moment, each entity must assess its local conditions, and thus decide its primitive behavior. In doing so, a local fitness function (here, the term "fitness function" is synonymous with objective function) is essential for evaluating the specific states of local entities. However, a natural question arising is which types of local fitness function are capable of guiding the computational systems to evolve toward the global optimum. In order to answer this question, let us first take a view on an elementary case in Figure 3.1, which is abstracted from the fitness landscape of epistasis (Duch and Arenas, 2005).

In Figure 3.1, it illustrates the requirements of building the local fitness function with three tentative fitness landscapes, in which each path evolves from the initial solution "ab" to the global optimum "AB" through updating each agent once, and the solid paths are regarded as the favored evolutionary paths. For selecting the favored trajectories, Figure 3.1a indicates that the local fitness function for each

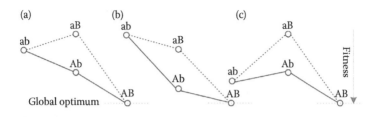

Figure 3.1 **(a–c) Illustration of accessible evolutionary paths. (With kind permission from Springer Science+Business Media:** *Artificial Intelligence Review,* **Toward understanding the optimization of complex systems, 38, 2012, 313–324, Liu, J. and Chen, Y. W.)**

agent should reflect the sign of the global fitness changes; seemingly, in the intermediate process, updating "a" to "A" yields the opposite fitness effect for updating "b" to "B"; Figure 3.1b and c show that the local fitness function should be capable of measuring the magnitude of the global fitness changes, for example, updating "a" to "A" has a more positive or less negative fitness effect than updating "b" to "B." With these two premises, reasonable local fitness functions can be defined for different computational systems. For example, in the Ising computing model, the local fitness of the spin σ_i can be represented as

$$f_\sigma(\sigma_i) = \sigma_i \left(h_i + \sum_{j \in N(i)} J_{ij} \sigma_j \right) \tag{3.1}$$

where $N(i)$ is the set of spins directly interacted with the spin σ_i. And then, the global fitness changes can be consistently calculated by the local fitness function when flipping the state of spins. With the definition of the consistent local fitness function, the emergent computing methods can now know how to apply the local behavioral rules in the presence of natural or priority-based selection mechanisms.

Next, we will take the TSP as a walk-through example of COPs. The TSP has attracted much interest in computing communities in the past decades due to the fact that it provides insights into a wide range of theoretical questions in discrete mathematics, theoretical computer science, computational biology, among others, and offers models and solutions to many real-world practical problems, ranging from scheduling, transportation, and to genome mapping. The TSP is usually stated as an optimization problem of finding the shortest closed tour that visits each city only once. In other words, given a set of n cities and the distance measure, d_{ij}, for all pairs i and j, the optimization goal is to find a permutation π of these n cities that minimizes the following global fitness function:

$$F(s) = \sum_{i=1}^{n-1} d_{\pi(i),\pi(i+1)} + d_{\pi(n),\pi(1)} \tag{3.2}$$

For a TSP solution s in the solution space S, it can be exclusively represented by the discrete states of all cities. As discussed in the previous chapter, if city i, which can be represented as entity x_i, is connected to its kth ($1 \leq k \leq n-1$) nearest neighbor, the discrete state variable of city i is defined as

$$s(x_i) = k, \quad i = 1, 2, \ldots, n-1 \tag{3.3}$$

In an ideal computational system, all cities will be connected to their first nearest neighbors, that is, $s(x_i) = 1$ for all i ($1 \leq i \leq n$). However, such ideal microstates are often frustrated by the interacting effects between cities, that is, competitions on the selection of forward-directed edges. As a result, a city may not always be connected to its first nearest neighbor, and possibly dwells on a specific energy state except the ground state, that is, $s(x_i) > 1$. Correspondingly, a local fitness function can be formulated to evaluate the microscopic dynamical characteristics of all cities.

Now, let us assume d_i to be the length of the forward-directed edge starting from city i; in a feasible solution s, the local fitness of city i can be defined as

$$f_s(x_i) = d_i - \min_{j \neq i} d_{ij}, \quad i = 1, 2, \ldots, n \tag{3.4}$$

Correspondingly, the global fitness function for any possible solution s can be represented as

$$F(s) = \sum_{i=1}^{n} \min_{j \neq i} d_{ij} + \sum_{i=1}^{n} f_s(x_i) \tag{3.5}$$

Obviously, the first term in Equation 3.5 is a constant, which can serve as a lower bound for a given TSP instance. The second term is the sum of the local fitness for all entities. That is to say, the global fitness of a combinatorial solution can be represented as a function of the distributed local fitness, and the optimal solution is equivalent to the microstate with the minimum sum of local fitness for all entities. Intuitively, the computational system can be optimized through updating the states of those entities with a worse local fitness. In a complex computational system, the relationship between discrete states and local fitness is usually nonlinear, and the value of local fitness is not necessarily equal even if two entities have the same state value.

Since local fitness is not as explicit as global fitness, as discussed in this chapter (Chen and Lu, 2007), the following questions will inevitably arise:

1. Can the optimization guided by local fitness achieve the global optimum of computational systems?
2. What will be the relationship between local fitness and global fitness?

To answer the above questions, the consistency and equivalence conditions between global fitness and local fitness are investigated.

Definition 3.1

Suppose that solution s' is constructed from s ($\forall s \in S$) by updating the state of entity i using a neighbor rule, local fitness is *consistent* with global fitness, *if* it is true that

$$\text{sgn}\{F(s') - F(s)\} = \text{sgn}\left\{\sum_{x \in X(s,s',i)} [f_{s'}(x) - f_s(x)]\right\} \quad (3.6)$$

where sgn{·} is a symbolic function, and $X(s, s', i)$ denotes the set of entities whose states are changed as a result of the interacting effects of updating the state of entity i. If the neighbor rule is a reversible move class, and $X(s, s', i) = X(s', s, i)$, then, the solution space can be represented as an undirected graph, otherwise a directed graph.

Definition 3.2

Local fitness is *equivalent* to global fitness *if*
   i. Local fitness is consistent with global fitness and
   ii. $\exists \alpha \in R^+, \beta \in R$ such that it is true that for all $s \in S$

$$F(s) = \alpha \sum_{x \in X} f_s(x) + \beta \quad (3.7)$$

According to Equation 3.6, it is evident that if local fitness is consistent with global fitness, improving one or some entities' local fitness will also optimize the global fitness of the whole computational system, and the process of self-organization will be effective in solving hard computational problems. Equivalence is a special case of consistency. These definitions make it easier for us to understand what types of local fitness function should be defined in designing self-organized computing methods.

3.2.3 Microscopic Analysis of Optimal Solutions

In a computational system, searching the optimal microstate will no doubt become simpler if we have a definite object in view. Therefore, in this section, the microscopic analysis of optimal solutions is investigated based on the preceding definitions of the discrete-state variable and local fitness function.

As we know, the optimal solutions for a large number of benchmark TSP problems have been given in Reinelt's TSPLIB (Reinelt, 1991; TSPLIB). Therefore, for the sake of illustration and testing, the TSP instance, called kroA100 (100 cities) is used as an example. In order to find a specific TSP tour, the discrete-state value and local fitness of each city are calculated sequentially. As a result, the microscopic characteristics of the optimal tour and a random tour can be found and have been comparatively shown in Figure 3.2.

As can be noted in Figure 3.2, in the optimal solution, the values of the state variables for all cities as well as their local fitness are far lower than the upper bound of the variables, for example, $s(x_i) \ll n - 1$ for all i ($1 \leq i \leq n$). This implies that those cities with a high local fitness in an initial solution should update their states, until the computational system self-organizes itself into a well-organized microscopic state, that is, the optimal solution.

In order to demonstrate the universality of this microscopic distribution, the statistical properties of the optimal TSP solutions can be further characterized by the kth nearest-neighbor distributions (NNDs) (Chen and Zhang, 2006). Without loss of generality, in this chapter, the kth-NND for any possible TSP tour s is defined as follows:

$$p(k) = \frac{r(k)}{n}, \quad k = 1, 2, \ldots, n-1 \tag{3.8}$$

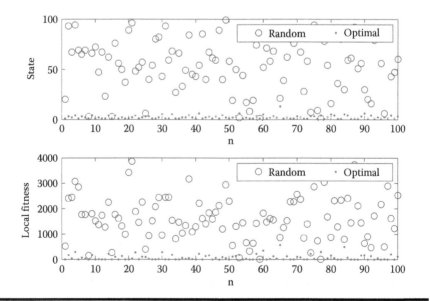

Figure 3.2 **Microscopic characteristics of the optimal tour and a random tour (kroA100). (Reprinted from *Expert Systems with Applications*, 38, Liu, J. et al. Self-organized combinatorial optimization. 10532–10540, Copyright 2011, with permission from Elsevier.)**

where $r(k)$ is the total number of the kth nearest neighbors for all forward-directed edges in a feasible tour, that is, $r(k) = |X(k)|$ if $X(k) = \{x_i | s(x_i) = k\}$. Obviously, $p(k) \in [0, 1]$, and its sum will be equal to one. Figure 3.3 shows the kth-NNDs $p(k)$ of optimal tours on a set of 10 benchmark Euclidean TSP instances, with sizes ranging from 51 to 2392 nodes.

As shown in Figure 3.3, each kth-NND $p(k)$ is approximately an exponential decay function of the neighboring rank k. Numerous experiments have indicated that the optimal solutions of almost all benchmark TSP instances in TSPLIB conform to this qualitative characterization of microscopic exponential distribution.

It is worth pointing out that the above analytical results of microscopic distributions are very useful both to mathematicians and to computer scientists. From the viewpoint of mathematics, the maximum neighboring rank $K_{max} = \max\{s(x_i)\}$ for all i ($1 \leq i \leq n$) in the optimal solution is very useful to reduce the dimension of the state space. Taking pr2392 as an example, if its maximum neighboring rank $K_{max} = 20$ can be approximately estimated, the solution space represented by discrete-state variables can be significantly reduced from $(n - 1)^n$ to 20^n, and thus, a low-dimensional state space can be obtained before being fed into mathematical tools. From the viewpoint of computer science, these analysis results are also of great value for the design of an effective self-organized computing method.

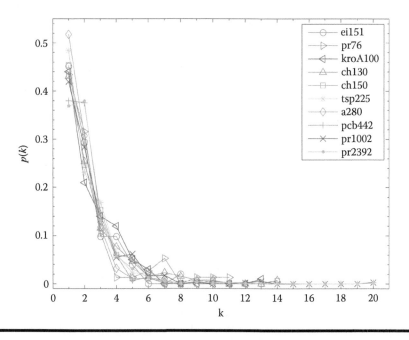

Figure 3.3 The kth NNDs $p(k)$ of optimal tours from eil51 to pr2392. (Reprinted from *Expert Systems with Applications*, 38, Liu, J. et al. Self-organized combinatorial optimization. 10532–10540, Copyright 2011, with permission from Elsevier.)

3.2.4 Neighborhood and Fitness Network

After unveiling the microscopic characteristics of optimal solutions, the solution space of COPs can now be analyzed further, and a reasonable searching dynamics can be designed to reach the optimal solution. In a fundamental sense, the effectiveness of combinatorial optimization methods has a close relationship with the underlying neighborhood definition. The neighborhood mapping, also called move class, is usually used to construct better solutions from the current solution. For any possible solution $s \in S$, if $N(s) \subseteq S$ is defined as a neighborhood function, all those feasible solutions in $N(s)$ are called neighbors of the solution s. With respect to a specific neighborhood $N(s)$, we call s' a strict locally optimal solution if the objective function $F(s') \leq F(s)$, $\forall s \in N(s')$. Obviously, the globally optimal solution s^*, $F(s^*) \leq F(s)$, $\forall s \in S$, belongs to the set of locally optimal solutions. If we can find out all the locally optimal solutions, the global optimum will be immediately obtained.

In order to analyze the computational complexity of search in a solution space, the concept of fitness landscape has been extensively applied to COPs (Altenberg, 1997; Hordijk, 1997; Merz and Freisleben, 2000; Reidys and Stadler, 2002). Generally speaking, if the dimension of search space is δ, that is, the cardinality of neighborhood $|N(s)| = \delta$, by adding an extra dimension that represents the fitness of each solution, the dimension of the fitness landscape will be equal to $\delta + 1$. Consequently, a hyperdimensional landscape with peaks and valleys can be used to characterize the search space of optimization problems. Its ruggedness directly reflects the search complexity from an initial solution to the global optimum. If a fitness landscape is highly rugged, it will be difficult to search it for a better solution due to the lower correlation among neighboring solutions (Hordijk, 1997; Merz, 2004). To study the properties of fitness landscapes, Kauffman (1993) developed a problem-independent *NK*-model with a tunable parameter. In *NK*-model, the symbol N represents the number of entities of a computational system, for example, cities in a TSP instance or genes in a genotype, and K reflects the degree of interactions among entities. More specifically, the local fitness contribution of each entity i is decided by the binary states of itself and K other interacting entities $\{i_1, ..., i_K\}$ $\subset \{1, ..., i-1, i+1, ..., N\}$. Thus, the global fitness of a solution $s = x_1, ..., x_N$ can be mathematically represented as follows:

$$F(s) = \frac{1}{N} \sum_{i=1}^{N} f_i(x_i; x_{i_1}, ..., x_{i_K}) \tag{3.9}$$

The local fitness f_i of entity i depends on its value x_i and the values of K other entities $x_{i_1}, ..., x_{i_K}$. In simulation, the local fitness function $f_i: \{0, 1\}^{K+1} \to I\!R$ assigns a random number from a uniform distribution between 0 and 1 to each of its inputs (Merz, 2004). With this *NK*-model, the ruggedness of fitness landscapes can be tuned from smooth to rugged by increasing the value of K from 0 to $N - 1$. When

K is small, the global fitness difference between neighboring solutions will be relatively small, and heuristics can easily find better solutions using correlated gradient information. When K is large, a large number of entities between neighboring solutions will possess different states that will greatly influence the global fitness. When $K = N - 1$, the landscape will be completely random, and the local fitness of all entities will be changed, even if we update only one entity's state. In this extreme case, no algorithm is substantially more efficient than exhaustive search. However, in practice, almost all computational systems are probably not as complex as the worst-case condition of $K = N - 1$ in the NK-model, for example, in the TSP, the maximum neighboring rank $K_{max} \ll N$. Therefore, it leaves us sufficient space to develop reasonable and efficient optimization methods, and hence to tackle those theoretically intractable computational problems.

Although the notion of the fitness landscape is useful to characterize the complexity of search space, it is impossible to visualize a solution space when the dimension of the neighborhood is higher than 2. Thus, in this chapter, we utilize the concept of the fitness network to represent the structure of solution spaces. For a given COP, the fitness network can be defined by a triple (S, F, N):

- A set of nodes (solutions) s on solution space S
- A fitness function $F: s \to R$
- A neighborhood function $N(s)$, which decides the structure of the fitness network

Thus, the fitness network can be interpreted as graph $G = (V, E)$ with vertex set $V = S$ and edge set $E = \{(s, s') \in S \times S | s' \in N(s)\}$. If the move class is a reversible, that is, if $s' \in N(s)$, then also $s \in N(s')$, and the fitness network is an undirected network; otherwise it is a directed network. On this complex network of search space, the out-degree of each nodes s is the cardinality of its neighborhood $|N(s)|$.

Let us take the illustrative fitness network of Figure 3.4 as an example. The hypothetical fitness network corresponds to a solution space with $(0, 1)$ binary sequences of length 4. In this fitness network, we have the Hamming distances $d(s, s') = 1$; for all neighboring solutions s and s', the height in the vertical direction reflects the fitness value, that is, the height of a node represents the fitness of the solution associated with it. Thus, the objective of optimization is to find the globally optimal solution "0011" in the bottom of the fitness network.

Based on the representation of the fitness network, the optimization dynamics can be described in terms of a searching trajectory navigating through the network in order to find the lowest-fitness node. Therefore, a neighborhood-based algorithm for combinatorial optimization can be essentially characterized by the neighborhood structure $N(s)$ and its searching dynamics in the fitness network.

In order to discuss whether or not an algorithmic solution can obtain the optimal solution from an arbitrary initial solution, let us first consider some properties of the fitness network.

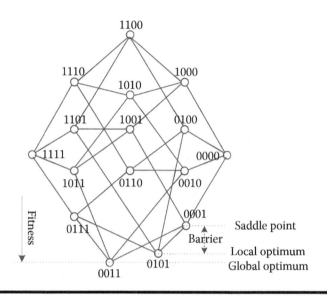

Figure 3.4 A hypothetical fitness network on the Hamming graph of binary sequences of length 4. (Reprinted from *Expert Systems with Applications*, **38**, Liu, J. et al. Self-organized combinatorial optimization. 10532–10540, Copyright 2011, with permission from Elsevier.)

Definition 3.3

In a fitness network, solution s' is *reachable* from s, if there exists at least a searching trajectory, $s = s_0, \dots, s_k, s_{k+1}, \dots, s_m = s'$, satisfying that $s_{k+1} \in N(s_k)$, $k = 0, \dots, m - 1$.

If a node is reachable from another node, then, there exists a set of consecutive nodes along a sequence of edges between them. Also, if node s_2 is reachable from node s_1, and node s_3 is reachable from node s_2, then, it follows that node s_3 is reachable from node s_1. The definition of reachability is applicable to both directed and undirected fitness networks.

Definition 3.4

A fitness network is called *ergodic*, if all other solutions are reachable from an arbitrary solution s ($\forall s \in S$).

Here, the term "ergodic," as adopted from physics, refers to all microstates that are accessible. Theoretically speaking, a probabilistic searching method should guarantee that the optimal solution is reachable from its initial solution s. Since we usually construct initial solutions by nondeterministic methods, the ergodicity of a fitness network is a necessary condition for finding the optimal solution.

In a fitness network as discussed above, local optima (such as solution "0101" in Figure 3.4) are undoubtedly barriers on the way to the optimal solution. Suppose that solution s_1 and s_2 are two local optima, and S^t denotes the solution set in a reachable trajectory from s_1 to s_2, the fitness barrier separating s_1 and s_2, as discussed by Reidys and Stadler (2002), can be defined as follows:

$$F_b[s_1, s_2] = \min\{\max[F(s)|s \in S^t]|S^t : reachable\, trajectories\, from\, s_1\, to\, s_2\} \quad (3.10)$$

A critical point $s \in S$ satisfying the min–max condition (3.10) is called a saddle point of the fitness network (e.g., solution "0001" in Figure 3.4). Then, the fitness barrier enclosing a local minimum s_1 can be evaluated by the height of the lowest saddle point between s_1 and a more favorable local minimum s_2 (Kern, 1993; Reidys and Stadler, 2002). Mathematically, it can be represented as follows:

$$F_b(s_1) = \min\{F[s_1, s_2] - F(s_1)|s_2 : F(s_2) < F(s_1)\} \quad (3.11)$$

This depth indirectly reflects the difficulty of escaping from the basins of attraction of the current local optimum s_1. As a rule of thumb, the quality of local optimal solutions will be improved, if we increase the size of the neighborhood. It can also be reflected by Equations 3.10 and 3.11 that an increased set of reachable trajectories may also decrease the depth of a local optimum. However, a larger neighborhood does not necessarily produce a more effective optimization method due to computational complexity, unless one can search the larger neighborhood in a more efficient manner.

3.2.5 Computational Complexity and Phase Transition

As mentioned before, NP-complete COPs theoretically require exponential time to find a globally optimal solution. However, it is possible for us to solve a typical case in a more efficient manner. In order to characterize the typical-case computational complexity, the concept of phase transition in statistical physics has been successfully applied to the analysis of COPs (Cheesman et al., 1991; Monasson et al., 1999; Achlioptas et al., 2005). Phase transition refers to a drastic change in some macroscopic properties, when a control parameter crosses a critical value, for example, water changing from ice (solid phase), to water (liquid phase), and to steam (gas phase) abruptly, as the temperature increases. Analogous to physical systems, it has been shown that many combinatorial decision problems also possess the phenomena of phase transition. Typical examples include the graph-coloring problem (Cheesman et al., 1991), K-satisfiability problem (Martin et al., 2001), and TSP (Gent and Walsh, 1996; Zhang, 2004), among others.

As a manifestation of phase-transition theory in combinatorial optimization, here, we discuss how the fundamentals of phase transition can be used to statistically analyze the solution space of COPs. As an illustration, we take a random two-dimensional Euclidean TSP instance, for example, with placing $n = 10$ cities on a

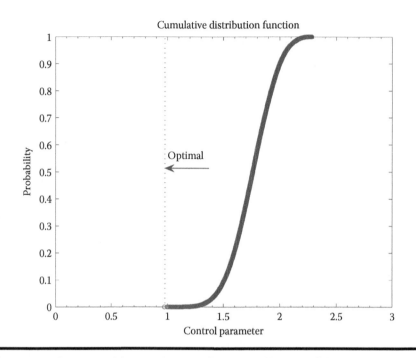

Figure 3.5 Phase transition on the cumulative distribution of dimensionless control parameters. (Reprinted from *Expert Systems with Applications*, 38, Liu, J. et al. Self-organized combinatorial optimization. 10532–10540, Copyright 2011, with permission from Elsevier.)

square of area A, and enumerate the length L of all $(n-1)!/2$ possible tours. For each solution, we calculate its dimensionless control parameter $L/\sqrt{n \cdot A}$ (Gent and Walsh, 1996). Figure 3.5 shows the phase transition on the probability distribution of all dimensionless control parameters. In the figure, the cumulative distribution function directly reflects the probability that the solution space takes on solutions less than or equal to a specific control parameter.

The observation of the phenomenon of phase transition can be shown in Figure 3.5, and we can approximately divide the solution space into two regions: (i) one region is underconstrained among entities, in which the density of solutions is high, thus making it relatively easy to find a solution; (ii) another region is overconstrained, in which the probability of the existence of a solution is toward zero, and it is very difficult to obtain a near-optimal solution, especially the optimal solution, which is located on the boundary of phase transition. Since the cumulative distribution is analogous to Gaussian distribution, it means that most of candidate solutions are located in the middle level of the fitness network.

Based on the above theoretical modeling, simulation, and analysis, we can note that the computational complexity of COPs always occurs on the boundary of phase transition, and therefore, a majority of solution samplings in a searching trajectory should focus on the bottom of the fitness network. A reasonable and effective optimization process for addressing hard COPs should consist of two types of steps: one for finding a new-optimal solution as efficiently as possible, and another for repeatedly escaping from one local optimum to another with enough fluctuated explorations.

3.3 Self-Organized Optimization

Following the line of analysis and discussions in the preceding section, a self-organized computing method for solving COPs is presented in this section. This optimization algorithm is inspired by the characteristics of collective emergent behavior in a dynamical system through the self-organization of its interacting entities.

3.3.1 Self-Organized Optimization Algorithm

In recent years, self-organization has been recognized as one of the most important research themes in studying the evolving process of complex dynamical systems. In self-organizing systems, the global behaviors seem to emerge spontaneously from local interactions among entities. A simple self-organization model was proposed by Bak and Sneppen (1993) for studying biological evolution. In the BS model, the evolution of a multientity system is driven by its extremal dynamics. At each iterative step, the entity with the worst fitness is selected to update its state, and simultaneously, the states of their neighbors will also be changed due to the triggered local interactions. After a number of update steps, the local fitness of almost all entities will transcend a threshold value, and this system will reach a highly correlated critical state, in which a little change of one entity may result in chain reactions and reorganize major parts of the complex system (Bak, 1996).

As analyzed in Sections 3.2.2 and 3.2.3, the combinatorial optimization solution can also be optimized by improving extremal local variables. Thus, the self-organized process driven by extremal dynamics can be used as an effective optimization dynamics. In the case of the TSP, the self-organized optimization algorithm under study can be given as follows (Liu et al., 2011):

Self-Organized Optimization Algorithm

   Step 1. Initialization
      Randomly construct an initial TSP tour s, which can be naturally represented as a permutation of n cities (path representation). Let s_{best} represent the best solution found so far, and set $s_{best} \leftarrow s$.

Step 2. For the "current" solution s:
   i. Calculate the local fitness f_i according to Equation 3.3 for each city i $(1 \leq i \leq n)$
   ii. Determine the order of all cities $\{\pi(1), \ldots, \pi(k), \ldots, \pi(n)\}$, in which $f_{\pi(1)} \geq \cdots \geq f_{\pi(k)} \geq \cdots \geq f_{\pi(n)}$, that is, the rank k is going from $k = 1$ for the city with the maximum local fitness to $k = n$ for the city with the minimum local fitness
   iii. Select a city $\pi(k)$ with the kth "high" local fitness by a scale-invariance probability distribution $P_k \propto k^{-\alpha}$ $(1 \leq k \leq n)$, where α is an adjustable parameter
   iv. Update the state of the selected city $\pi(k)$ by the 2-*opt* move class (Gutin and Punnen, 2002), and then, a neighborhood $N(s)$ of the current solution s is constructed
   v. Calculate the global fitness $F(s)$ of all solutions in the neighborhood $N(s)$ and sequence all solutions by the descending order of global fitness. If the best solution s' in the neighborhood is better than the current solution, that is, $F(s') - F(s) < 0$, accept $s \leftarrow s'$, and set $s_{best} \leftarrow s'$; otherwise, select a neighbor s' having the lth "low" global fitness by another power-law distribution $P_l \propto l^{-\beta}$ $(1 \leq l \leq |N(s)|)$ to replace the current solution, where β is another power parameter, and $|N(s)|$ is the cardinality of the neighbor $N(s)$.

Step 3. Repeat at step 2 until the termination criterion is satisfied
Step 4. Return s_{best} and $F(s_{best})$

Generally speaking, the power parameter α is used to control the self-organized optimization process. For $\alpha = 0$, the updated entities are randomly selected, and for $\alpha \to \infty$, only the extremal entity with the worst local fitness is forced to change at each step. Numerous experiments have shown that a value of $\alpha \approx 1 + 1/\ln(n)$ seems to provide the best results on self-organized systems (Boettcher and Percus, 2003; Chen et al., 2007). Another parameter β is used to adjust the hill-climbing ability on the basin of local optima. It is worth noting that calculating the global fitness of a solution at step 2 (v) can be done more efficiently by catching and computing those entities being updated rather than computing all of them from scratch. Usually, a predefined iterative number or *CPU* time can be selected as a termination criterion.

By applying the above-proposed self-organized optimization algorithm to the TSP instance of kroA100, where the parameters $\alpha = 1.25$, $\beta = 2.75$, the typical searching dynamics of the self-organized optimization algorithm has been plotted in Figure 3.6 (Liu et al., 2011).

It can be seen from Figure 3.6 that a near-optimal solution can be obtained sufficiently fast (as shown in the inset of Figure 3.6) by the initial greedy self-organizing process, and then, the enough fluctuated search near the global optimum can help to escape from the basin of the current local optimum and explore new local optima in the fitness network.

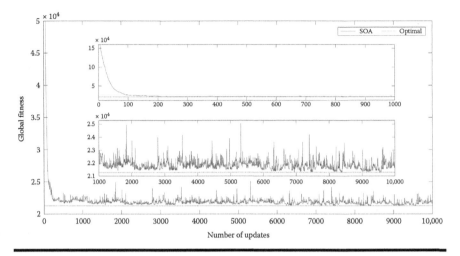

Figure 3.6 The searching dynamics produced by the proposed self-organized optimization algorithm. (Reprinted from *Expert Systems with Applications*, 38, Liu, J. et al. Self-organized combinatorial optimization. 10532–10540, Copyright 2011, with permission from Elsevier.)

The extent of fluctuations, which reflect the ability of hill climbing near the bottom of the fitness network, can be adjusted by the parameter β. As we increase the value of β, the fluctuations become gradually reduced. While $\beta \rightarrow \infty$, there are no fluctuations, and the algorithm will directly converge into a local optimum. The experiments also show that the proposed algorithm provides superior performance when $\beta \approx 2.75 \pm 0.25$ for the example problems under simulation.

According to the analytical results in Section 3.2, the self-organized optimization dynamics combining the greedy self-organizing process and the fluctuated explorations is an effective search process in complex fitness networks. Furthermore, in order to validate the self-organized behavior of the proposed algorithm, we have randomly generated a large-scale Euclidean TSP instance, with $n = 2048$, and then statistically analyzed the optimized results of our optimization algorithm. The kth-NND of an optimized result is presented in Figure 3.7 (Liu et al., 2011).

As shown in Figure 3.7, the kth-NND for the optimized result obtained by the self-organized optimization algorithm is a nearly perfect exponential distribution. In the exponentially fitted function $p(k) = ae^{-bk}$, the coefficients $a = 0.6791 \pm 0.0142$, $b = 0.5173 \pm 0.0115$, with 99% confidence bounds.

3.3.2 Comparison with Related Methods

In the past few decades, different nature-inspired or stochastic methods have been proposed for tackling COPs. In the following sections, three classical stochastic search algorithms are revisited for purposes of comparison and analysis.

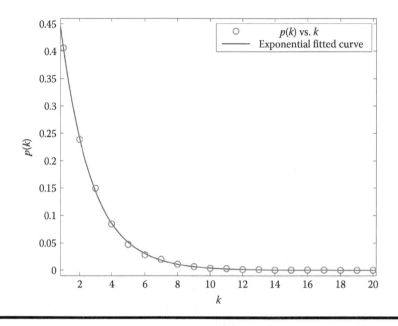

Figure 3.7 Optimization result from the self-organized optimization algorithm. (Reprinted from *Expert Systems with Applications*, 38, Liu, J. et al. Self-organized combinatorial optimization. 10532–10540, Copyright 2011, with permission from Elsevier.)

3.3.2.1 Simulated Annealing

SA is a stochastic search technique that simulates the thermodynamical process of annealing in metallurgy. At each step, the SA algorithm decides whether to replace the current solution according to the Boltzmann probability $\exp(-\Delta E/T)$, where ΔE is the fitness difference between the current solution and a neighboring solution. When the control temperature T is large, a system configuration with worsening energy may be accepted with high probability. As the parameter is sufficiently close to zero by an exponential cooling schedule, the SA algorithm will no longer accept the uphill moves, and then an approximate solution is assumed to be obtained. Since numerous microstates are sampled far from the optimal solution, where the probability of solutions is relatively high, and a feasible solution is easily obtained, as discussed in Section 2.3.2, the SA optimization dynamics is probably not as efficient as the currently studied self-organized optimization dynamics.

3.3.2.2 Genetic Algorithm

GA is a population-based searching algorithm based on the evolutionary theory of natural selection. Generally speaking, the algorithm starts from a population of candidate solutions, and each individual is evaluated in terms of global fitness. At each generation, a set of individuals are probabilistically selected from the current population according to their global fitness, and updated through crossover and mutation operators for creating a new population of candidate solutions. As compared with our self-organized optimization algorithm, the GA optimization dynamics seems to be a purposeless generation-test process since it mimics evolution at the global system-level solution, and does not go deep into the microscopic entities of the computational system.

3.3.2.3 Extremal Optimization

Although both EO and our proposed algorithm are inspired by the self-organized process of the BS evolution model, two significant differences should be pointed out:

   i. The definition of local fitness differs. For example, in a TSP tour, the local fitness of city i is defined as $f_i = 3/(p + q)$, if it is connected to its pth and qth nearest neighbors, respectively (Boettcher and Percus, 2000). Obviously, it is not consistent with the global fitness, as discussed above, and it is hardly possible to obtain the global optimum.
   ii. The method for selecting a new solution s' to replace the current solution s is different. It is more like a physical process rather than optimization.

3.3.3 Experimental Validation

In order to further validate the effectiveness and efficiency of the proposed self-organized optimization algorithm, a set of 10 benchmark Euclidean TSP instances is selected, with sizes ranging from 51 to 2392 nodes, as the test benchmarks. All of them are available from Reinelt's TSPLIB (Reinelt, 1991).

In our computational tests, SA, τ-EO, and the proposed self-organized algorithm (SOA) use the same 2-opt-move class at each iterative step. SA usually requires to tune several parameters case by case for obtaining a proper decreasing-temperature schedule. The parameters of τ-EO and SOA are approximately set as $\tau = \alpha \approx 1 + 1/\ln(n)$. GA employs the superior edge recombination crossover operator and displacement mutation operator as concluded in Potvin (1996) and Larrañaga (1999). All these algorithms are encoded in C++ and implemented on a Pentium IV (2.4 GHz).

Table 3.1 shows the comparative computational results. The mean value $\bar{F}$ and standard deviation σ of error (%) = $100 \times (F(s) - F(s^*))/F(s^*)$, where $F(s)$ denotes the solution found by a specific algorithm in each of the 10 trials, are used to measure the performance and effectiveness of optimization algorithms. The computation time (t_{CPU}) is used to evaluate the efficiency of optimization algorithms.

It can be seen from Table 3.1 (Liu et al., 2011) that the proposed self-organized optimization method provides superior optimization performances in both

Table 3.1 Comparison of the Computational Results from SA, GA, τ-EO, and SOA

Problem	$F(s^*)$	SA			GA			τ-EO			SOA		
		$\bar{F}$	σ	t_{CPU}	$\bar{F}$	σ	t_{CPU}	$\bar{F}$	σ	t_{CPU}	$\bar{F}$	Σ	t_{CPU}
eil51	426	0.00	0.00	2	0.00	0.00	25	0.00	0.00	2	0.00	0.00	2
pr76	108159	0.21	0.24	3	0.00	0.00	30	0.00	0.00	3	0.00	0.00	3
kroA100	21282	0.69	0.45	5	0.05	0.07	65	0.12	0.09	5	0.02	0.04	5
ch130	6110	0.74	0.54	10	0.29	0.25	102	0.31	0.34	10	0.13	0.15	10
ch150	6528	0.92	0.61	12	0.38	0.40	149	0.51	0.42	12	0.22	0.20	12
tsp225	3916	1.42	0.77	21	0.66	0.35	212	0.65	0.49	21	0.45	0.34	21
a280	2579	1.45	0.71	45	0.64	0.51	362	0.53	0.46	45	0.32	0.31	45
pcb442	50778	2.07	1.05	125	1.53	0.62	1011	1.52	0.68	125	1.15	0.53	125
pr1002	259045	3.16	1.24	300	1.89	0.80	2957	1.95	0.76	300	1.77	0.69	300
pr2392	378032	4.45	2.23	900	3.83	1.51	5326	3.55	1.54	900	3.21	1.25	900

Source: Reprinted from *Expert Systems with Applications*, 38, Liu, J. et al. Self-organized combinatorial optimization. 10532–10540, Copyright 2011, with permission from Elsevier.

effectiveness and efficiency, and outperforms the state-of-the-art SA, GA, and EO for all 10 test instances.

3.4 Summary

In this chapter, an analytic characterization for COPs is developed from the self-organizing system's point of view. Based on the definitions of the discrete-state variable and local fitness, the empirical observation of the microscopic distribution is discussed with respect to an optimal solution. A notion of fitness network is also introduced in order to characterize the structure of the solution space, and the searching complexity of hard optimization problems is statistically analyzed from the characteristics of phase transition.

The performance of a self-organized optimization method was presented as well as demonstrated for solving hard computational systems. For different optimization problems, only the definitions of a consistent local fitness function and the neighboring rules are required.

Based on our analytic and algorithmic discussions, it is noted that this chapter offers new insights into, as well as may inspire, further studies on the systematic and microscopic analysis of NP-complete problems. The chapter paves the way for utilizing self-organization in solving COPs. The future work will involve in-depth studies on the optimization dynamics of computational systems by means of introducing the fundamentals of self-organizing systems.

MODIFIED EO AND INTEGRATION OF EO WITH OTHER SOLUTIONS TO COMPUTATIONAL INTELLIGENCE

II

Chapter 4

Modified Extremal Optimization

4.1 Introduction

To improve the performance of the original τ-EO algorithm and extend its application area, this chapter presenting some modified versions is organized as follows: In Section 4.2, the modified EO with extended evolutionary probability distribution as proposed by Zeng et al. (2010b) is discussed. Section 4.3 presents a multistage EO (MSEO) with dynamical evolutionary mechanism (Zeng et al., 2010c). In Section 4.4, another modified version called the backbone-guided EO algorithm proposed by Zeng et al. (2012), which utilizes the backbone information that guides the search process of EO approaching the optimal region more efficiently, is described. Furthermore, Section 4.5 presents PEO algorithm (Chen et al., 2006). Finally, the summary of this chapter is given in Section 4.6.

4.2 Modified EO with Extended Evolutionary Probability Distribution

Similar to other popular EAs, for example, GA, evolutionary probability distributions are used to select those variables with worst fitness for mutation or the solutions from the neighborhood. The update of the current solution is critical for the performances of τ-EO and its modified versions, such as SOA (Chen et al., 2007), and reference energy extremal optimization (REEO) (Zhang and Zeng, 2008). In most of the research works regarding EO algorithms and their applications, the power-law distribution is widely accepted as an effective evolutionary probability distribution in the original τ-EO algorithm and its modified versions. Hoffmann et al. (2004) have proved that using any linear performance measure would make a strictly monotonic distribution over fitness ranks k, such as the power-law distribution advocated by Boettcher and Percus (2000, 2001a). It will be optimal only if all selection distributions perform equally well. Motivated by this viewpoint and the optimization scheme of SOA, a more general modified EO framework with extended evolutionary probability distributions was proposed by Zeng et al. (2010b) to explore complex configuration spaces of hard COPs. In addition, this modified EO framework is also motivated by the degree distributions of complex networks in the emergent network science research domain during the past decade (Albert and Barabási, 2000; Strogatz, 2001; Barabási and Oltvai, 2004; Barabási, 2007; Clauset et al., 2009). Generally speaking, there are three main types of networks, including "random," "scale-free," and "hierarchical" networks, of which degree distributions correspond to exponential (or Poisson), power-law, and power law with exponential cutoff distributions, respectively. The basic idea behind the proposed method is the utilization of exponential distributions or hybrid ones (e.g., power laws with exponential cutoff) to replace the power laws in SOA.

4.2.1 Evolutionary Probability Distribution

Originally developed from the fundamentals of statistical physics, EO is based on the extremal dynamics of SOC, which describes a class of systems that have a critical point as an attractor. Their macroscopic behavior exhibits the spatial and temporal scale-invariance characteristics of the critical point of a phase transition. The inherent nonequilibrium dynamics of EO is the result of the fitness rank ordering and power-law distribution-based updating method adopted by τ-EO. The choice of a power-law distribution for $P(k)$ ensures that no regime of fitness is excluded from further evolution, since $P(k)$ varies in a gradual, scale-free manner over ranks k (Boettcher and Percus, 1999). Clearly, the evolutionary mechanism including the rank ordering and updating rules can be viewed as a "memory," which allows EO to retain well-adapted pieces of a solution. In this sense, EO mirrors one of the crucial properties noted in a well-known coevolutionary model called the BS model. From the perspective of optimization, EO successively replaces extremely undesirable variables of a single suboptimal solution with new ones randomly or according to power-law distribution.

In fact, the power-law distribution has attracted increasing attention over the years for the following two main reasons. The first one is its mathematical properties which sometimes lead to surprising physical consequences. Second, a variety of experimental studies show that the distributions occur in a diverse range of natural and man-made phenomena (Clauset et al., 2009), for example, the intensities of earthquakes, the populations of cities, and the sizes of power outages. In Boettcher and Frank (2006), τ-EO is characterized by a power-law distribution over the fitness ranks k,

$$P_\tau(k) = \frac{\tau - 1}{1 - n^{1-\tau}} k^{-\tau} \quad (1 \le k \le n) \tag{4.1}$$

where τ is a positive constant, called critical exponent of power laws and often $\tau > 1$. For a fixed dimension n in a specified problem, it is obvious that the coefficient $\tau - 1/1 - n^{1-\tau}$ is also a positive constant. Therefore, Equation 4.1 can be represented as the following general form:

$$Pp(k) \propto k^{-\tau} \quad (1 \le k \le n) \tag{4.2}$$

The scale-free property is common but not universal (Strogatz, 2001). This statement motivates us to explore other probability distributions to replace the power-laws-based evolution rules in EO-similar methods.

As a negative example, μ-EO (Boettcher and Frank, 2006) was introduced to demonstrate the usefulness of power-law distributions of τ-EO. In detail, μ-EO is characterized by an exponential distribution over the fitness ranks k,

$$P_\mu(k) = \frac{e^\mu - 1}{1 - e^{-\mu n}} e^{-\mu k} \quad (1 \le k \le n) \tag{4.3}$$

where μ is a positive constant. For a fixed dimension n in a specified problem, it is obvious that the coefficient $e^\mu - 1/1 - e^{-\mu n}$ is also a positive constant, so Equation 4.3 can be represented as the following general form:

$$Pe(k) \propto e^{-\mu k} \quad (1 \le k \le n) \tag{4.4}$$

In general, complex networks are categorized into three types: random, scale-free, and hierarchical networks (Barabási and Oltvai, 2004), in which the probability $P(k)$ that a node is connected to k other nodes is bounded, following exponential, power-law, and power law with exponential cutoff distributions, respectively.

Empirical measurements, however, indicate that real networks deviate from simple power-law behavior (Barabási, 2007). The most typical deviation is the flattening of the degree distribution at small values of k, while a less typical deviation is the exponential cutoff for high values of k. Thus, a proper fit to the degree distribution of real networks has the form

$$P_h(k) \propto (k + k_0)^{-\gamma} e^{-k/k_x} \quad (1 \le k \le n) \tag{4.5}$$

where k_0, k_x represent the small-degree cutoff and the length scale of the high-degree exponential cutoff, respectively, and γ is a positive constant. The scale-free behavior of real networks is therefore evident only between k_0 and k_x. This distribution can be reformulated as the following form termed "hybrid distribution" here:

$$Ph(k) \propto k^{-h} e^{-hk} \quad (1 \le k \le n) \tag{4.6}$$

where h is a positive constant.

The cumulative probabilities of these different types of distributions, including power-law, exponential, and hybrid ones, are given as follows (Zeng et al., 2010b):

$$Qp(K) = \frac{\sum_{k=1}^{K} k^{-\tau}}{\sum_{k=1}^{n} k^{-\tau}} \quad (1 \le K \le n) \tag{4.7}$$

$$Qe(K) = \frac{\sum_{k=1}^{K} e^{-\mu k}}{\sum_{k=1}^{n} e^{-\mu k}} \quad (1 \le K \le n) \tag{4.8}$$

$$Qh(K) = \frac{\sum_{k=1}^{K} e^{-hk} k^{-h}}{\sum_{k=1}^{n} e^{-hk} k^{-h}} \quad (1 \le K \le n) \tag{4.9}$$

In fact, for the specific finite size problems, for example, $n = 532$, these probability distributions and their cumulative ones are shown in Figure 4.1 (Zeng et al., 2010b). When the parameters μ, τ, and h are assigned small values, for example, $\mu = 0.037$, $\tau = 1.15$, and $h = 0.05$, the probabilities $Pe(k)$, $Pp(k)$, and $Ph(k)$ for $n = 532$ decrease significantly as k varies in [1,100] while unobviously as k in [100, 532], but the specific distributions are different, which are shown in Figure 4.1a. By contrast, when the parameters are assigned large values, for example, $\mu = 0.52$, $\tau = 2.20$, and $h = 0.35$, $Pe(k)$, $Pp(k)$, and $Ph(k)$ for $n = 532$ decrease significantly as k varies in [1,10] while the changes are unobvious as k in [10,532] and the different specific distribution are shown in Figure 4.1b.

4.2.2 Modified EO Algorithm with Extended Evolutionary Probability Distribution

In TSP, the local fitness λ_i for city i and global fitness $C(S)$ for any possible solution S are defined as follows (Zeng et al., 2010b):

$$\lambda_i = d_i - \min_{i \ne j} d_{ij}, \quad i = 1, 2, \dots, n \tag{4.10}$$

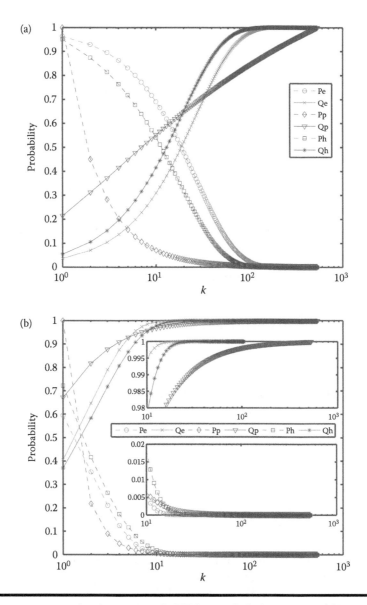

Figure 4.1 (See color insert.) Probabilities and their corresponding cumulative ones of power-law, exponential, and hybrid distributions for $n = 532$, of which (a) $\mu = 0.037$, $\tau = 1.15$, and $h = 0.05$; (b) $\mu = 0.52$, $\tau = 2.20$, and $h = 0.325$. (Reprinted from *Physica A*, 389 (9), Zeng, G. Q. et al. Study on probability distributions for evolution in modified extremal optimization, 1922–1930, Copyright 2010c, with permission from Elsevier.)

$$C(S) = \sum_{i=1}^{n} \min_{i \neq j} d_{ij} + \sum_{i=1}^{n} \lambda_i \qquad (4.11)$$

where d_i represents the length of the forward-directed edge starting from city i in a feasible solution S, and d_{ij} is the intercity distance between ith and jth cities.

Inherited with the optimization scheme of SOA (Chen et al., 2007), a modified EO framework with extended evolutionary probability distributions (Zeng et al., 2010b) was proposed in this section by replacing the original power-law distributions used in SOA with other probability distributions, such as exponential or hybrid ones. The details of this framework are described as follows:

1. Initialize a random configuration S and set $S_{best} = S$.
2. For the current configuration S,
   a. Evaluate local fitness value λ_i for each variable i and rank all the variables according to λ_i, that is, find a permutation Π_1 of the labels i such that $\lambda_{\Pi_1(1)} \geq \lambda_{\Pi_1(2)} \geq \cdots \geq \lambda_{\Pi_1(n)}$.
   b. Select a rank $\Pi_1(k_1)$ according to a probability distribution $P_1(k_1)$, $1 \leq k_1 \leq n$ and update the state of the selected city $\Pi_1(k_1)$ by 2-opt move (Helsgaun, 2000), and then the neighborhood $N(S)$ of the current solution S is constructed.
   c. Calculate the global fitness $C(S)$ of all the configurations in $N(S)$ and rank them according to $C(S)$, that is, find a permutation Π_2 of the labels j such that $C_{\Pi_2(1)} \geq C_{\Pi_2(2)} \geq \cdots \geq$.
   d. $C_{\Pi_2(|N(S)|)}$, where $|N(S)|$ is the cardinality of the neighbor $N(S)$.
   e. Choose a new configuration $S' \in N(S)$ according to another probability distribution $P_2(k_2)$, $1 \leq k_2 \leq |N(S)|$.
   f. Accept $S := S'$ unconditionally.
   g. If $C(S) < C(S_{best})$, store $S_{best} = S$.
3. Repeat step 2 as long as desired.
4. Return S_{best} and $C(S_{best})$.

The modified EO algorithms under the above framework with different evolutionary probability distributions are shown in Table 4.1 (Zeng et al., 2010b). Here, power-law, exponential, and hybrid distributions are abbreviated as "P," "E," and "H," respectively. Note that the postfix of each algorithm defined in Table 4.1 denotes the types of probability distributions used in $P_1(k_1)$ and $P_2(k_2)$. For example, the MEO-HH algorithm represents a modified EO algorithm in which the hybrid distributions "H" are used both in $P_1(k_1)$ and $P_2(k_2)$. It is obvious that this framework can be viewed as a generalization of SOA since the updating rules are changed by using different probability distributions. In particular, when $P_1(k_1)$ and $P_2(k_2)$ follow power-law distributions, the proposed algorithm is SOA.

Table 4.1   Modified EO Algorithms with Different Evolutionary Probability Distributions

Algorithm	$P_1(k_1)$	$P_2(k_2)$
τ-EO	$k_1^{-\tau_1}$	–
SOA	$k_1^{-\tau_1}$	$k_2^{-\tau_2}$
MEO-EE	$e^{-\mu_1 k_1}$	$e^{-\mu_2 k_2}$
MEO-EP	$e^{-\mu_1 k_1}$	$k_2^{-\tau_2}$
MEO-EH	$e^{-\mu_1 k_1}$	$e^{-h_2 k_2} k_2^{-h_2}$
MEO-HE	$e^{-h_1 k_1} k_1^{-h_1}$	$e^{-\mu_2 k_2}$
MEO-HP	$e^{-h_1 k_1} k_1^{-h_1}$	$k_2^{-\tau_2}$
MEO-HH	$e^{-h_1 k_1} k_1^{-h_1}$	$e^{-h_2 k_2} k_2^{-h_2}$

Source: Reprinted from *Physica A*, 389 (21), Zeng, G. Q. et al. Multistage extremal optimization for hard travelling salesman problem. 5037–5044, Copyright 2010b, with permission from Elsevier.

4.2.3 *Experimental Results*

To demonstrate the effectiveness of the proposed MEO algorithms shown in Table 4.1, Zeng et al. (2010a,b) chose some well-known TSP instances from TSPLIB95 (available at http://www.iwr.uni-heidelberg.de/groups/comopt/software /TSPLIB95/) as a test bed.

These instances such as bier127, pcb442, and att532 (Schneider et al., 1996) have highly degenerated ground states, which means there are a very large amount of nearly optimal solutions quite close to global optima, a fact that leads to large difficulties in terms of many existing optimization methods. Table 4.2 (Zeng, 2011) shows the results obtained by the proposed algorithms and other physics-inspired optimization methods. f_b, f_m, and f_w are the best, mean, and worst fitness value of 10 runs, respectively, and f_o is the optimal value of the specific instance. The performance of these algorithms is evaluated by the following indices: the best error $e_b = (f_o - f_b)/f_o \times 100\%$, the average error $e_m = (f_o - f_m)/f_o \times 100\%$, and the worst error $e_w = (f_o - f_w)/f_o \times 100\%$. Clearly, the proposed MEO algorithms with different evolutionary probability distributions provide better performance than SOA, τ-EO, and SA under the same computational time. Furthermore, the comparative performance of MEO algorithms with other popular algorithms, such as SOA,

Table 4.2 Comparative Performance of MEO Algorithms, SOA, τ-EO, and SA for TSP Benchmark Problems with Highly Degenerated Ground States

Problem	bier127			pcb442			att532		
fo	118,293			50,778			86,729		
$t_{CPU}(s)$	14			115			165		
Algorithm	e_b (%)	e_m (%)	e_w (%)	e_b (%)	e_m (%)	e_w (%)	e_b (%)	e_m (%)	e_w (%)
SA	0.670	1.180	1.500	0.971	2.070	2.999	1.727	2.960	4.000
τ-EO	0.448	0.740	1.000	0.510	1.520	2.275	1.254	2.330	3.195
SOA	0.000	0.167	0.462	0.440	1.150	1.800	0.993	1.862	2.735
MEO-EE	0.000	0.065	0.200	0.097	0.869	1.607	0.396	1.057	1.614
MEO-EP	0.000	0.031	0.101	0.399	0.721	0.975	0.364	1.060	1.615
MEO-EH	0.000	0.042	0.175	0.037	0.760	1.150	0.512	0.985	1.859
MEO-HE	0.000	0.063	0.194	0.272	0.986	1.233	0.667	1.115	1.642
MEO-HP	0.000	0.032	0.100	0.435	0.814	1.140	0.615	0.995	1.623
MEO-HH	0.000	0.060	0.186	0.293	0.863	1.112	0.646	1.087	1.607

Source: Zeng, G. Q. 2011. Research on modified extremal optimization algorithms and their applications in combinatorial optimization problems. Doctoral dissertation, Zhejiang University, Hangzhou, China.

τ-EO, GA, and SA for other TSP benchmark instances from TSPLIB95 is shown in Table 4.3.

Although power-law-based evolutionary probability distribution is one of the main characteristics of SOA, the above experimental results (Zeng et al., 2010a,b; Zeng, 2011) on a variety of TSP benchmark instances have shown that the proposed MEO algorithms with exponential or hybrid distributions is superior to SOA, τ-EO, and SA. Furthermore, this study appears to demonstrate that the μ-EO with exponential distribution (Boettcher and Frank, 2006) can provide better performance than τ-EO at least for hard TSP, which can dispel the misconception of Boettcher and Frank (2006) that μ-EO with exponential distributions fails to perform well for hard optimization problems. From an optimization point of view, our results indicate that the power law is not the only proper probability distribution for EO-similar methods, the exponential and hybrid distributions may be other choices. In fact, the key idea behind the proposed algorithm (Zeng et al., 2010b) has extended to solve other optimization problems, for example, MAX-SAT (Zeng et al., 2011). The research results further demonstrate the effectiveness of the proposed MEO algorithms with extended evolutionary probability distributions.

Table 4.3 Comparison of MEO Algorithms with Other Popular Algorithms for TSP Benchmark Instances from TSPLIB95

(a) Part 1								
Problem	kroA100		ch130		ch150		tsp225	
f_o	21,282		6110		6528		3859	
$t_{CPU}(s)$	4		9		11		19	
Algorithm	e_m (%)	σ	e_m (%)	Σ	e_m (%)	Σ	e_m (%)	σ
SA	0.691	0.453	0.744	0.542	0.920	0.611	1.423	0.770
GA	0.255	0.107	0.486	0.310	0.552	0.429	0.991	0.452
τ-EO	0.124	0.090	0.312	0.343	0.509	0.420	0.647	0.491
SOA	0.023	0.044	0.134	0.151	0.223	0.202	0.478	0.355
MEO-EE	0.014	0.036	0.077	0.125	0.154	0.177	0.297	0.283
MEO-EP	0.010	0.032	0.070	0.118	0.142	0.165	0.286	0.249
MEO-EH	0.016	0.037	0.089	0.132	0.169	0.183	0.322	0.303
MEO-HE	0.013	0.034	0.085	0.135	0.166	0.180	0.311	0.314
MEO-HP	0.011	0.035	0.072	0.127	0.155	0.179	0.290	0.292
MEO-HH	0.015	0.037	0.078	0.129	0.147	0.167	0.303	0.286
(b) Part 2								
Problem	a280		lin318		rat783		pr1002	
f_o	2579		42,029		8806		259,045	
$t_{CPU}(s)$	41		69		380		710	
Algorithm	e_m (%)	σ	e_m (%)	σ	e_m (%)	σ	e_m (%)	σ
SA	1.454	0.708	1.799	0.813	2.941	1.010	3.162	1.246
GA	0.755	0.546	1.592	0.625	2.556	0.822	2.677	1.033
τ-EO	0.532	0.461	1.545	0.618	2.656	0.697	1.953	0.762
SOA	0.325	0.317	1.083	0.457	2.499	0.563	1.770	0.691
MEO-EE	0.212	0.245	0.919	0.372	2.011	0.492	1.244	0.468
MEO-EP	0.200	0.234	0.892	0.367	1.924	0.483	1.089	0.437

(Continued)

Table 4.3 (*Continued*)   Comparison of MEO Algorithms with Other Popular Algorithms for TSP Benchmark Instances from TSPLIB95

(b) Part 2 (Continued)								
Problem	a280		lin318		rat783		pr1002	
f_o	2579		42,029		8806		259,045	
$t_{CPU}(s)$	41		69		380		710	
Algorithm	e_m (%)	σ	e_m (%)	σ	e_m (%)	σ	e_m (%)	σ
MEO-EH	0.214	0.243	0.987	0.395	1.995	0.498	1.203	0.476
MEO-HE	0.211	0.238	1.032	0.403	2.016	0.501	1.231	0.432
MEO-HP	0.246	0.251	0.903	0.369	1.938	0.496	1.137	0.503
MEO-HH	0.282	0.277	1.066	0.421	2.000	0.505	1.711	0.617

Source: Zeng, G. Q. 2011. Research on modified extremal optimization algorithms and their applications in combinatorial optimization problems. Doctoral dissertation, Zhejiang University, Hangzhou, China.

4.3 Multistage EO

In all previous work concerning EO except the theoretical analysis (Hoffmann et al., 2004; Heilmann et al., 2004), the selection over the ranks of the degree of freedom for updating depends on a time-independent probability distribution. More specifically, in the power-law distribution adopted by all existing EO versions, the value of critical exponent τ is always fixed as a constant in the whole search process. In this section, Zeng et al. (2010c) proposed a novel method called MSEO by adopting different values of the control parameters in different stages. The main goal of this study is attempting to demonstrate that a dynamical probability distribution-based evolutionary mechanism is more effective than the traditional static strategy.

4.3.1 Motivations

In fact, there are two motivations behind the proposed MSEO method. One is the effect of the control parameter τ on the search dynamics of τ-EO. Specifically, when τ is small, the search is similar to a random walk. Conversely, it approaches to a deterministic local search, only updating those worst variables for very large values of τ. It is clear that the adjustable parameters are crucial to control the search fluctuations. Once the search gets trapped into a local optimum, it seems to be difficult to bring it to a better one if the control parameters are not adjusted. To illustrate the necessity of modifying the previous time-independent strategy of the control parameters, we design a simple experimental study on the performances

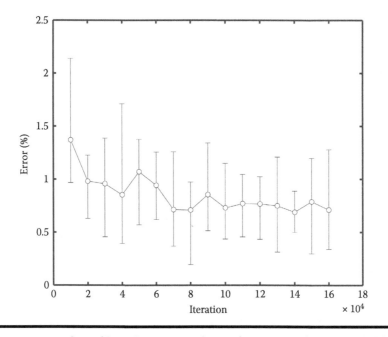

Figure 4.2 Number of iterations versus the performances of MEO-HH with the same initial configuration and the same control parameters for pcb442 instance. The error bars represent the best and worst performances over 10 independent runs. (Reprinted from *Physica A*, 389 (21), Zeng, G. Q. et al. Multistage extremal optimization for hard travelling salesman problem. 5037–5044, Copyright 2010b, with permission from Elsevier.)

of MEO-HH with the same initial configuration and the same value of control parameters but different runtime, that is, the number of iterations. The resulting performances are obtained by 10 independent runs and is shown in Figure 4.2 (Zeng et al., 2010c). Obviously, for the case with the same initial configurations and the same control parameters, the average and the best performances are the best at some critical number of iterations, that is, 80,000 iterations in this experiment. In other words, even though the runtime is extended, it may be difficult to further improve the performance of this algorithm after the critical runtime. In fact, similar phenomena can be observed in other experiments under this case.

From the above experiments, the intuition in improving the performances is to design a dynamical evolutionary mechanism. As a consequence, a possible improved method is to adjust the values of control parameters after some critical runtime. The question arising naturally is how to determine the critical values of runtime. Here, we will determine them empirically, for example, presetting them as some constant values in practice.

Another motivation of the proposed MSEO is our recent study (Zeng et al., 2010a) concerning the effects of initial configurations on the performances of modified EO. More specifically, for the same evolutionary mechanism, the algorithm with the initial configurations constructed by some heuristics is generally superior to that starting from random ones. This indicates that the quality of initial configurations plays an important role in governing the performances of modified EO algorithms. It should be noted that the optimal values of the control parameters applied to the algorithm starting from a biased initial configuration are different from those with a randomly selected one. Thus, varying the values of the control parameters during the different search processes is likely to improve the performances of a modified EO algorithm starting from different-quality initial configurations in different stages.

4.3.2 MSEO Algorithm

The key idea behind MSEO is to perform optimization by using different values of the parameters in different search stages. Generally speaking, optimization starts with randomly collecting some local optima by multistart techniques. In all later stages, it always selects the best configuration generated from the previous stage as the initial one for optimization in the current stage. The process repeats until the stopping criterion is satisfied. The scheme of the proposed method is as follows (Zeng et al., 2010c):

1. In the first stage, that is, $m = 1$, use a modified EO algorithm with the same evolutionary probability distribution parameters (p_{11}, p_{12}) starting from random or nearest-neighbor search (NNS)-based initial configurations (Zeng et al., 2010a) by N_1 different independent runs, where the number of iterations in each run is set as I_1 and obtain N_1 configurations.
2. For m from 2 to M, where M denotes the total number of stages.
3. Select the best one $s_{(m-1)b}$ from the stage $m - 1$ and set $s_{(m-1)b}$ as the initial configuration of the mth stage, that is, $s_{m0} = s_{(m-1)b}$.
4. In mth stage, use a modified EO with the evolutionary probability distribution parameters (p_{m1}, p_{m2}) starting from s_{m0} by N_m different independent runs, and the number of iterations in each run is set as I_m and obtain N_m configurations.
5. *End for* and output the finial N_M configurations and the corresponding global fitness.

Note that the modified EO algorithm used in each stage has several choices from Table 4.1. Of course, the quality of the selected algorithm in each stage has a direct influence on the final performances of MSEO. It should be also emphasized that we focus on illustrating how MSEO works in this chapter and the comparison of MSEO for different modified EO algorithms in each stage is an open subject in future research. Thus, for the sake of convenience, the

NNMEO-HH and MEO-HH are selected in the first stage and the rest of the stages, respectively, for tests in Section 4.3.3. Note that NNMEO-HH (Zeng et al., 2010a) is a modified EO algorithm with NNS-based initial configurations ("NN") and a hybrid distributions-based evolutionary mechanism in $P_1(k_1)$ and $P_2(k_2)$.

Obviously, the performances of the proposed method are governed by a set of control parameters, including M, N_m, I_m, p_{m1}, and p_{m2}. For simplicity, M, N_m, and I_m are predefined as constants here. Our study focuses on the method of varying the values of control parameters to enhance the solutions. It is evident that the control parameters h_{m1}, h_{m2} play a critical role in the performance of MSEO. In this sense, the key to MSEO is to change the values of the control parameters (p_{m1}, p_{m2}) in different stages. In general, the determination of control parameters used in the first stage is similar to that of the normal EO-similar method in the research works (Chen et al., 2007a; Zeng et al., 2010b). For the other stages, the values of the parameters used in the current stage are always larger than those in the last stage. This may be explained from the perspective of the "backbone" idea (Schneider et al., 1996). Specifically speaking, we expect some good components of the local minimum generated by the last stage can be frozen and the others are optimized in this stage, which indicates that the original problem reduces to a smaller size one. To achieve this goal, the larger values of the parameters in the current stages than the previous one are needed. As a result, adjusting control parameters in each stage to make the search approach the ground states as deeply as possible is similar to a robust backbone-based guided search method.

4.3.3 Experimental Results

4.3.3.1 The Simplest Case: Two-Stage EO

In order to demonstrate the effectiveness of the proposed method, Zeng et al. (2010c) chose some well-known TSP instances from TSPLIB95, for example, lin318, pcb442, att532, rat783, and pr1002 as the test bed. The studies by Schneider et al. (1996) show that pcb442 and att532 instances have highly degenerated ground stages, which indicate that the existence of a huge number of suboptimal solutions quite close to the global optima result leads to large difficulties in terms of many optimization algorithms. First, we consider the simplest case of the proposed method, that is, two-stage EO (TSEO). Here, MEO-HH is utilized in each stage. For simplicity, each stage in TSEO goes through 10 independent runs and each run iterates 90,000 and 10,000 steps in the first and second stage, respectively. For comparison, Zeng et al. (2010c) performed other physics-inspired algorithms, including classical SA, τ-EO, and a series of modified EO within the same runtime. Note that all the algorithms were implemented in Microsoft Visual Studio on a Pentium 1.86 GHz PC with dual-core processor T2390 and 2 GB RAM running

Windows Vista Basic systems. Here, τ-EO is implemented with a new definition of fitness rather than the original definition. In fact, τ-EO here outperforms the original one with nonlinear function of fitness (Boettcher and Percus, 2000). The resulting performances of these algorithms are shown in Table 4.4. Clearly, even TSEO being the simplest case of MSEO can provide much better performances than classical SA, τ-EO, and single-stage modified EO called NNMEO-HH under fine-tuning within the same runtime.

Figure 4.3 (Zeng et al., 2010c) shows the typical search dynamics of the algorithms including NNMEO-HH and TSEO for pcb442 instances. Obviously, the modified EO algorithms including NNMEO-HH and TSEO descend fast near the optimum but with different fluctuations. By comparing these search processes, we can observe that changing the values of the control parameters after some number of iterations makes the search of TSEO as far down in the energy landscape as possible while the static strategy adopted by NNMEO-HH do not. From the respective histograms, the fluctuation characteristics can be analyzed easily. The fluctuated dynamics depend on the updating probability distributions adopted by the proposed optimization algorithms. More specifically, NNMEO-HH follows a bell-shaped distribution while TSEO has a "good" cutoff. In other words, TSEO is more likely to approach lower states than NNMEO-HH under the same runtime. It is clear that the more the frequency of this cutoff, the better the performance that is obtained. From the viewpoint of mathematics, the search dynamics of these algorithms can be described and analyzed by the Markov process.

4.3.3.2 Complex Case

In general, given more stages, the better performance of an algorithm may be obtained, but longer runtime is required. Therefore, a reasonable approach is the tradeoff between the number of stages and the resulting performances. Here, Zeng et al. (2010c) studied more complex cases than the simple TSEO, for example, three-stage EO. In each stage, MEO-HH is responsible for the corresponding optimization. The number of iterations and the number of independent runs are prefixed as constants. The detailed results are shown in Table 4.5 (Zeng et al., 2010c). It is remarked that MSEO does not guarantee the search to reach the ground states (optimum) whereas it always makes the search more deeply than the corresponding single-stage algorithm. More importantly, the experimental results illustrate a simple dynamical probability distribution strategy used in MSEO may be more appropriate than the previous static mechanism.

In conclusion, the experimental results on the well-known hard TSP instances demonstrate the proposed MSEO may provide much better performances than other physics-inspired algorithms such as classical SA, τ-EO, and a single-stage modified EO. In other words, the simple dynamical probability distribution evolutionary mechanism adopted in MSEO appears to be more appropriate and effective

Table 4.4   Comparison of TSEO with Other Physics-Inspired Algorithms Including SA, τ-EO, MEO-HH, and SOA for Some Hard TSP Instances

Problem	Optimum	Algorithm	Best	Average	Worst	t_{CPU} (s)
lin318	42,029	SA	42,449	42,785	43,120	69
		τ-EO	42,274	42,679	43,015	
		SOA	42,215	42,484	42,869	
		MEO-HH	42,200	42,482	42,642	
		TSEO	42,082	42,105	42,133	
pcb442	50,778	SA	51,271	51,829	52,301	115
		τ-EO	51,037	51,550	51,933	
		SOA	51,001	51,362	51,692	
		MEO-HH	50,945	51,171	51,375	
		TSEO	50,809	50,824	50,828	
att532	86,729	SA	88,227	89,296	90,198	165
		τ-EO	87,830	88,750	89,591	
		SOA	87,590	88,344	89,101	
		MEO-HH	87,078	87,764	88,475	
		TSEO	86,954	87,023	87,098	
rat783	8806	SA	9005	9065	9105	380
		τ-EO	8985	9040	9090	
		SOA	8974	9026	9075	
		MEO-HH	8952	8982	9023	
		TSEO	8885	8900	8910	
pr1002	2,59,045	SA	2,64,018	2,67,230	2,69,406	710
		τ-EO	2,62,930	2,64,096	2,66,557	
		SOA	2,62,548	2,63,630	2,65,443	
		MEO-HH	2,62,300	2,63,478	2,65,217	
		TSEO	2,61,341	2,61,696	2,61,990	

Source: Reprinted from *Physica A*, 389 (21), Zeng, G. Q. et al. Multistage extremal optimization for hard travelling salesman problem. 5037–5044, Copyright 2010b, with permission from Elsevier.

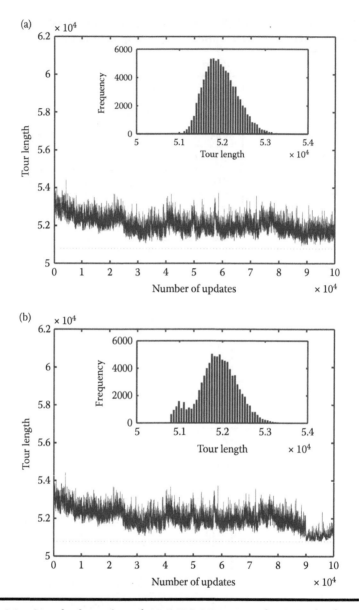

Figure 4.3 Search dynamics of NNMEO-HH (a) and TSEO (b) for pcb442 instance. The insets are the respective histograms of the frequency with which a particular tour length is obtained during the fluctuation process. (Reprinted from *Physica A*, 389 (21), Zeng, G. Q. et al. Multistage extremal optimization for hard travelling salesman problem. 5037–5044, Copyright 2010b, with permission from Elsevier.)

Table 4.5 Performances of MSEO for TSP Instances, Where Each Stage Goes through 10 Independent Runs

Problem	Optimum	Stages	Best	Average	Worst	t_{CPU} (s)
lin318	42,029	2	42,082	42,105	42,133	69
	42,029	3	42,053	42,062	42,077	77
pcb442	50,778	2	50,809	50,824	50,828	115
	50,778	3	50,787	50,790	50,795	127
att532	86,729	2	86,954	87,023	87,098	165
	86,729	3	86,876	86,900	86,950	182
rat783	8806	2	8885	8900	8910	380
	8806	3	8860	8873	8881	456
pr1002	2,59,045	2	2,61,341	2,61,696	2,61,990	710
	2,59,045	3	2,61,090	2,61,239	2,61,325	860

Source: Reprinted from *Physica A*, 389 (21), Zeng, G. Q. et al. Multistage extremal optimization for hard travelling salesman problem. 5037–5044, Copyright 2010b, with permission from Elsevier.

than the traditional static strategy. Of course, a more adaptive schedule of the control parameters may be devised to cross the energy barriers effectively and efficiently.

4.3.4 Adjustable Parameters versus Performance

As analyzed in Section 4.3.2, the control parameters of the probability distributions adopted in MSEO are crucial for its performance. Although the systematic method of determining the optimal values of these parameters is still an open issue, here we study it numerically based on some important rules. Generally speaking, the determination of the control parameters used in the first stage is similar to our recent analysis and numerical method (Zeng et al., 2010b). For the algorithms defined in Table 4.1, the control parameter of $P_1(k_1)$ always chooses small values while that of $P_2(k_2)$ chooses large values. For example, the optimal values of τ_1 used in τ-EO and SOA approximately satisfy the equation $\tau_1 = 1 + 1/\ln(n)$ (Boettcher and Percus, 2001a) while those of τ_2 used in SOA and MEO-EP (Zeng et al., 2010b) range from 1.8 to 3.0. In fact, the control parameters h_1 and h_2 used in the standard MEO-HH play analogous roles to τ_1 and τ_2 used in SOA (Chen et al., 2007a), respectively. Similar to τ_1 in τ-EO and SOA, the optimal values of h_1 used in the standard MEO-HH generally range from 0.03 to 0.07 so that most of the bad variables (or viewed as microparticles) have a chance to be selected to mutate and the worse ones have more probabilities. The optimal values of h_2 always range from 0.30 to 0.50 to guarantee that the better configuration (or viewed as macro-state) has more probability to be selected as a new one than unconditionally accepted. This is also consistent with the statistical property of the kth-nearest-neighbor distribution of optimal tours found in many TSP instances (Chen and Zhang, 2006). For the other stages, the values of the parameters used in $P_1(k_1)$ and $P_2(k_2)$ of the current stage are always larger than those in the last stage according to the "backbone-similar" idea.

Based on the above analysis, Zeng et al. (2010c) chose the pcb442 instance to illustrate how to determine those optimal values of the control parameters used in the two-stage EO algorithm TSEO. For the pcb442 instance, the optimal values of the control parameters h_{11} and h_{21} of $P_1(k_1)$ used in the first and second stage, respectively, are determined as $h_{11} \approx 0.056 \pm 0.005$ and $h_{21} \approx 0.060 \pm 0.015$ according to their similar effect to $\tau_1 \approx 1 + 1/\ln(442) = 1.15$ in τ-EO and SOA, and aforementioned rules. Figure 4.4 illustrates the effects of the control parameters h_{12} and h_{22} of $P_2(k_2)$ used in the first and second stage, respectively, on the corresponding performances of TSEO. These performances are measured by the best, average, and worst errors (%), which are defined as $100 \times$ (best $-$ optimum)/optimum, $100 \times$ (average $-$ optimum)/optimum, $100 \times$ (worst $-$ optimum)/optimum, respectively, over 10 independent runs when varying the values of h_{12} and h_{22}. Due to the operation that the best configuration obtained in the first stage is selected as the initial one in the second stage, one should focus on evaluating the best errors to determine the optimal value of h_{12}. From Figure 4.4, it is clear that the best configuration is obtained in the first stage when $h_{12} \approx 0.365 \pm 0.005$. The optimal value of h_{22} is determined in terms of the final comprehensive performances obtained in the second stage. From Figure 4.4, the optimal value of h_{22} used in TSEO is approximately from 0.720 to 0.730. By the similar method, these optimal values of the control parameters used in TSEO can be determined for other tested TSP instances.

For the three-stage algorithm MSEO, the optimal values of the control parameters h_{11}, h_{12}, and h_{21} are the same as those in TSEO, yet those of h_{22} should be determined according to the best errors obtained in the second stage. From Figure 4.4, we can observe that the best configuration is obtained in the second stage when $h_{22} \approx 0.565 \pm 0.005$ for the pcb442 instance. Similar to the determination method of the parameters h_{21} and h_{22} used in TSEO, the optimal values of h_{31} and h_{32} used in the third stage can be also determined.

4.4 Backbone-Guided EO

A large deal of research works (Schneider et al., 1996; Monasson et al., 1999; Singer et al., 2000; Dubolis and Dequen, 2001; Slaney and Walsh, 2001; Zhang, 2001, 2002, 2004; Telelis and Stamatopoulos, 2002; Schneider, 2003; Kilby et al., 2005; Zhang and Looks, 2005; Menaï and Batouche, 2006) have shown that the computational complexity of an optimization problem depends not only on its dimension,

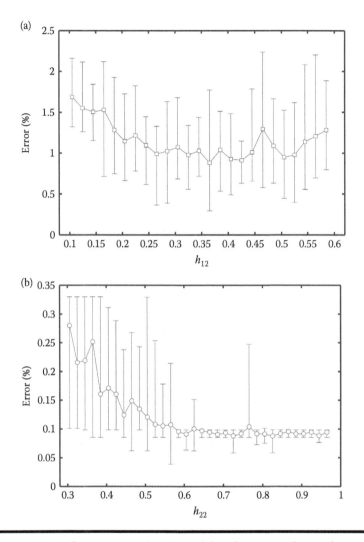

Figure 4.4 Control parameters h_{12} (a) and h_{22} (b) versus the performances of the first and second stage, respectively, in the TSEO algorithm. (Reprinted from *Physica A*, 389 (21), Zeng, G. Q. et al. Multistage extremal optimization for hard travelling salesman problem. 5037–5044, Copyright 2010b, with permission from Elsevier.)

but also on some inherent structural properties, for example, backbone. As one of the most interesting and important structures, backbone has been used to explain the difficulty of problem instances (Monasson et al., 1999; Singer, 2000; Slaney and Walsh, 2001; Zhang, 2001, 2002; Kilby et al., 2005). The problems with larger backbone are generally harder for local search algorithms to solve because the clustered solutions in these problems often result in these algorithms making mistakes more easily and wasting time searching empty subspaces before correcting the bad assignments (Slaney and Walsh, 2001). On the other hand, the utilization of the backbone information may help the design of effective and efficient optimization algorithms (Schneider et al., 1996; Dubolis and Dequen, 2001; Telelis and Stamatopoulos, 2002; Schneider, 2003; Zhang, 2004; Zhang and Looks, 2005; Menaï and Batouche, 2006). For example, Schneider et al. (1996) and Schneider (2003) have developed a powerful parallel algorithm for TSP by using its backbone information. Dubolis and Dequen (2001) incorporated the backbone information in a DPL-type algorithm for random 3-SAT problem. Telelis and Stamatopoulos (2002) developed a heuristic backbone sampling method to generate initial solutions for a local search algorithm based on the concept of backbone. Zhang (2004) proposed a backbone-guided WALKSAT method where the backbone information is embedded in a popular local search algorithm, such as WALKSAT. Furthermore, the basic idea has been extended to TSP (Zhang and Looks, 2005), and the partial MAX-SAT problem (Menaï and Batouche, 2006). The experimental results have shown that these backbone-based methods provide better performance than the pure local search ones. Nevertheless, almost all existing EO-based algorithms have overlooked the inherent structural properties behind the optimization problems, for example, backbone information.

This section presents another method called backbone-guided extremal optimization (BGEO) (Zeng et al., 2012) for the hard MAX-SAT problem. Menaï and Batouche (2006) developed a modified EO algorithm called Bose–Einstein-EO (BE-EO) to solve the MAX-SAT problem. The basic idea behind BE-EO is to sample initial configurations set based on Bose–Einstein distribution to the original τ-EO search process. The experimental results on both random and structured MAX-SAT instances demonstrate BE-EO's superiority to more elaborate stochastic optimization methods such as SA (Hansen and Jaumard, 1990), GSAT (Selman and Kautz, 1993), WALKSAT (Selman et al., 1994), and Tabu search (Szedmak, 2001). In Zeng et al. (2011), a more generalized EO framework termed as EOSAT was proposed to solve the MAX-SAT problem. The modified algorithms, such as BE-EEO and BE-HEO, provide better performance than BE-EO. Therefore, by incorporating the backbone information into the EOSAT framework, the BGEO method proposed in the work of Zeng et al. (2012) is possible to guide the search approach to the optimal solutions, and to further improve the performance of the original EO algorithms.

4.4.1 Definitions of Fitness and Backbones

According to the seminal work (Boettcher and Percus, 2000, 2001a), one of the most important issues for designing EO-based algorithms is the appropriate definition of local and global fitness. More specifically, the global fitness (i.e., objective function) of an optimization problem should be decomposed into the local fitness (i.e., the contribution from the decision variables).

For a given configuration S of a weighted MAX-SAT problem, the local fitness λ_i of each variable x_i is defined as follows (Zeng et al., 2012):

$$\lambda_i = \frac{-\sum_{x_i \in C_j \text{ and } C_j(S)=0} w_j}{\sum_{x_i \in C_k} w_k} \tag{4.12}$$

In other words, the local fitness is defined as the fraction of the sum of weights of unsatisfied clauses in which the variable x_i appears by the sum of weights of clauses connected to this variable.

The global fitness $C(S)$ is defined as the sum of the contribution from each variable (Zeng et al., 2012), that is,

$$C(S) = -\sum_{i=1}^{n} \left(\lambda_i \sum_{x_i \in C_k} w_k \right) = -\sum_{i=1}^{n} (c_i \lambda_i), \quad \text{where } c_i = \sum_{x_i \in C_k} w_k \tag{4.13}$$

where c_i is a constant for a given problem.

For the MAX-SAT problem, the exact backbone **B** is the set of variables having the same assignments in the set $\mathbf{S}_{global}$ of global optimal solutions. The formal definition is given as follows (Zeng et al., 2012):

$$\mathbf{B} = \{x_i | \forall s_j, s_k \in \mathbf{S}_{global}, s_j(x_i) = s_k(x_i)\} \tag{4.14}$$

Nevertheless, the exact backbone information of a given problem instance are even more difficult to obtain than actual problem solutions (Menaï and Batouche, 2006). An approximate approach to estimate the backbone information is considering the local minima as "real" optimal ones.

The quasi-backbone $\mathbf{X_B}$ is the set of variables having the same assignments in the set $\mathbf{S}_{local}$ of some local optimal solutions. Its formal definition is given as follows (Zeng et al., 2012):

$$\mathbf{X_B} = \{x_i | \forall s_j, s_k \in \mathbf{S}_{local}, s_j(x_i) = s_k(x_i)\} \tag{4.15}$$

4.4.2 BGEO Algorithm

It has shown that for the MAX-SAT problem, only the BE distribution can guarantee that an initial assignment set is generated with an arbitrary proportion of 1s and 0s (Szedmak, 2001). Moreover, the experimental results (Menaï and Batouche, 2006) on random and structured MAX-SAT instances have shown that the BE-EO algorithm starting from BE-based initial configurations outperforms τ-EO from uniformly random ones. Therefore, a BE-based assignment called "BE-based Initial

---

Backbone-Guided Extremal Optimization

Input: a MAX-SAT instance; *MI*: the maximum iterations; R_l: the maximum independent runs of the *l*th iteration; SS_l: the maximum sample size in the *l*th iteration; MS_l: the maximum steps of EO algorithm in the *l*th iteration; p_l: the adjustable parameter for evolutionary probability distribution of EO algorithm in the *l*th iteration;

Output: S_B: the best configurations found; $C(S_B)$: the total weights of unsatisfied clauses.

---

1. Initialization: set the backbone set $\mathbf{X_B} = \mathbf{0}$, nonbackbone set $\mathbf{X_{NB}} = \mathbf{X}$,
2.   *for l* = 1: *MI*
3.   *for j* = 1: R_k
4.   *for i* = 1: SS_k
5.   Fix the values of $\mathbf{X_B}$, initialize $\mathbf{X_{NB}} = \mathbf{X} - \mathbf{X_B}$ by BEICG, and construct the initial solution S_i, set $S_{best} = S_i$
6.   For the current solution S_i
                    a. Evaluate λ_i for each variable x_i and rank all the variables according to λ_i, that is, find a permutation Π_1 of the labels *i* such that $\lambda_{\Pi_1(1)} \geq \lambda_{\Pi_1(2)} \geq \cdots \geq_{\Pi_1(n)}$;
                    b. Select a rank $\Pi(k)$ according to a probability distribution $P_l(k_1), 1 \leq k_1 \leq n$ and denotethe corresponding variable as x_j;
                    c. Flip the value of x_j and set $S_{new} = S_i$ in which x_j value is flipped;
                    d. Accept $S_i = S_{new}$ unconditionally;
                    e. If $C(S_{new}) \leq C(S_{best})$, then $S_{best} = S_{new}$;
7.   Repeat the step 6 until the maximum steps MS_l, and obtain the best solution $S_{if} = S_{best}$
8.   *end for*
9.   Choose the best solution S_{bj} from the solution set$\{S_{if}\}$
10.   *end for*
11.   Obtain the solution set $\mathbf{S}_l = \{S_{bj}\}$, extract the backbone information from $\mathbf{S}_l$, update $\mathbf{X_B}$ and $\mathbf{X_{NB}}$
12.   *end for*
13.   Choose the best solution S_B from $\mathbf{S} = \bigcup_{l=1}^{MI} \mathbf{S}_l$, and obtain the corresponding cost $C(S_B)$

Figure 4.5 Framework of BGEO. (From Zeng, G. Q. et al., 2012. *International Journal of Innovative Computing, Information and Control* **8 (12): 8355–8366. With permission.)**

Configuration Generator (BEICG)" is used as the initial configuration of the proposed framework in this section.

In this section, Zeng et al. (2012) incorporated the quasi-backbone information into the EO algorithm, and obtained the framework of BGEO, which is described in Figure 4.5.

The BGEO framework can be viewed as an iterative process, which consists of the backbone estimation phase and backbone-guided optimization phase. In the first iteration, that is, $l = 1$, the BGEO collects R_l local optimum solutions starting from pure randomly generated initial solutions without any backbone information. In the following iterations, BGEO explores the complex landscape by utilizing the backbone information obtained in the last iteration. When the evolutionary probability distribution $P_l(k_1)$ adopted in BGEO algorithm is chosen as power-law, exponential, and hybrid distribution, respectively, the corresponding algorithm is called BG-PEO, BG-EEO, and BG-HEO, respectively, so the corresponding parameter p_l is τ_l, μ_l, and h_l, respectively.

Obviously, the performance of BGEO depends on these parameters, including MI, R_l, SS_l, MS_l, and p_l. Therefore, determining the appropriate values of these parameters to make BGEO achieve the best performance is a critical issue. Here, MI and R_l are all positive constants. According to the study on BE-EO (Menaï and Batouche, 2006), the parameters SS_l, MS_l are as follows:

$$SS_l = C_{l1} \times |\mathbf{X}_{NB}(l)| \tag{4.16}$$

$$MS_l = C_{l2} \times |\mathbf{X}_{NB}(l)| \tag{4.17}$$

where C_{l1}, C_{l2} are all positive constants and $\leq|\mathbf{X}_{NB}(l)|$ is the number of nonbackbone variables in the lth iteration.

Obviously, p_l plays an analogous role in the proportion p of random and greedy moves in WALKSAT (Selman and Kautz, 1993), and the noise parameter η in FMS (Seitz et al., 2005). Due to the different features of the first and the remaining iterations, p_l is given in the following form:

$$p_l = \begin{cases} p_c, & l = 1 \\ p_c + d* |\mathbf{X}_B(l-1)|, & 2 \leq l \leq MI \end{cases} \tag{4.18}$$

where p_c is the initial value of the parameter p_l and d is a positive constant.

From the BGEO framework, it is clear that the optimization of the $(l+1)$th iteration always starts from the initial solutions where all are based on the backbone information extracted in the lth iteration, so the size of the backbone extracted in $(l+1)$th iteration is generally more than at least equal to that in lth iteration, that is, $n \geq |\mathbf{X}_B(MI)| \geq \cdots \geq |\mathbf{X}_B(l+1)| \geq |\mathbf{X}_B(l)| \geq \cdots \geq |\mathbf{X}_B(l)|$. In other words, the size of the remaining problem that needs to be optimized will be smaller and smaller as the number of iterations increase. As a consequence, there must exist a finite constant MI_{max} such that $\leq|\mathbf{X}_B(MI_{max})| \to n$. To illustrate this observation, Figure 4.6 shows the dynamics of the pseduo backbone size during the search process of BGEO for some uniform satisfiable MAX-3-SAT instances "uf-n:m," in which n is the number of the variables and m is the number of the clauses. The backbone size increases as the number of iteration increases.

For the uf-100:430 instance, the search dynamics of the best global fitness in BGEO are shown in the top of Figure 4.7. For the hard MAX-3-SAT instances, BGEO can reach high-quality solutions in finite iterations. The bottom of Figure 4.7 gives the performance comparison of BE-EEO and BGEO algorithms. Obviously, BGEO performs better than BE-EEO in the same runtime.

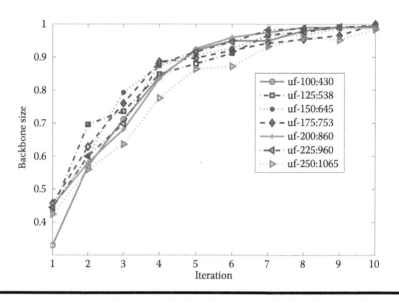

Figure 4.6 Dynamics of the pseudo backbone size during the search process of BGEO. (From Zeng, G. Q. et al., 2012. *International Journal of Innovative Computing, Information and Control* 8 (12): 8355–8366. With permission.)

Furthermore, the superiority of the BGEO algorithm is demonstrated by the experimental results in Section 4.4.3.

4.4.3 Experimental Results

In order to demonstrate the effectiveness of the BGEO, we choose the hard MAX-SAT problem instances from SATLIB (available at http://www.cs.ubc.ca/~hoos/SATLIB/benchm.html) as the test bed. The tested problems include random unweighted MAX-3-SAT instances, MAX-3-SAT instances near the phase transition, and MAX-3-SAT instances with controlled backbone size (CBS). Note that all algorithms are implemented in MATLAB® 7.6 on a Pentium 1.86 GHz PC with dual-core processor T2390 and 2 GB RAM running Windows Vista Basic systems. The performances of these algorithms are measured by the best, mean, and worst errors denoted as e_b, e_m, and e_w, respectively. The errors are defined as $e_b(\%) = 100 \times (m_b - m_o)/m$, $e_m(\%) = 100 \times (m_m - m_o)/m$, and $e_w(\%) = 100 \times (m_w - m_o)/m$, respectively, where m_b, m_m, and m_w are the minimal, average, and maximal number of unsatisfied clauses over 10 independent runs, respectively, and m_o is the optimal solution.

The experimental results (Menaï and Batouche, 2006) on random and structured MAX-SAT instances have shown that BE-EO can provide better or at least competitive performance than more elaborate stochastic optimization methods, such as SA (Hansen and Jaumard, 1990), GSAT (Selman and Kautz, 1993),

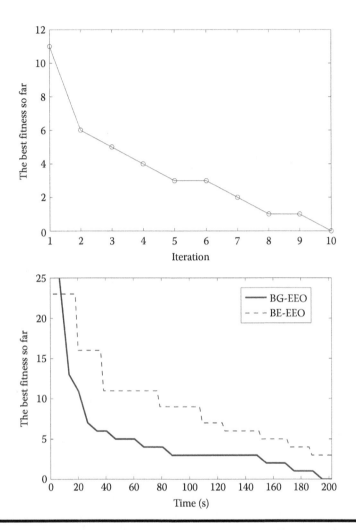

Figure 4.7 For uf-100:430, the top is the search dynamics of the best global fitness in BGEO and the bottom is the comparison of BG-EEO and BE-EEO. (From Zeng, G. Q. et al., 2012. *International Journal of Innovative Computing, Information and Control* 8 (12): 8355–8366. With permission.)

WALKSAT (Selman et al., 1994), and TS-CSP (Szedmak, 2001). Furthermore, the superiority of the BE-EEO and BE-HEO algorithms under the EOSAT framework to the BE-EO algorithm is demonstrated by our recent research (Zeng et al., 2011). Consequently, this section concentrates on comparing BG-EEO with these reported algorithms by the experiments on the random unweighted MAX-3-SAT instances. The results on the random unweighted MAX-3-SAT instances are shown in Table 4.6, where the performances of these algorithms are measured by the average error (%). It is obvious that BG-EEO is superior to these reported algorithms.

Table 4.6 Comparsion of BG-EEO and the Reported Algorithms on Random Unweighted MAX-3-SAT Instances

Variables (n)	100	100	300	300	500
Clauses (m)	500	700	1500	2000	5000
SA	1.64	2.587	2.000	2.900	4.528
TS-CSP	0.453	1.755	0.523	1.595	3.328
GWSAT	0.556	1.914	0.551	1.597	3.279
WSAT	0.552	1.914	0.541	1.614	3.340
τ-EO	0.800	1.880	0.600	1.900	3.352
BE-EO	0.632	1.860	0.500	1.550	3.100
BE-EEO	0.524	1.732	0.486	1.479	3.003
BG-EEO	0.400	1.553	0.264	1.245	2.875

Source: From Zeng, G. Q. et al., 2012. *International Journal of Innovative Computing, Information and Control* 8 (12): 8355–8366. With permission.

The tested MAX-3-SAT satisfiable and unsatisfiable instances are represented as "uf-n:m" and "uuf-n:m" here, in which n is the number of the variables and m is the number of the clauses. For example, "uuf-50:218" represents the unsatisfiable instance that has 50 variables and 218 clauses. These instances with $\alpha = m/n$ ranges from 4.260 to 4.360, which are close to the critical threshold of the phase transition $\alpha_c \approx 4.267$. For each unsatisfiable instance, the optimal number of unsatisfied clauses is 1, that is, $m_o = 1$. Therefore, we focus on the optimization problem, MAX-3-SAT, that is to find an assignment to maximize the number of satisfied clauses. In other words, MAX-3-SAT is equivalent to minimize the number of unsatisfied clauses. The experimental results on these satisfiable and unsatisfiable instances near the phase transition are shown in Tables 4.7 and 4.8, respectively. Especially, the BG-EEO algorithm reaches the optimal solutions for some instances shown in bold. It is obvious that BG-EEO (Zeng et al., 2012) provides better performance than BE-EO (Menaï and Batouche, 2006) and BE-EEO (Zeng et al., 2011) algorithms for these satisfiable and unsatisfiable instances near the phase transition.

Table 4.9 gives the comparison of BG-EEO and BE-EEO for some large instances. Clearly, BG-EEO outperforms the reported BE-EEO algorithms for these hard instances. MAX-3-SAT instances with CBS are different from those given in Tables 4.6 through 4.9 in that they have some backbone variables, where the backbone size is defined as b. Singer et al. (2000) have shown that the search cost is very high even for small size problems but with large backbone size. Therefore, these CBS instances from SATLIB are chosen for testing the superiority of the

Table 4.7 Comparison of BG-EEO and the Reported BE-EO, BE-EEO Algorithms for MAX-3-SAT Satisfiable Instances Near Phase Transition

Problem	α	BE-EO			BE-EEO			BG-EEO		
		e_b	e_m	e_w	e_b	e_m	e_w	e_b	e_m	e_w
uf-50:218	4.360	0.92	2.20	2.75	0.46	1.74	2.29	**0.00**	0.00	0.00
uf-75:325	4.333	1.54	2.62	3.07	0.62	1.45	2.15	**0.00**	0.15	0.31
uf-100:430	4.300	1.63	2.47	2.80	0.70	1.83	2.33	**0.00**	0.23	0.46
uf-125:538	4.304	2.04	2.68	3.35	1.11	1.99	2.23	0.19	0.28	0.37
uf-150:645	4.300	2.17	2.53	2.95	1.40	1.95	2.33	0.16	0.31	0.47
uf-175:753	4.303	2.39	2.76	2.92	1.59	1.91	2.26	0.13	0.33	0.40
uf-200:860	4.300	2.67	3.13	3.49	1.86	2.29	2.56	0.12	0.17	0.23
uf-225:960	4.267	2.08	2.79	3.23	1.67	1.99	2.29	0.10	0.31	0.62
uf-250:1065	4.260	1.88	2.76	3.29	1.60	1.94	2.07	0.09	0.35	0.66

Source: From Zeng, G. Q. et al., 2012. *International Journal of Innovative Computing, Information and Control* 8 (12): 8355–8366. With permission.

Table 4.8 Comparison of BG-EEO and the Reported BE-EO, BE-EEO Algorithms for MAX-3-SAT Unsatisfiable Instances Near Phase Transition

Problem	α	BE-EO			BE-EEO			BG-EEO		
		e_b	e_m	e_w	e_b	e_m	e_w	e_b	e_m	e_w
uuf-50:218	4.360	1.38	2.38	3.21	0.46	1.88	2.75	**0.00**	0.00	0.00
uuf-75:325	4.333	1.85	2.65	3.37	1.23	1.94	2.46	**0.00**	0.17	0.31
uuf-100:430	4.300	1.86	2.63	2.80	1.16	1.88	2.56	**0.00**	0.26	0.46
uuf-125:538	4.304	1.86	2.70	3.35	1.30	2.08	3.16	0.19	0.30	0.37
uuf-150:645	4.300	2.17	2.71	3.41	1.40	2.05	2.64	0.16	0.32	0.47
uuf-175:753	4.303	2.92	3.33	3.98	1.73	2.30	2.67	0.13	0.34	0.40
uuf-200:860	4.300	3.49	3.85	4.30	2.44	2.72	3.14	0.12	0.16	0.23
uuf-225:960	4.267	2.81	3.48	4.17	2.19	2.64	3.44	0.10	0.44	0.62
uuf-250:1065	4.260	3.09	3.51	4.38	1.78	2.28	2.72	0.09	0.45	0.66

Source: From Zeng, G. Q. et al., 2012. *International Journal of Innovative Computing, Information and Control* 8 (12): 8355–8366. With permission.

Table 4.9   BG-EEO versus BE-EEO Algorithm for the
Large MAX-SAT Instances

Problem	A	BE-EEO			BG-EEO		
		e_b	e_m	e_w	e_b	e_m	e_w
f600_2550	4.25	1.73	2.96	3.89	0.52	0.96	1.57
f1000_4250	4.25	2.08	3.35	4.74	0.67	1.21	1.92
f2000_8500	4.25	2.87	4.01	4.92	0.83	1.50	2.33

Source: From Zeng, G. Q. et al., 2012. *International Journal of Innovative Computing, Information and Control* 8 (12): 8355–8366. With permission.

proposed BGEO method. The control parameter α of these instances ranging from 4.03 to 4.49 is near the critical threshold of the phase transition $\alpha_c \approx 4.267$. Moreover, the values of b in these instances range from 10 to 90 at each α value. The comparison of BG-EEO and the reported BE-EO and BE-EEO algorithms for these CBS instances is shown in Table 4.10. It is clear that BG-EEO (Zeng et al., 2012) performs much better than BE-EO (Menaï and Batouche, 2006) and BE-EEO (Zeng et al., 2011) for these hard CBS instances. Especially, the BG-EEO algorithm reaches the optimal solutions for some instances shown inbold.

4.5 Population-Based EO

Many real-world optimization problems involve complicated constraints. What constitute the difficulties of the constrained optimization problem are various limits on the decision variables, the constraints involved, the interference among constraints, and the interrelationship between the constraints, objective functions, and decision variables. This has motivated the development of a considerable number of approaches to tackling constrained optimization problems such as Stochastic Ranking (SR) (Runarsson and Yao, 2000), Adaptive Segregational Constraint Handling Evolutionary Algorithm (ASCHEA) (Hamida and Schoenauer, 2002), Simple Multimembered Evolution Strategy (SMES) (Mezura-Montes and Coello, 2005), etc. In this section, EO will be applied to solving numerical constrained optimization problems. To enhance and improve the search performance and efficiency of EO, Chen et al. (2006) developed a novel EO strategy with population-based search, called PEO. In addition, Chen et al. (2006) adopted the adaptive Lévy mutation operator, which makes PEO able to carry out not only coarse-grained but also fine-grained search. It is worth noting that there exists no adjustable parameter in PEO, which makes PEO more charming than other methods. Finally, PEO is successfully applied in solving six popular benchmark problems and

Table 4.10   BG-EEO versus the BE-EO, BE-EEO Algorithms for CBS Instances

Problem	α	B	BE-EO			BE-EEO			BG-EEO		
			e_b	e_m	e_w	e_b	e_m	e_w	e_b	e_m	e_w
CBS_100_403	4.03	10	1.99	2.68	3.23	0.99	1.91	2.48	**0.00**	0.07	0.25
		30	1.99	2.46	2.98	1.24	1.76	2.23	0.25	0.45	0.74
		50	1.49	2.06	2.48	1.24	1.61	1.98	0.50	0.74	0.99
		70	2.43	2.75	3.16	1.24	1.74	1.98	0.50	0.74	0.99
		90	2.23	2.73	3.47	1.74	2.13	2.48	0.74	0.99	1.24
CBS_100_411	4.11	10	1.46	2.12	2.68	0.97	1.58	1.95	0.24	0.49	0.73
		30	1.22	2.19	2.92	1.22	1.68	2.68	0.24	0.49	0.97
		50	1.95	2.68	3.16	1.70	2.09	2.92	0.24	0.61	0.97
		70	1.95	2.70	3.89	1.70	2.12	2.19	0.49	0.78	0.97
		90	1.95	2.53	3.41	1.46	1.85	2.19	0.24	0.68	0.97
CBS_100_418	4.18	10	1.67	2.34	3.11	1.22	1.75	2.19	0.24	0.38	0.72
		30	2.39	2.68	3.11	1.44	2.08	2.39	0.24	0.36	0.72
		50	2.63	2.73	3.11	1.44	2.03	2.39	0.48	0.72	0.96
		70	0.96	2.44	3.11	0.96	1.48	2.15	0.48	0.62	0.72
		90	2.15	2.68	3.11	1.20	1.87	2.39	0.48	0.72	0.96
CBS_100_423	4.23	10	1.65	2.13	2.60	0.71	1.47	2.13	**0.00**	0.47	0.95
		30	1.42	2.48	3.55	1.18	1.84	2.13	0.24	0.47	0.71
		50	1.65	2.53	3.07	1.18	1.80	2.36	0.24	0.57	0.71
		70	1.42	2.41	2.84	0.95	1.77	2.36	0.24	0.54	0.71
		90	1.89	2.55	3.31	1.42	2.00	2.13	0.24	0.80	0.95
CBS_100_429	4.29	10	1.17	2.45	3.03	0.47	1.52	2.10	**0.00**	0.65	0.93
		30	1.63	2.17	2.56	1.17	1.75	2.10	**0.00**	0.47	0.70
		50	2.10	2.66	3.26	0.93	1.70	2.33	**0.00**	0.47	0.70
		70	1.40	2.24	2.80	0.70	1.70	2.33	0.23	0.58	0.93
		90	2.10	2.77	3.26	1.86	2.24	2.56	0.23	0.70	0.93

(Continued)

Table 4.10 (*Continued*)   BG-EEO versus the BE-EO, BE-EEO Algorithms for CBS Instances

Problem	α	B	BE-EO			BE-EEO			BG-EEO		
			e_b	e_m	e_w	e_b	e_m	e_w	e_b	e_m	e_w
CBS_100_435	4.35	10	1.15	1.91	2.53	0.69	1.22	1.61	**0.00**	0.28	0.46
		30	1.38	1.91	2.53	0.92	1.47	1.84	**0.00**	0.32	0.46
		50	2.30	2.67	3.45	1.61	2.05	2.30	0.46	0.69	0.92
		70	1.61	2.46	2.99	1.15	1.89	2.30	0.23	0.64	0.92
		90	2.30	2.57	2.76	0.92	1.66	2.30	0.23	0.60	0.92
CBS_100_441	4.41	10	1.36	2.12	2.49	0.68	1.45	2.04	0.23	0.45	0.68
		30	1.59	2.15	2.95	0.45	1.16	2.04	**0.00**	0.23	0.45
		50	1.81	2.59	3.40	1.13	1.88	2.72	0.23	0.45	0.91
		70	1.81	2.49	2.95	1.59	1.90	2.49	0.45	0.68	0.91
		90	2.49	2.90	3.63	1.36	2.06	2.72	0.45	0.73	0.91
CBS_100_449	4.49	10	1.56	2.27	3.12	0.67	1.44	2.00	0.22	0.45	0.67
		30	1.78	2.45	3.12	1.34	1.87	2.23	0.45	0.58	0.89
		50	1.78	2.05	2.90	0.89	1.27	1.78	0.22	0.33	0.67
		70	2.45	2.92	3.34	1.34	2.16	2.67	0.22	0.56	0.89
		90	2.67	3.27	3.79	1.56	2.32	2.90	0.67	0.89	1.11

Source: From Zeng, G. Q. et al., 2012. *International Journal of Innovative Computing, Information and Control* 8 (12): 8355–8366. With permission.

shows competitive performance when compared with three state-of-the-art search methods, that is, SR, ASCHEA, and SMES.

4.5.1 Problem Formulation of Numerical Constrained Optimization Problems

A general nonlinear programming problem can be formulated as follows (Chen et al., 2006):

$$\text{minimize} \quad f(X), \quad X = [x_1, \ldots, x_n]^T \in R^n \tag{4.19}$$

$$\text{subject to } g_i(X) \le 0, \quad i = 1, \ldots, q \tag{4.20}$$

$$h_j(X) = 0, \quad j = q+1, \ldots, r \tag{4.21}$$

where $f(X)$ is the objective function, $X \in S \cap F$. $S \subseteq R^n$ is defined as the whole search space which is an n-dimensional space bounded by the following parametric constraints:

$$l_j \le x_j \le u_j, \quad j = 1, \ldots, n \tag{4.22}$$

where l_j and u_j are the lower and upper bound of x_j, respectively, and $F \subseteq R^n$ is defined as the feasible region. It is clear that $F \subseteq S$.

In this section, the methods for handling constrained nonlinear programming problems are based on the concept of penalty functions, which penalize unfeasible solutions. A set of functions $P_i(X)$ $(1 \le i \le r)$ is used to construct the penalty. The function $P_i(X)$ measures the violation of the ith constraint in the following way:

$$P_i(x) = \begin{cases} \max\{0, g_i(X)\}_i^2 & \text{if } 1 \le i \le q \\ |h_i(X)|_i^2 & \text{if } q+1 \le i \le r \end{cases} \tag{4.23}$$

4.5.2 PEO Algorithm

It is worth recalling that EO performs a search through sequential changes on a single solution, namely, the point-to-point search rather than the population-based search applied in GA. In order to accelerate the convergence speed, Chen et al. (2006) developed a novel real-coded EO search algorithm, the so-called PEO, through introducing the population search strategies being popularly used in EAs to EO. Similar to the EAs, PEO operates on the evolution of solutions generation after generation. By uniformly placing the population of initial random solutions on the search space, PEO can explore the wide search space, avoiding getting trapped into local optima. On the other hand, similar to EO, PEO performs only one operation, that is, mutation, on each variable. Each solution evolves to its SOC state by always forcing the worst variable to change.

Inspired by De Sousa and Ramos (2002) and De Sousa et al. (2004), Chen et al. (2006) defined the fitness of each variable for the constrained optimization problems as follows. For the minimization problems without equality and inequality constraints, the fitness λ_i of variable x_i means the mutation cost, that is, $OBJ(S_i) - OBJ(S_{best})$, where S_i is the new solution after performing mutation only on x_i and leaving all other variables fixed, $OBJ(S_i)$ is the objective value of S_i and $OBJ(S_{best})$ is the best objective value found so far. For the minimization problem with equality and inequality constraints, the sum of all the penalties $Q(S_i) = \Sigma_{j=1}^{r} P_j(S_i)$ should be incorporated into the fitness λ_i, that is, $\lambda_i = OBJ(S_i) - OBJ(S_{best}) + Q(S_i)$. It is worth pointing out that those variables which meet the constraints are considered as badly adapted individuals and thus low fitness will be assigned to them.

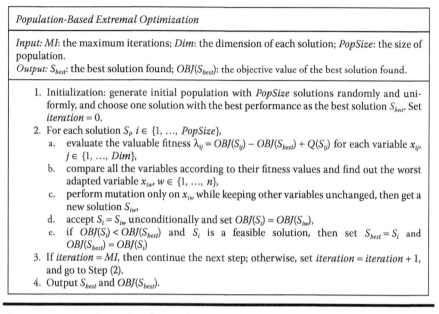

Population-Based Extremal Optimization

Input: MI: the maximum iterations; *Dim:* the dimension of each solution; *PopSize:* the size of population.
Output: S_{best}: the best solution found; *OBJ(S_{best}):* the objective value of the best solution found.

1. Initialization: generate initial population with *PopSize* solutions randomly and uniformly, and choose one solution with the best performance as the best solution S_{best}. Set *iteration* = 0.
2. For each solution S_i, $i \in \{1, ..., PopSize\}$,
   a. evaluate the valuable fitness $\lambda_{ij} = OBJ(S_{ij}) - OBJ(S_{best}) + Q(S_{ij})$ for each variable x_{ij}, $j \in \{1, ..., Dim\}$,
   b. compare all the variables according to their fitness values and find out the worst adapted variable x_{iw}, $w \in \{1, ..., n\}$,
   c. perform mutation only on x_{iw} while keeping other variables unchanged, then get a new solution S_{iw},
   d. accept $S_i = S_{iw}$ unconditionally and set $OBJ(S_i) = OBJ(S_{iw})$,
   e. if $OBJ(S_i) < OBJ(S_{best})$ and S_i is a feasible solution, then set $S_{best} = S_i$ and $OBJ(S_{best}) = OBJ(S_i)$
3. If *iteration = MI*, then continue the next step; otherwise, set *iteration = iteration + 1*, and go to Step (2).
4. Output S_{best} and $OBJ(S_{best})$.

Figure 4.8 Pseudo-code of PEO for numerical constrained minimization problem. (With kind permission from Springer Science + Business Media: *Proceedings of the 2006 International Conference on Computational Intelligence and Security (CIS'2006),* Population-based extremal optimization with adaptive Lévy mutation for constrained optimization. 2006, pp. 258–261, Chen, M. R. et al.)

On the contrary, those variables which do not satisfy the constraints will be considered as well-adapted species and be assigned high fitness.

For a numerical constrained minimization problem, PEO proceeds as shown in Figure 4.8.

4.5.3 Mutation Operator

Note that there is merely a mutation operator in PEO. Therefore, the mutation plays a key role in the PEO search that generates new solutions. Many mutation operators have been proposed in the past two decades, such as Gaussian mutation, Cauchy mutation, and so on. Yao et al. (1999) have pointed out that Cauchy mutation performs better when the current search point is far away from the global optimum, while Gaussian mutation is better at finding a local optimum in a good region. It would be ideal if Cauchy mutation is used when search points are far away from the global optimum and Gaussian mutation is adopted when search points are in the neighborhood of the global optimum. Unfortunately, the global optimum is usually unknown in practice, making the ideal switch from Cauchy to Gaussian mutation very difficult. In the PEO algorithm, Chen et al. (2006)

adopted the adaptive Lévy mutation which was proposed by Lee and Yao (2001), to easily switch Cauchy mutation to Gaussian mutation. Lévy mutation is, in a sense, a generalization of Cauchy mutation since Cauchy distribution is a special case of Lévy distribution. By adjusting the parameter α in Lévy distribution, one can tune the shape of the probability density function, which in turn yields adjustable variation in mutation step sizes. In addition, Lévy mutation provides an opportunity for mutating a parent using a distribution which is neither Cauchy nor Gaussian. The Lévy probability distribution has the following form (Mantegna, 1994):

$$L_{\alpha,\gamma}(y) = \frac{1}{\pi} \int_0^\infty e^{-\gamma q^\alpha} \cos(qy) dq \qquad (4.24)$$

As can be easily seen from Equation 4.24, the distribution is symmetric with respect to $y = 0$ and has two parameters, γ and α. γ is the scaling factor satisfying $\gamma > 0$ and α satisfies $0 < \alpha < 2$. The analytic form of the integral is not known for general α except for a few cases. In particular, for $\alpha = 1$, the integral can be carried out analytically and is known as the Cauchy probability distribution. In the limit of $\alpha \to 2$, the distribution approaches the Gaussian distribution. The parameter α controls the shape of the probability distribution in such a way, that one can obtain different shapes of probability distribution. In Chen et al. (2006), Lévy mutation performs with the following representation:

$$x_k^{t+1} = x_k^t + L_k(\alpha) \qquad (4.25)$$

where $L_k(\alpha)$ is a Lévy random variable with the scaling factor $\gamma = 1$ for the kth variable. To generate a Lévy random number, Chen et al. (2006) used an effective algorithm presented by Mantegna (1994). It is known that Gaussian mutation ($\alpha = 2$) works better for searching a small local neighborhood, whereas Cauchy mutation ($\alpha = 1$) is good at searching a large area of the search space. By adding additional two candidate offspring ($\alpha = 1.4$ and 1.7), one is not fixed to the two extremes. It must be indicated that, unlike the method in Lee and Yao (2001), the mutation in PEO does not compare the anticipated outcomes of different values of α due to the characteristics of EO. In PEO, the Lévy mutation with $\alpha = 1$ (i.e., Cauchy mutation) is first adopted. It means the large step size will be taken first at each mutation. If the new generated variable after mutation goes beyond the intervals of the decision variables, the Lévy mutation with $\alpha = 1.4, 1.7, 2$ will be carried out in turn, that is, the step size will become smaller than before. Thus, PEO combines the advantages of coarse-grained search and fine-grained search. The above analysis shows that the adaptive Lévy mutation is very simple yet effective. Unlike some switching algorithms which have to decide when to switch between different mutations during search, the adaptive Lévy mutation does not need to make such decisions and introduces no adjustable parameters.

4.5.4 Experimental Results

To testify the performance of PEO, Chen et al. (2006) selected 6 (g04, g05, g07, g09, g10, and g12) out of 13 benchmark functions published in Runarsson and Yao (2000) as test functions (note that all the six test functions are minimization task), since the characteristics of those functions contain the "difficulties" in having global optimization problems by using an EA. For more details about the expressions of those benchmark problems, readers can refer to Runarsson and Yao (2000). In Chen et al. (2006), all the algorithms were encoded in the floating point representation. The source codes of all experiments were coded in JAVA. Besides, inequality constraints can be incorporated into the fitness via the relevant penalty items. All equality constraints can be converted into inequality constraints, $|h(X)| - \varepsilon \leq 0$, using the degree of violation ε. The value of ε for function g05 was set to 0.0001. In all the algorithms, the population size was 100 and the maximum number of generations was 5000. Thirty independent runs were carried out for each test function. Figure 4.9a–f shows the simulation results of PEO on the six test problems, respectively. The average of the best results of every 100 generations found in 30 independent runs are shown in Figure 4.9. Table 4.11 summarizes the experimental results when the PEO with adaptive Lévy mutation (for simplicity, we call it PEOAL) is used. Table 4.11 also shows the known "optimal" solution for each problem and statistics. These include the best objective value found, mean, standard deviation, and worst found. Furthermore, Chen et al. (2006) compared PEO against three state-of-the-art approaches: SR, ASCHEA, and SMES. The best, mean, and worst results obtained by each approach are shown in Tables 4.12 through 4.14. "NA" in all tables means the results are "not available." The results provided by these approaches were taken from the original references for each method.

As can be seen from Table 4.11, PEO was capable of finding the global optimum in two test functions (g05 and g12). It is interesting to note that PEO also found solutions very close to the global optima in the remaining four functions (g04, g07, g09, g10). Furthermore, as observed from Figure 4.9, PEO was able to approach the global optimum quickly. Thus, PEO possesses good performance in accuracy and convergence speed. When compared with respect to the three state-of-the-art techniques previously indicated, we found the following (see Tables 4.12 through 4.14):

1. Compared with SR: PEO found better "best," "mean," and "worst" solutions in two functions (g5 and g10). It also provided similar "best," "mean," and "worst" solutions in function g12. Slightly better "best" results were found by SR in the remaining functions (g04, g07, g09).
2. Compared with ASCHEA: PEO was able to find better "best" and "mean" results in two functions (g05, g10). ASCHEA surpassed our mean results in three functions (g04, g07, g09). We did not compare the worst results due to the

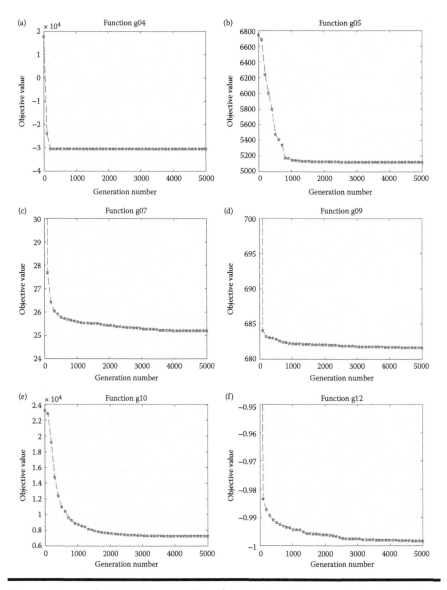

Figure 4.9 (a–f) Simulation results of PEO algorithm on six test functions. (With kind permission from Springer Science+Business Media: *Proceedings of the 2006 International Conference on Computational Intelligence and Security (CIS'2006)*, Population-based extremal optimization with adaptive Lévy mutation for constrained optimization. 2006, pp. 258–261, Chen, M. R. et al.)

Table 4.11   Experimental Results of PEO on Six Test Functions

Problem	Optimal	Best	Mean	Worst	St. Dev.
g04	−30665.539	−30652.146	−30641.177	−30629.763	5.45E + 0
g05	5126.498	5126.498	5126.527	5126.585	2.5E − 2
g07	24.306	24.798	25.130	25.325	1.18E − 1
g09	680.630	680.706	681.498	682.228	3.36E − 1
g10	7049.331	7051.573	7160.620	7294.895	5.82E + 1
g12	−1.000000	−1.0000	−1.000	−1.00	9.8E − 4

Source: With kind permission from Springer Science+Business Media: *Proceedings of the 2006 International Conference on Computational Intelligence and Security (CIS'2006)*, Population-based extremal optimization with adaptive Lévy mutation for constrained optimization. 2006, pp. 258–261, Chen, M. R. et al.

Table 4.12   Comparison of the Best Results Obtained

Problem	Optimal	PEOAL	SR	ASCHEA	SMES
g04	−30665.539	−30652.146	−30665.539	−30665.5	−30665.539
g05	5126.498	5126.498	5126.497	5126.5	5126.599
g07	24.306	24.798	24.307	24.3323	24.327
g09	680.630	680.706	680.630	680.630	680.632
g10	7049.331	7051.573	7054.316	7061.13	7051.903
g12	−1.000000	−1.0000	−1.000000	NA	−1.000

Source: With kind permission from Springer Science+Business Media: *Proceedings of the 2006 International Conference on Computational Intelligence and Security (CIS'2006)*, Population-based extremal optimization with adaptive Lévy mutation for constrained optimization. 2006, pp. 258–261, Chen, M. R. et al.

fact that they were not available for ASCHEA. In addition, we did not perform comparisons with respect to ASCHEA using function g12 for the same reason.
3. Compared with SMES: PEO found better "best," "mean," and "worst" results in two functions (g05, g10) and similar "best," "mean," and "worst" results in function g12. SMES outperformed PEO in the remaining functions.

From the aforementioned comparisons, it is obvious that PEO shows very competitive performance with respect to those three state-of-the-art approaches.

Table 4.13 Comparison of the Mean Results Obtained

Problem	Optimal	PEOAL	SR	ASCHEA	SMES
g04	−30665.539	−30641.177	−30665.539	−30665.5	−30665.539
g05	5126.498	5126.527	5128.881	5141.65	5174.492
g07	24.306	25.130	24.372	24.66	24.475
g09	680.630	681.498	680.665	680.641	680.643
g10	7049.331	7160.620	7559.192	7193.11	7253.047
g12	−1.000000	−1.000	−1.000000	NA	−1.000

Source: With kind permission from Springer Science+Business Media: *Proceedings of the 2006 International Conference on Computational Intelligence and Security (CIS'2006)*, Population-based extremal optimization with adaptive Lévy mutation for constrained optimization. 2006, pp. 258–261, Chen, M. R. et al.

Table 4.14 Comparison of the Worst Results Obtained

Problem	Optimal	PEOAL	SR	ASCHEA	SMES
g04	−30665.539	−30629.763	−30665.539	NA	−30665.539
g05	5126.498	5126.585	5142.472	NA	5304.167
g07	24.306	25.325	24.642	NA	24.843
g09	680.630	682.228	680.763	NA	680.719
g10	7049.331	7294.895	8835.655	NA	7638.366
g12	−1.000000	−1.00	−1.000000	NA	−1.000

Source: With kind permission from Springer Science+Business Media: *Proceedings of the 2006 International Conference on Computational Intelligence and Security (CIS'2006)*, Population-based extremal optimization with adaptive Lévy mutation for constrained optimization. 2006, pp. 258–261, Chen, M. R. et al.

4.5.5 Advantages of PEO

From the above experimental results, we can see that PEO with the adaptive Lévy mutation has the following advantages:

1. There is no adjustable parameter in PEO. This makes PEO more charming than other state-of-the-art methods.
2. Only one operator, that is, mutation operator, exists in PEO, which makes PEO simple and convenient.

3. PEO possesses good performance in accuracy and convergence speed.
4. By incorporating the adaptive Lévy mutation, PEO can perform globally and search locally.

4.6 Summary

This chapter introduces some modified EO versions, such as modified EO with extended evolutionary probability distributions (Zeng et al., 2010b), MSEO with dynamical evolutionary mechanism (Zeng et al., 2010c), backbone-guided EO algorithm (Zeng et al., 2011), PEO algorithm (Chen et al., 2006), which are to improve the performance of the original EO algorithm and extend the application area. The experimental results have demonstrated the effectiveness of these modified EO algorithms. The main topics studied in this chapter are summarized as follows:

1. This chapter introduces a modified EO framework with extended evolutionary probability distributions (Zeng et al., 2010b) which are not restricted to power laws. The numerical test on a variety of TSP benchmark instance (Zeng et al., 2010a, b) show that the modified algorithms are superior to other statistical physics-oriented methods, such as SA, τ-EO, and SOA. This also indicates that τ-EO with exponential distributions can provide better performance than τ-EO at least for hard TSP, which can dispel the misconception of Boettcher and Frank (2006) that τ-EO with exponential distributions fails to perform well for hard optimization problems. Second, from the optimization point of view, power laws may be not the best evolution probability distributions at least for some class of hard problems, for example, TSP.

2. A MSEO method (Zeng et al., 2010c) for COPs is also introduced in this chapter. The basic idea behind this method is repeating the process that consists of choosing the best configuration obtained by the previous stage as an initial configuration for the current stage and adjusting the values of control parameters in a modified EO until obtaining high-quality configurations. The experimental results on the well-known hard TSP instances demonstrate that the proposed MSEO may provide much better performances than other physics-inspired algorithms such as classical SA, τ-EO, and a single-stage modified EO. In other words, the simple dynamical probability distribution evolutionary mechanism adopted in MSEO appears to be more appropriate and effective than the traditional static strategy.

3. Furthermore, this chapter also introduces the BGEO (Zeng et al., 2012) for hard SAT and MAX-SAT problems. The basic idea behind BGEO is to incorporate the backbone information extracted from the history of search process into EO to guide the entire search process to approach the optimal or at

least high-quality solutions. The BGEO is essentially a biased local search method that exploits the "big valley" structure of the configuration space. Also, it is similar to the population learning the large-scale structure of the fitness landscape. The experimental results on a variety of hard MAX-SAT problem instances have shown that BGEO outperforms the reported BE-EO algorithm and BE-EEO algorithm without backbone information.

4. Finally, PEO (Chen et al., 2006) with adaptive Lévy mutation is introduced in detail in this chapter, which has many advantages, such as no adjustable parameter, only one mutation operator, and good performance in terms of accuracy and convergence speed, abilities of coarse-grained search and fine-grained search. Experimental results indicate that PEO has very competitive performance with respect to those state-of-the-art approaches.

Chapter 5

Memetic Algorithms with Extremal Optimization

5.1 Introduction to MAs

In the past few decades, the new theory and methods of CI have been developed from fundamentals to practical applications in terms of the combination of computer sciences and control theories. Although the CI methods don't have a solid theoretical foundation, they have been successfully applied in many real-world problems, particularly for those complex systems optimization with global and/ or LS. Among them, the evolutionary computation methods have been great successes in control systems analysis, model identification, and design. However, due to the inherent shortcomings of Darwin's theory, the evolutionary computations could suffer from a lower searching efficiency and inaccuracy to provide real-time solutions.

In recent years, a particular class of global–LS hybrids named MAs are proposed (Moscato, 1989), which are motivated by Dawkins's theory (Dawkins, 1976). MAs are a class of stochastic heuristics for global optimization that combine the global- search nature of EA with LS to improve individual solutions (Hart et al., 2004; Krasnogor and Smith, 2005). They have been successfully applied to hundreds of real-world problems such as optimization of combinatorial optimization (Merz, 2000), multiobjective optimization (MOO) (Knowles and Come, 2001), bioinformatics (Krasnogor, 2004), etc.

As mentioned above, conventional optimization techniques using the deterministic rule-based search often fail or get trapped in local optimum when solving complex problems. On the other hand, compared with deterministic optimization techniques, some CI methods are inefficient and imprecise in fine-tuned LS although they are good at global search, especially when they approach a local region near the global optimum. According to the so-called "No-Free-Lunch" Theorem by Wolpert and Macready (1997), a search algorithm strictly performs in accordance with the quantity and quality of the problem knowledge they incorporate. This fact clearly underpins the exploitation of problem knowledge intrinsic to MAs (Moscato and Cotta, 2003). Under the framework of MAs, the stochastic global-search heuristics approaches are combined with problem-specific solvers, which are a combination of Neo-Darwinian's natural evolution principles and Dawkins' concept of meme (Dawkins, 1976), defined as a unit of cultural evolution that is capable of performing individual learning (local refinement). In MAs, the global character of the search is given by the evolutionary nature of the CI approach while the LS aspect is usually performed by means of constructive methods, intelligent LS heuristics, or other search techniques (Hart et al., 2004). The hybrid algorithms can combine the global explorative power of the CI method with the local exploitation behaviors of conventional optimization techniques, complementing their individual weak points, and thus outperforming either one used alone. MAs have also been named genetic local searchers (Merz, 2000), hybrid GAs (He and Mort, 2000), Lamarckian GAs (Ong and Keane, 2004), Baldwinian GAs (Ku and Mak, 1998), etc.

5.2 Design Principle of MAs

As the theoretical foundation of the evolutionary computation, such as GA, evolutionary strategy, evolutionary programming, genetic programming, etc., the evolution mechanism derived from the natural process of life evolution has been generally recognized and widely used. However, some existing evolutionary models can't reveal a common fact in real natural life evolution: In most cases, the adaptive evolutionary process of the whole complex system is composed of the coevolution process of local interaction between multiple subsystems and the system itself. The bionic algorithms based on the above-mentioned evolutionary model mainly have the following shortcomings and deficiencies:

1. The traditional EAs use a predefined fitness function; according to the theory of coevolution, the fitness of individuals is naturally formed in the environment when struggling for survival and will change as the environment changes.
2. Without considering the possibility of cooperation between organisms, the traditional EAs only consider the competition between organisms, but the truth is that the coexistence of competition and cooperation really exists, which is called coevolution.

Coevolution refers to the phenomenon that two interacting species adapt to each other in the process of evolution, and one species changes in genetic evolution because of another species' influence. This interaction behavior is quite common in nature, as trees grow higher in order to avoid giraffes feeding; on the other hand, this change will lead to giraffes evolving with longer necks.

Ehrlich and Raven put forward the concept of cooperative coevolution when they noticed that the plants and plant hoppers (butterflies) will interact with each other's evolution (Ehrlich and Raven, 1964). In 1977, Haken proposed the "coordination theory," which describes the cooperative behavior between internal elements of the system in the process of evolution and interaction, and states that cooperative behavior is the necessary condition for system evolution (Haken, 1977).

Coevolutionary theory believes that the process of biological evolution in nature is the individual selection pressures in the environment, while the environment includes not only the nonbiological factors but also biological factors. So, the evolution of a species is bound to change the selection pressure effects on other organisms, causing other creatures to also change; these changes in turn lead to further changes of related species. In many cases, two or more separated species often influence each other during the evolution process to form a collaborative interaction of adaptive systems. Considering ecosystem stability and biological diversity, just the coevolutionary strategy, not the competition, will help the components in an ecosystem to have higher energy conversion efficiency, strengthen the self-organizing ability of the system itself, maintain the diversity of ecosystems, and benefit more.

Cooperative coevolution may be interpreted as an evolutionary mechanism, different species interact with each other and coevolve, and this mechanism of biological evolution has an important significance. This can also be understood as a kind of evolution result, because the cooperative coevolution instance is a kind of collaborative relationship, from which instances coordination evolution theory is summarized. Generalized coevolution refers to some kinds of interdependent relationship between biology and biology, biology and environment in the evolution process, which occurs at different biological levels, that is, gene level, cellular level and individual level, level of population, and ecosystem. This can be reflected in the molecular level of deoxyribonucleic acid (DNA) and protein sequences synergy mutation, or the molecular level of DNA and protein sequence mutation, and even in the coevolution of the macro-level species morphological traits, behavior, etc. In a natural ecosystem, the phenomenon of population-based coevolution is very common. In the long-term evolution process, the relation of interactive populations developed from unilateral dependence to dependencies on both sides, is indispensable between populations and has always been marked by interaction, interdependence, mutual promotion, and complementary behavior.

To be clear, coevolution is not a specific algorithm, but an optimization idea derived from biological evolution. Coevolution is the evolution mechanism that imitates cooperation among species in a natural ecosystem. It draws on the ecology group synergy theory, using the principle of automatic adjustment and automatic adaptation between populations. There is a mutual influence and mutual restriction between the population in evolution, driving each other to improve their own and global performance. Therefore, considering the characteristic of the problem to be solved, making combinations of different optimization algorithms based on the coevolution mechanism, can improve the solution efficiency and precision.

Considering the above characteristics of coevolution, it is easy for us to associate it with the popular MAs. The MAs combine the global-search ability of a random algorithm with the high-searching efficiency and precision of the LS algorithm, and realize the collaboration between different search methods, to achieve the unity of the global-search ability and LS efficiency. In recent years, researchers found that the choice of LS operator will significantly influence the performance of the MAs. So, a kind of new collaborative MAs occur, called coevolving memetic algorithm (COMA), which consists of two species: one is based on the gene encoding the candidate solution of the population, and another is based on the LS of meme-coding operator populations (also called a meme population). As the candidate solutions evolve, the meme population itself will evolve too, which leads the LS operator to gradually change into the most suitable form. Smith summarized the development and application of COMA algorithms in 2007 (Smith, 2007).

From the above description, we can see that both the coevolutionary algorithm and MAs are derived from the mechanisms of nature and human society. Coevolution plays more important roles in the cooperative model from the perspective of the theory and the coordination mechanism between different populations at same/different levels in the evolution process, whereas MAs focus on the algorithm realization and applications of the macroscopic behavior and the collaborative optimization between different levels and different evolutionary behaviors (operators). These two algorithms have much in common, such as communication, influence, and relations between multiple populations/levels; there are also some differences, such as the coordination mechanism. With the development of theoretical research, the difference between them is becoming increasingly smaller.

From the description and analysis of EO dynamics in Section I, we can see that the EO algorithm reflects coevolutionary thought to some extent. That is, the ability of a component depends not only on the environment around it, but also on the changes of its adjacent components. Those components will not only compete but also coordinate for common development. However, the BS model, as the foundation of EO, is the simple and rough approximation of biological evolution in nature. The described coordination and evolution are components based on the links at the gene level, while the actual coevolution occurs at different levels and between different populations, even different evolution modes. Comprehensive coevolution thought is well reflected by the proposed MAs. In the MAs, the global search and LS operators were effectively combined to realize the cooperation among different evolution levels and modes. Therefore, under the instruction of coevolution between different levels/modes, this chapter proposes two novel hybrid optimization methods with the combination of recently proposed heuristic EO and popular deterministic methods under the conceptual umbrella of MAs. The proposed methods balance both aspects through the hybridization of heuristic EO as the global-search scheme and the deterministic method as the LS scheme. Owing to the inherited high-searching efficiency/accuracy from deterministic methods and the global-searching capability from the EO method, the proposed method shows considerable performance improvement over traditional methods.

In the following sections, we will introduce the authors' research and developments on integration of EO with LM (Levenberg–Marquardt gradient search [GS]), SQP (sequential quadratic programming), PSO (particle swarm optimization), ABC (artificial bee colony), and GA to form the hybrid EO–LM (Chen and Lu, 2010a), EO–SQP (Chen and Lu, 2011a), PSO–EO (Chen et al., 2010b), ABC–EO (Chen et al., 2014a), and GA–EO, respectively.

5.3 EO–LM Integration

5.3.1 Introduction

As mentioned above, the efficiency of optimization search can be improved significantly by incorporating a LS procedure into optimization; the LS algorithm could be gradient-based methods such as LM GS or other methods. In this section, a hybrid EO–LM algorithm is introduced and applied in NN (neural network) training. The structure of the hybrid EO–LM algorithm is based on standard EO, the characteristic of GS is added by propagating the individual solution with the LM algorithm during EO evolution. The proposed EO–LM solution has the abilities to avoid local minimum and perform detailed LS with both efficiency and robustness. The incorporation of the stochastic EO method with the conventional deterministic LM algorithm can combine the global explorative power of EO with the local exploitation behaviors of LM, complementing their individual weak points, and thus make multilayer perceptron (MLP) network training superior in generalization, computation efficiency, and avoiding local minima.

The properties of the feed-forward MLP network are governed by the activation functions of neurons and the synaptic connections between the layered neurons, as shown in Figure 5.1. The associative memories from input space to output space are built up and stored in the synaptic weights through supervised learning from learning samples. The performance under its working environment measures the generalization capability of an MLP network (Haykin, 1994). After introduced by Werbos (1974) and popularized by Rumelhart et al. (1986a, b), the GS-based back propagation (BP) algorithm has been the most popular learning technique in MLP network training due to the simplicity and applicability of its implementation. However, in view of the drawbacks of GS in nature, such as easily trapping into local minima, sensitivity to initial weights, poor generalization (Haykin, 1994; Salomon, 1998), etc., there have been a variety of well-known attempts to improve

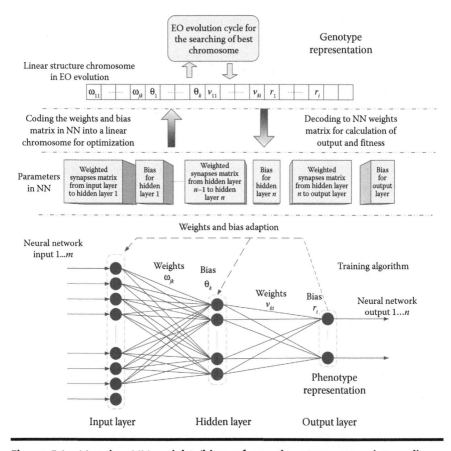

Figure 5.1 Mapping NN weights/biases from phenotype space into a linear chromosome of genotype space. (From Chen, P. et al. *International Journal of Computational Intelligence Systems* 3: 622–631, 2010. With permission.)

the original BP algorithm (Jacobs, 1988; Rigler et al., 1991; Fukuoka et al., 1998). The applications of these approaches may result in better solutions, but require higher computation cost (Dengiz et al., 2009).

On the other hand, Hush has proved that the parameter optimization for an MLP network with sigmoid function is an NP-hard problem (Hush, 1999). The recent research results in bioinspired CI (Engelbrecht, 2007) (e.g., EAs, EO, and ant colony optimization [ACO]) and their superior capabilities in solving NP-hard and complex optimization problems have motivated researchers to use CI methods for the training of the MLP network. One way to overcome the drawbacks of BP learning is to formulate the training process as CI-based evolution of the MLP network structure, synaptic weights, learning rule, input features (Arifovic and Gencay, 2001; Li et al., 2007; Yao and Islam, 2008; Dengiz et al., 2009; Fasih et al., 2009; Reyaz-Ahmed et al., 2009; Sedkia et al., 2009), etc. In fact, the NN evolution with CI methods may significantly enlarge its search space and provide better performance than BP algorithms. However, most CI methods are rather inefficient in fine-tuned LS although they are good at global search, especially when the searching solutions approach a local region near the global optimum, this will result in high computation cost.

Based on the complexity of nonlinear optimization involved in NN learning, P. Chen et al. (2010) presented the development of a novel MA-based hybrid method called the "EO–LM" learning algorithm, which combines the recently proposed heuristic EO (Boettcher and Percus, 1999) with the popular LM GS algorithm (Hagan and Menhaj, 1994). In this section, we will first give the math formulation for the problem under study, then illustrate the EO–LM fundamentals and algorithms, and finally show the comparison results between the EO–LM and standard LM algorithms on three experimental problems.

5.3.2 Problem Statement and Math Formulation

A feed-forward MLP network with a single hidden layer is shown in Figure 5.1, if we select the tan-sigmoid and linear function as the activation functions of hidden and output layers, respectively; the map from jth input x_j ($j = 1, \ldots, m$) to ith output $\hat{y}_i$ ($i = 1, \ldots, n$) can be written as (Haykin, 1994)

$$\hat{y}_i = f(X, w, v, \theta, r) = \sum_{k=1}^{p} (v_{ki} z_k + r_i)$$

$$= \sum_{k=1}^{p} \left(v_{ki} \log \left(\sum_{j=1}^{m} \omega_{jk} x_j + \theta_k \right) + r_i \right) \quad i = 1, \ldots, n$$

(5.1)

z_k, the kth hidden layer variable ($k = 1, \ldots, p$)
ω_{jk}, the weight linking the jth input variable with the kth hidden layer variable
vki, the weight linking the kth hidden layer variable with the ith output variable
θ_k, the bias of the kth hidden layer variable
r_i, the bias of the ith output variable
log sig, the logistic transfer function log sig (a) = $1/[(1 + \exp(-a))]$

In NN training, the learning samples are often divided into a training dataset and a validation dataset, and the former is used for updating the network weights and biases. The error on the validation set is monitored during the training process, which will guarantee the generalization of the NN. The aim of this study is to develop a novel MA-based NN-learning algorithm for an MLP network that may provide good performance in generalization and robustness with the minimum output error during the optimization process of the synaptic weights

$$\begin{cases} \text{Min} \quad E(w,v,\theta,r) = \sum_{i=1}^{n} \sum_{l=1}^{n\_Train} \left[y_i^l - \hat{y}_i^l \right]^2 \\ s.t \quad w \in R^{m^*p}, v \in R^{p^*n}, \theta \in R^p, r \in R^n \end{cases} \tag{5.2}$$

where *n_Train* represents the training data number. *w*, *v*, θ, and *r* are bounded by the searching space of the optimization algorithm. y_i represents the *i*th desired output.

5.3.3 Introduction of LM GS

The LM GS algorithm was introduced to feed-forward network training to provide better performance (Hagan and Menhaj, 1994). Generally, the LM algorithm is a Hessian-based algorithm for nonlinear least-squares optimization (Nocedal and Stephen, 2006). Similar to the quasi-Newton methods, the LM algorithm was designed to approach second-order training speed without having to compute the Hessian matrix. Under the assumption that the error function is some kind of a squared sum, the Hessian matrix can be approximated as

$$H = J^T J \tag{5.3}$$

and the gradient can be computed as

$$g = J^T e \tag{5.4}$$

where *J* is the Jacobian matrix that contains first derivatives of the network errors with respect to weights and biases, and *e* is an error vector. The Jacobian matrix can be computed through a standard BP technique that is much less complex than computing the Hessian matrix (Hagan and Menhaj, 1994).

The LM algorithm uses this approximation to the Hessian matrix in the following Newton-like update:

$$x_{k+1} = x_k - [J^T J + \mu I]^{-1} J^T e \tag{5.5}$$

The parameter μ is a scalar controlling the behavior of the algorithm. The convergence behavior of the LM is similar to that of the Gauss–Newton method. Near a solution with a small residual, it performs well and keeps a very fast convergence rate; while for the large-residual case, the performance of the Gauss–Newton and LM algorithms is usually poor (Nocedal and Stephen, 2006).

5.3.4 MA-Based Hybrid EO–LM Algorithm

The fundamentals in NN learning have been an attractive fundamental research area in the physics community in the past two decades (Seung et al., 1992; Engel, 2001). The cooperative behaviors of interacting neurons and synapses have been widely studied in terms of statistical physics. The competition between training error and entropy may lead to discontinuous behavior of the NN (Kinzel, 1998). The corresponding research also discovers discontinuous properties as a function of model parameters or the number of presented examples (Kinzel, 1999; Biehl et al., 2000). By analogy with physical systems, a phase transition is defined as a qualitative change in network behavior that occurs when the network parameters cross a critical boundary. The learning curves in the NN training exhibit a diversity of behaviors such as phase transition, which has been observed and reported (Fukumizu and Amari, 2000; Hohmann et al., 2008). The problem caused by local minima and plateaus of a NN is studied and a set of critical points in parameter space have been found. In practical applications, the learning process often suffers extremely slow dynamics around some critical points, which causes a very slow decrease of the training error for a long time before a sudden exit from it (Fukumizu and Amari, 2000). This phenomenon can be interpreted as a typical phase transition in NN learning. In this study, we discover that even a small weight change might result in large generalization performance changes in the end-point–temperature prediction for a production- scale basic oxygen furnace in steelmaking, as shown in Figure 5.2a; namely, the training dynamic responses show a second-order phase transition from good to poor generalization. Unlike the equilibrium-based approaches, researchers have surprisingly found that heuristic algorithms inspired by nonequilibrium physical processes such as EO do not seem to experience difficulties at criticality. In this study, we not only find the phase transition in practice and its damage to the training performance using the LM algorithm (see Figure 5.2a), but also discover that the proposed EO–LM algorithm may provide good performance (see Figure 5.2b). This opens a new door for hybrid training with the EO–LM algorithm.

As mentioned before, EO and its derivatives have been extensively applied in solving numerous NP-hard optimization problems. In this study, an EO–LM algorithm is developed and applied in NN training (Chen and Lu, 2010a, reproduced here in full, with permission from *Journal of Zhejiang University, Science A*). The proposed hybrid "EO–LM" solution has the abilities in avoiding local minimum, dealing with phase transition, and performing a detailed search when approaching the near-optimal region. The EO–LM learning is executed between two phases in parallel: the genotype phase for EO–LM and the phenotype phase for NN as shown in Figure 5.1. The synaptic weights and bias are coded as a real-valued chromosome, that is, a data string, to be evolved during EO–LM learning iterations.

Here, we illustrate the workflow of the algorithm and introduce three mutation operators adopted in this study: standard EO mutation, LM mutation, and multistart Gaussian mutation. To utilize the advantages of each mutation operator, one or more

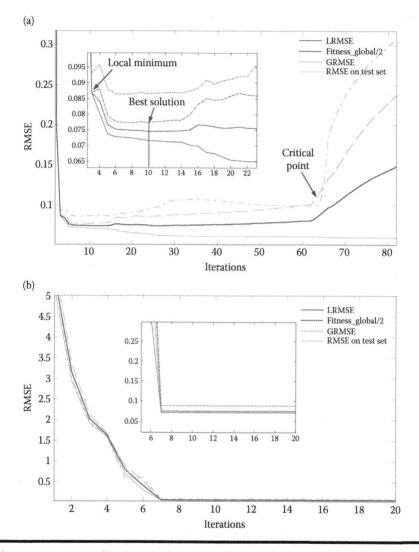

Figure 5.2 Generalization ability on training and validation data set and the phase transition in NN training. (a) Learning curve using LM and (b) learning curve using EO–LM.

phases of LS (mutation operator) are applied to the best solution S so far based on a probability parameter p_m in each generation. In contrast to the standard EO mutation, when LM mutation or multistart Gaussian mutation is adopted, we use the "GEO_{var}" (De Sousa et al., 2004) strategy to evolve the current solution by improving all variables simultaneously, as an attempt to speed up the process of searching the local minimum. There are two evolutionary levels during the proposed EO–LM optimization: on one hand, evolution takes place at the "chromosome level" as in any other EA; chromosomes (genes) represent solutions and features of the problem to be solved. On the other hand, evolution also happens at the "meme level," that is, the behaviors that individuals will use to alter the survival value of their chromosomes (Krasnogor and Gustafson, 2004). Accordingly, the solutions are evaluated by fitness functions of two different levels: the fitness of the respective gene itself (global fitness), the interaction fitness between the associated gene and meme (local fitness). Thus, both genetic and meme materials are coevolved, the evolutionary changes at the gene level are expected to influence the evolution at the meme level, and vice versa. The proposed EO–LM is able to self-assemble different mutation operators and coevolve the behaviors it needs to successfully solve the NN supervised learning problem. The flowchart of the proposed EO–LM algorithm to optimize parameters (the connection weights and the biases) of MLP network is shown in Figure 5.3.

The work steps of the proposed EO–LM-based MLP-training algorithm in this study can be described as below:

1. Define the number of hidden layers, the numbers of input neurons, output neurons, and the control parameters to be used in EO–LM algorithm.
2. Initialize the NN with randomly generated weights and biases based on the predefined structure in Step (1).

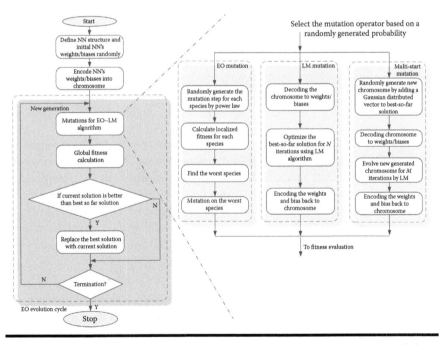

Figure 5.3 Flowchart of the EO–LM-based NN learning. (Reproduced from Chen, P. and Lu, Y. Z., *Journal of Zhejiang University, Science A* **11: 841–848, 2010a. With permission.)**

3. Map the weights/biases matrices of the NN from the problem oriented phenotype space into a chromosome, as shown in Figure 5.1.
4. For the first iteration of EO, decode the initial chromosome S back to weights/biases matrices and calculate the object fitness function, set $S_{best} = S$.
5. Decide what kind of mutation operators should be imposed to the current chromosome S based on randomly generated probability parameters p_m, if $P_m \leq P_{m\_basic}$, go to (i); else if $P_{m\_basic} < P_m \leq P_{m\_LM}$, go to (ii); else if $P_m > P_{m\_LM}$, go to (iii).

a. Perform the standard EO mutation on the best so-far solution S.
   i. Change the value of each component in the current S and get a set of new solutions S'_k, $k \in [1, 2, ..., n]$.
   ii. Sequentially evaluate the localized fitness λ_k specified in Equation 5.10 for every S'_k, and rank them according to their fitness values.
   iii. Choose the best solution S' from the new solutions set [S'], which is a neighbor subspace of the best so far solution S.

b. Perform the LM mutation on the current chromosome S.
   i. Decode the chromosome S back to weights/biases matrices in MLP networks.
   ii. The weight vector is updated for N iterations by

$$S' = S - [J^T J + \mu I]^{-1} J^T e \qquad (5.6)$$

   where J is the Jacobian matrix, e is a vector of network errors defined in the following equation:

$$e(x) = \sum_{i=1}^{n} \sum_{l=1}^{n\_Train} (y_i^l - \hat{y}_i^l)^2 \qquad (5.7)$$

   iii. Encode the updated weights/biases matrices to the chromosome S'.

c. Perform the multistart Gaussian mutation on the current chromosome S. Multistart methods have their main objective to increase diversity, whereby larger parts of the search space are explored. This strategy is often adopted in MAs to explore the neighborhood of the current solution.
   i. Generate a new chromosome S'_0 by adding a Gaussian distribution random vector with n dimensions to the best so far chromosome S.

$$S'_0 = S + Scale * N(0,1) \qquad (5.8)$$

   where n is the length of chromosome, Scale is the mutation step size.
   ii. Decode the chromosome S'_0 back to weights/biases matrices in MLP networks.
   iii. Training the MLP network based on the LM algorithm in Equation 5.6 for M iterations.
   iv. Encode the updated weights/biases matrices to the chromosome S'.

6. Decode the chromosome S' back to weights/biases matrices and calculate the global object fitness function. If $F(S') < F(S_{best})$, Set $S_{best} = S'$.
7. If the termination criteria are not satisfied, go to Step (5), else go to the next step.
8. Return S_{best}.

5.3.5 Fitness Function

The fitness function measures how fit an individual (i.e., solution) is, and the "fittest" one has more chance to be inherited into the next generation. A "global fitness" must be defined to evaluate how good a solution is. The errors on training set and validation set are often used to control and monitor the NN training process. Overfitting usually occurs during the NN training with descending training error and ascending prediction error. It greatly debilitates the generalization ability of a network. Consequently, in this study, the global fitness is defined as the sum of root mean square error ($RMSE$) on training set ($LRMSE$—root mean square error on training dataset [which reflects the learning performance of the model]) and validation set ($GRMSE$—root mean square error on validation dataset [which reflects the generalization ability of the model]), as defined below

$$Fitness_{global}(S) = LRMSE_{S(w,v,\theta,r)} + GRMSE_{S(w,v,\theta,r)}$$

$$= \sqrt{\frac{\sum_{i=1}^{n}\sum_{l=1}^{n\_Train}(y_i^l - \hat{y}_i^l)^2}{n * n\_Train}} + \sqrt{\frac{\sum_{i=1}^{n}\sum_{l=1}^{n\_Valid}(y_i^l - \hat{y}_i^l)^2}{n * n\_Valid}} \qquad (5.9)$$

Unlike GA, which works with a population of candidate solutions, EO depends on a single individual (i.e., chromosome)-based evolution. Through performing mutation on the worst component and its neighbors successively, the individual in EO can evolve itself toward the global optimal solution generation by generation. This requires a suitable representation which permits each component to be assigned a quality measure (i.e., fitness) called "local fitness." In this study, the local fitness λ_k is defined as an improvement in $LRMSE$ made by the mutation imposed on the kth component of best-so-far chromosome S:

$$\lambda_k = Fitness_{local}(k) = \Delta LRMSE(k)$$

$$= LRMSE_{S(w,v,\theta,r)} - LRMSE_{Sk(w,v,\theta,r)} \qquad (5.10)$$

As long as the NN is not overfitting, the improvement on the local fitness λ_k will also improve the global fitness described in Equation 5.9.

5.3.6 Experimental Tests on Benchmark Problems

This section presents the experimental tests of the EO–LM algorithm on several benchmark problems. Without loss of generality, input/output data used are normalized to the range [0.1, 0.9].

5.3.6.1 A Multi-Input, Single-Output Static Nonlinear Function

In this example, the EO–LM algorithm is applied to establish a multi-input, single-output (MISO) mapping specified by a highly nonlinear function as follows:

$$Y = 0.2(x_1^{0.5} + 2x_1x_2^{0.5} + x_2x_3 + x_3) \qquad (5.11)$$

First we randomly generate 300 I/O observation pairs, for which the input variables are generated randomly in region I: $x_1 \in [0, 1]$, $x_2 \in [0, 1]$, $x_3 \in [0, 1]$, use 200 of them as learning data and the other 100 as interpolation testing data to measure the generalization performance, we randomly generate another 100 I/O data pairs within region II: $x_1 \in [1, 2]$, $x_2 \in [1, 2]$, $x_3 \in [1, 2]$, they are beyond region I and will be used as an extrapolation testing data set. In this case, the structure of the network is {3, 2, 1}, namely three, two, and one nodes in input, hidden, and output layers, respectively, totally 11 parameters need to be optimized in this example.

For the purpose of fair comparison, this test was repeated 10 times using the Monte Carlo method for the EO–LM and standard LM algorithms. The comparison results between these two algorithms are listed in Table 5.1. In addition to the popular criteria of RMSE and mean error (ME), the efficiency coefficients R^2 or R, which measures the proportion of the variation of the observations around the mean (usually explained by the fitted regression model), is also be used as an additional measure. The value of R^2 or R falls between 0 and 1. When the R^2 or R reaches to 1, the model's outputs perfectly agree with its system's actual outputs. Table 5.1 and Figure 5.4 show the comparison between EO–LM and LM based on the statistical data.

Table 5.1 shows the performance of each algorithm over 10 runs, giving an indication of robustness and generalization ability of each algorithm. As mentioned above, the better solution has smaller RMSE over "train," "interpolation test," and "extrapolation test" datasets, and lower standard deviation as well. We can see that the EO–LM algorithm performs much better than the standard LM algorithm. Figure 5.4 shows comparison of RMSE distribution between EO–LM and LM on training and test data. From Table 5.1 and Figure 5.4, we can see that the solution is more robust and consistent than that evolved by the standard LM.

Figure 5.5 gives the comparisons in generalization performance between the EO–LM and LM algorithms on training, interpolation, and extrapolation test

Table 5.1   Comparison between the EO–LM and LM

		EO–LM Algorithm			LM Algorithm		
		Train Data	Test Data (Interset)	Test Data (Extra set)	Train Data	Test Data (Interset)	Test Data (Extra set)
RMSE	Max	0.0077	0.0091	0.1857	0.0306	0.0328	0.4000
	Min	0.0075	0.0090	0.0688	0.0075	0.0091	0.1328
	Mean	0.0076	0.0090	0.1273	0.0141	0.0149	0.2212
	Standard deviation	6.76e − 05	4.56e − 05	0.0378	0.00997	0.0094	0.0977
R^2	Mean	0.9985	0.9970	0.7819	0.9924	0.9890	N/A
Training Time		61.44 s			74.53 s		

Source: Adapted from Chen, P. et al., *International Journal of Computational Intelligence Systems* 3: 622–631, 2010.

datasets. We can see both algorithms provide good results for training dataset and EO–LM performs slightly better on the interpolation test dataset, but much better on the extrapolation test dataset. The extrapolation test dataset is a most crucial test phase to tell which algorithm will have a better ability to learn, EO–LM performs pretty well in this case, while LM corrupts.

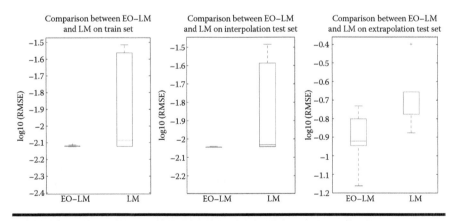

Figure 5.4   Comparison of RMSE distribution between EO–LM and LM on train/ interpolation test/extrapolation test set. (From Chen, P. et al., *International Journal of Computational Intelligence Systems* 3: 622–631, 2010. With permission.)

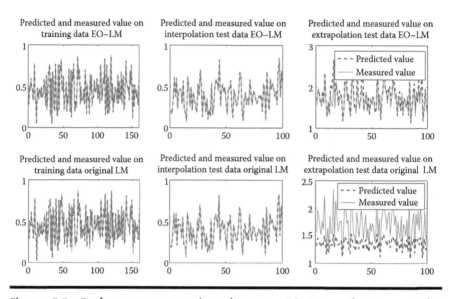

Figure 5.5 **Performance comparison between EO–LM and LM on train/ interpolation test/extrapolation test set. (From Chen, P. et al.,** *International Journal of Computational Intelligence Systems* **3: 622–631, 2010. With permission.)**

5.3.6.2 Five-Dimensional Ackley Function Regression

In this example, the EO–LM algorithm is applied to build a model for an "Ackley function" benchmark as shown in Equation 5.12. The Ackley function is a continuous, multimodal function obtained by modulating an exponential function with a cosine wave of moderate amplitude, as shown in Figure 5.6. Originally, it was formulated by Ackley only for the two-dimensional case (Ackley, 1987); then, the problem was generalized to n dimensions

$$f(X) = 20 + e - 20e^{\left(-0.2\sqrt{1/n\sum_{i=1}^{n} x_i^2}\right)} - e^{1/n\sum_{i=1}^{n} \cos(2\pi x_i)} \tag{5.12}$$

In this study, $n = 5$. The structure of the network is {5, 8, 1} and a total of 57 parameters need to be optimized. The training process was repeated 10 times for each algorithm randomly. The performance on training set and testing set of the proposed EO–LM and standard LM is listed in Table 5.2.

The prediction of the EO–LM algorithm simulated with test data in Figure 5.6 shows a better agreement with the target than the best results by the standard LM.

5.3.6.3 Dynamic Modeling for Continuously Stirred Tank Reactor

The system considered here is an isothermal continuous stirred tank reactor (CSTR) process with first-order reaction $A + B \rightarrow P$, in the presence of excess

Table 5.2 Comparison between the EO–LM and LM on Ackley Function

	Train Set				Test Set
	Mean_ RMSE	Max_ RMSE	Min_ RMSE	Std. Dev_RMSE	Mean_ RMSE
EO–LM	0.0379	0.0396	0.0365	8.7888e − 04	0.0276
LM	0.0394	0.0403	0.0369	0.0011	0.0308

concentration of A. The corresponding multi-input multioutput (MIMO) CSTR model is as follows (Martinsen et al., 2004):

$$
\begin{cases}
\dfrac{dx_1}{dt} = u_1 + u_2 - k_1\sqrt{x_1} \\[2ex]
\dfrac{dx_2}{dt} = (C_{B1} - x_2)\dfrac{u_1}{x_1} + (C_{B2} - x_2)\dfrac{u_2}{x_1} - \dfrac{k_2 x_2}{(1 + x_2)^2}
\end{cases}
\tag{5.13}
$$

$$
y = \begin{pmatrix} 1 & 0 \\ 0 & 1 \end{pmatrix}[x_1 \quad x_2]^T
$$

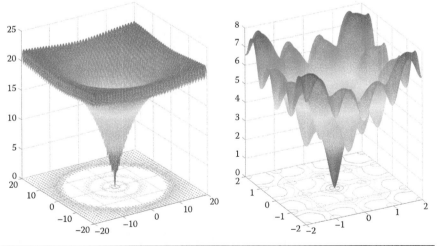

Landscape for two-dimensional Ackley function   Landscape for two-dimensional Ackley function

Figure 5.6 (See color insert.) Landscape for two-dimensional Ackley function; left: surface plot in an area from −20 to 20, right: focus around the area of the global optimum at [0, 0] in an area from −2 to 2.

Table 5.3 Comparison between the EO–LM and LM on CSTR Model

| | | EO–LM Algorithm | | | | | LM Algorithm | | | | | |
| | | Train Set | | Test Set | | | Train Set | | Test Data Set | | |
		Y1	Y2	Y1	Y2		Y1	Y2	Y1	Y2	
RMSE	Best	0.0023	0.0024	0.0043	0.0045	0.0025	0.0031	0.0048	0.0048			
	Mean	0.0024	0.0025	0.0046	0.0052	0.0027	0.0031	0.0051	0.0054			
	Standard deviation	8.05e − 05	1.81e − 04	2.41e − 04	5.98e − 04	3.01e − 04	3.38e − 04	3.53e − 04	4.13e − 04			

Source: From Chen, P. et al., *International Journal of Computational Intelligence Systems* 3: 622–631, 2010. With permission.

$$s.t \quad \begin{cases} 0.1 \le u_1 \le 2 \\ 0.1 \le u_2 \le 2 \end{cases} \tag{5.14}$$

where $k_1 = 0.2$, $k_2 = 1$, $C_{B1} = 24.9$, and $C_{B2} = 0.1$.

The plant has two inputs and two outputs, their relationship expresses strong nonlinearity. For $(u_1, u_2) = (1, 1)$, the CSTR has three equilibrium points at $x_1 = 100$, $x_2 = (0.633, 2.72, 7.07)$, with the middle equilibrium point being unstable, and the others stable.

Here, we use the nonlinear autoregressive model with exogenous inputs (NARX) to model the dynamic systems shown in Equations 5.13 and 5.14, it can be mathematically represented as follows:

$$Y(n) = f(u_1(n-1), u_2(n-1), y_1(n-1), y_2(n-1)) \tag{5.15}$$

where $Y(n) = [y_1(n), y_2(n)]^T = [x_1(n), x_2(n)]^T$, In this case, a MIMO MLP network is employed to build the nonlinear mapping $f(\bullet)$ between inputs and outputs. The structure of the network is {4, 10, 2}, totally 72 parameters need to be optimized. We randomly generate 500 observation pairs and use 400 of them as the learning data and the other 100 as the test data set. The example was run 10 times for each algorithm randomly. The performance by the EO–LM and standard LM is listed in Table 5.3.

Table 5.3 shows the comparison between predicted and measured values at training and test phases by the hybrid EO–LM and LM. We can see that the EO–LM performs better than the standard LM both on train data set and test data set.

It can be seen the EO–LM is better than the LM in both RMSE and generalization performance. We can see that the EO–LM can easily avoid the local minima, overfitting, or underfitting problems. The dynamic responses of the weights corresponding to Figure 5.2 are shown in Figure 5.7. It can be seen that the LM suffers a phase transition from the 62th iteration and the whole weights suddenly evolved large fluctuations simultaneously when the set of parameters of the network crosses its boundary. The LM takes a long time to converge a "metastable" state slowly with a minor improvement in *LRMSE*. In contrast, the EO–LM also has a phase transition or critical point at the fifth iteration, but the EO–LM is able to handle it quite well due to the natural capability of EO to deal with phase transitions.

5.4 EO–SQP Integration

5.4.1 Introduction

With high demand in decision and optimization for many real-world problems and the progress in computer science, the research on novel global optimization

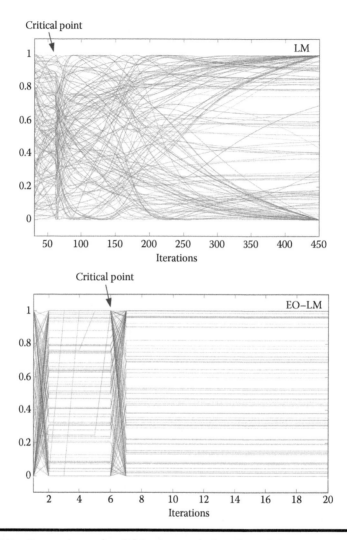

Figure 5.7 **Comparison of weights change during the training process using LM and EO–LM.**

solutions has been a challenge to academic and industrial societies. During the past few decades, various optimization techniques have been intensively studied; these techniques follow different approaches and can be divided roughly into three main categories, namely, the deterministic methods (Nocedal and Stephen, 2006), stochastic methods (Kall and Wallace, 1994), and bioinspired CI (Engelbrecht, 2007).

In general, most global optimization problems are intractable, especially when the optimization problem has complex landscape and the feasible region is concave and covers a very small part of the whole search space. Solution accuracy and global convergence are two important factors in the development of optimization techniques. Deterministic search methods are known to be very efficient with high accuracy. Unfortunately, they are easily trapped in local minima. It is hardly possible to design a deterministic algorithm that would outperform the exhaustive search in assuring the solution obtained to be the true global optimum. On the other hand, methods of CI (EAs, EO, ACO, etc.) are much more effective for traversing these complex surfaces and inherently better suited for avoiding local minima (Engelbrecht, 2007). However, CI has its weakness in slow convergence and providing a precise-enough solution because of the failure to exploit local information. Moreover, for constrained optimization problems, the given constraints must be met to have feasible solutions; CI methods often lack an explicit mechanism to bias the search in feasible regions.

This section proposes a novel hybrid EO–SQP method with the combination of EO and the popular deterministic SQP under the conceptual umbrella of MAs. EO is a general-purpose heuristic algorithm, with the superior features of SOC, nonequilibrium dynamics, coevolutions in statistical mechanics, and ecosystems, respectively. Inspired by far-from-equilibrium dynamics, this approach provides a new philosophy to optimize based on nonequilibrium statistical physics and the capability to elucidate the properties of phase transitions in complex optimization problems. SQP has been one of the most popular methods for nonlinear optimization because of its efficiency in solving medium- and small-size nonlinear programming (NLP) problems. It guarantees local optima as it follows a GS direction from starting point toward optimum point and has special advantages in dealing with various constraints. This will be particularly helpful for the hybrid EO–SQP algorithm when solving constrained optimization problems: the SQP can also serve as a means of "repairing" infeasible solutions during EO evolution. The proposed method balances both aspects through the hybridization of heuristic EO as the global-search scheme and deterministic SQP as the LS scheme. The performance of the proposed EO–SQP algorithm is tested on 12 benchmark numerical optimization problems and compared with some other state-of-the-art approaches. The experimental results show the EO–SQP method is capable of finding the optimal or near-optimal solutions for NLP problems effectively and efficiently.

5.4.2 Problem Formulation

Many real-world optimization problems can be mathematically modeled in terms of a desired objective function subject to a set of constraints as follows:

$$\text{Minimize } f(X), \quad X = [x_1, x_2, \ldots, x_n] \tag{5.16}$$

subject to

$$g_t(X) \leq 0; \quad t = 1, 2, \ldots, p \tag{5.17}$$

$$h_u(X) = 0; \quad u = 1, 2, \ldots, q \tag{5.18}$$

$$\underline{x}_v \le x_v \le \bar{x}_v; \quad v = 1, 2, \ldots, n \tag{5.19}$$

where $X \in R^n$ is an n-dimensional vector representing the solution of problems (5.16) through (5.19), $f(X)$ is the objective function, which needs to satisfy p-inequality constraints $g_t(X)$, q-equality constraints $h_u(X)$, x_v and $\bar{x}_v$ are the lower and upper bounds of the variable x_v. The objective function and the constraints may have properties such as linear or nonlinear, convex or nonconvex, differential or nondifferential, unimodal, or multimodal. This research is not restricted to any particular property.

The above formulation is an instance of the well-known NLP problem. In general, the global optimization of NLP is one of the toughest NP-hard problems. Solving this type of problem has become a challenge to computer science and operations research.

5.4.3 Introduction of SQP

After its initial proposal by Wilson (1963), the SQP method was popularized in the 1970s by Han (1976) and Powell (1978). SQP proves itself as the most successful method and outperforms the other NLP methods in terms of efficiency and accuracy to solve nonlinear optimization problems. The solution procedure is on the basis of formulating and solving a quadratic subproblem with iterative search. At each iteration, an approximation is made from the Hessian matrix of Lagrangian function; after that a quadratic programming (QP) is used to compute an optimal search direction with the updated Hessian matrix in a subregion; finally the optimum search length is determined by a single-variable optimization method using a line search strategy with an objective to minimize a merit function. The procedure will be repeated until the given termination criteria are satisfied.

SQP used in this study consists of three main stages as follows:

1. Update the Hessian matrix of Lagrangian function based on the most popular algorithms called "Broyden–Fletcher–Goldfarb–Shanno" (BFGS) as shown below

$$H_{k+1} = H_k + \frac{q_k q_k^T}{q_k^T s_k} - \frac{H_k^T s_k^T s_k H_k}{s_k^T H_k s_k} \tag{5.20}$$

where

$$s_k = x_{k+1} - x_k \tag{5.21}$$

$$q_k = \left(\nabla f(x_{k+1}) + \sum_{i=1}^{m} \lambda_i \cdot \nabla g_i(x_{k+1}) \right) - \left(\nabla f(x_k) + \sum_{i=1}^{m} \lambda_i \cdot \nabla g_i(x_{k+1}) \right) \quad (5.22)$$

and λ_i, $i = 1, \ldots, m$, is the estimation of Lagrangian multiplier.

2. In the second stage, a search direction d_k can be obtained by solving the QP subproblem

$$\min_{d \in \Re^n} q(d) = \frac{1}{2} d_k^T H_k d_k + \nabla f(x_k)^T d_k \quad (5.23)$$

subject to

$$\left[\nabla g(x_k) \right]^T d_k + g_i(x_k) = 0 \quad i = 1, \ldots, m_e$$
$$\left[\nabla g(x_k) \right]^T d_k + g_i(x_k) \leq 0 \quad i = m_e + 1, \ldots, m \quad (5.24)$$

where

H_k, Hessian matrix of Lagrangian function defined by $L(x, \lambda) = f(x) + \lambda^T g_i(x)$ at $x = x_k$,

d_k, Basis for a search direction at iteration k,

$f(x)$, Objective function in Equation 5.16,

$g(x)$, Constraints described in Equations 5.17 and 5.18,

m_e, Number of equality constraints,

m, Number of inequality constraints

3. The next iteration x_{k+1} is updated by

$$x_{k+1} = x_k + a_k \cdot d_k \quad (5.25)$$

The optimum search length parameter a_k is determined by an appropriate line search procedure based on the current point x_k and the search direction d_k, so that a sufficient decrease in a merit function is obtained. This ensures a balance between reducing the objective function and reducing infeasibility.

5.4.4 MA-Based Hybrid EO–SQP Algorithm

As mentioned above, the deterministic rule inspired conventional optimization techniques often fail or get trapped in local optimum when solving complex problems. In contrast to deterministic optimization techniques, many CI-based optimization methods are good at global search, but relatively poor in fine-tuned LS when the solutions approach a local region near the global optimum. According to the so-called "No-Free-Lunch" Theorem by Wolpert and Macready (1997), a search algorithm strictly performs in accordance with the amount and quality of the problem knowledge they incorporate. This fact clearly underpins the exploitation of problem knowledge intrinsic to MAs. Under the framework of MAs, the stochastic global- search heuristics work together with problem-specific solvers, in which Neo-Darwinian's natural evolution principles are combined with Dawkins' concept of a meme (Dawkins, 1976) defined as a unit of cultural evolution that is capable of performing individual learning (local refinement). The global character of the search is given by the evolutionary nature of CI approaches while the LS is usually performed by means of constructive methods, intelligent LS heuristics or other search techniques (Hart et al., 2004). The hybrid algorithms can combine the global explorative power of CI methods with the local exploitation behavior of conventional optimization techniques, complement their individual weak points, and thus outperform either one used alone.

Moreover, since the natural link between hard optimization and statistical physics, the dynamic properties and computational complexity of the optimization have been attractive fundamental research topics in the physics community in the past two decades. Murty and Kabadi (1987, p. 118) first used the technique of discrete combinatorial complexity theory to study the computational difficulty of continuous optimization problems and found that "Computing a global minimum, or checking whether a given feasible solution is a global minimum, for a smooth nonconvex NLP, may be hard problems in general." It has been recognized that one of the real complexities in optimization comes from the phase transition, for example, "easy–hard–easy" search path (Rogers et al., 2006). Phase transitions are found in many combinatorial optimization problems, and have been observed in the region of continuous parameter space containing the hardest instances (Monasson et al., 1999; Fukumizu and Amari, 2000; Ramos et al., 2005). It has been shown that many problems exhibit "critical boundaries," across which dramatic changes occur in the computational difficulty and solution character, the problems become easier to solve away from the boundary (De Sousa et al., 2004). Unlike the equilibrium approaches such as SA, EO as a general-purpose method inspired by nonequilibrium physical processes shows no signs of diminished performance near the critical point, which is deemed to be the origin of the hardest instances in terms of computational complexity. This opens a new door for the development of a high-performance solution with fast global convergence and good accuracy in terms of a hybrid EO–SQP algorithm proposed in this section.

In this section, an MA-based hybrid EO–SQP algorithm is developed and applied to NLP problems. The proposed algorithm is a hybridization of EO and SQP. We intend to make use of the capacity of both algorithms: the ability of EO to find a solution close to the global optimum and effectively dealing with phase transition; the ability of SQP to fine-tune a solution quickly by means of LS and repair infeasible solutions. To implement EO–SQP optimization, the following practical issues need to be addressed.

5.4.5 Fitness Function Definition

The fitness function measures how fit an individual (i.e., solution) is, and the "fittest" one has more chance to be inherited into the next generation. A "global fitness" must be defined to evaluate a solution in the proposed EO–SQP algorithm. To solve the NLP optimization problems, the global fitness is defined as the object function value in Equation 5.16 for the unconstrained benchmark problem:

$$Fitness_{global}(S) = f(S) \qquad (5.26)$$

For constrained NLP optimization problems, a popular penalization strategy is used in EO–SQP evolution in order to transform the constrained problem to unconstrained one. If a solution is infeasible, its fitness value is penalized according to the violations of constraints defined in Equations 5.17 and 5.18:

$$Fitness_{global}(S) = f(S) + \underset{\prod g_t, \prod h_u}{Penalty(S)} \qquad (5.27)$$

Unlike GA, which works with a population of candidate solutions, EO depends on a single individual (i.e., chromosome)-based evolution. Through always performing mutation on the worst component and its neighbors successively, the individual in EO can evolve itself toward the global optimal solution generation by generation. This requires a suitable representation which permits each component to be assigned with a quality measure (i.e., fitness) called "local fitness." In this study, the local fitness λ_k is defined as an improvement in global fitness $Fitness_{global}$ made by the mutation imposed on the kth component of best-so-far chromosome S:

$$\lambda_k = Fitness_{local}(k) = \Delta Fitness_{global}(k) = Fitness_{global}(S) - Fitness_{global}(S'_k) \quad (5.28)$$

5.4.6 Termination Criteria

The termination criteria are used for the detection of an appropriate time to stop the optimization runs. In this study, the termination criteria are designed based on two widely used rules. If the predefined maximum generation is exceeded; or an error measure in dependence on the known optimum is satisfied (we can assume that the algorithm has managed to discover the global minimum), the algorithm should terminate. Denote S_{best} as the best-so-far solution found by the algorithms and S^* as the optimum solution of the functions. The search is considered successful, or in other words, the near-optimal solution is found, if S_{best} satisfies that $|(F^* - F)/F^*| < 1e - 3$ (for the case optimum value $F^* \neq 0$) or $|(F^* - F)| < 1e - 3$ (for the case optimum value $F^* = 0$). These criteria are perfectly suitable for comparing the performance of different algorithms.

5.4.7 Workflow and Algorithm

The hybrid algorithm proposed in this study combines the EO and SQP method. The structure of the hybrid EO–SQP is based on the standard EO with which the characteristic of GS is added by propagating the individual solution with the SQP algorithm during EO evolution. In this section, we illustrate the workflow of the EO–SQP algorithm and introduce three mutation operators adopted in this study: the standard EO mutation, SQP mutation, and multistart Gaussian mutation; Gaussian mutation has its main objective to increase diversity, whereby larger parts of the search space are explored. This strategy is often adopted to explore the neighborhood of the current solution. To utilize the advantages of each mutation operator, one or more phases of LS (mutation operator) are applied to the best-so-far solution S based on a probability parameter p_m in each generation. In contrast to the standard EO mutation, when SQP mutation or multistart Gaussian mutation is adopted, we use the "GEO_{var}" (De Sousa et al., 2004) strategy to evolve the current solution by improving all variables simultaneously, as an attempt to speed up the process of LS. The flowchart of the proposed EO–SQP algorithm is shown in Figure 5.8.

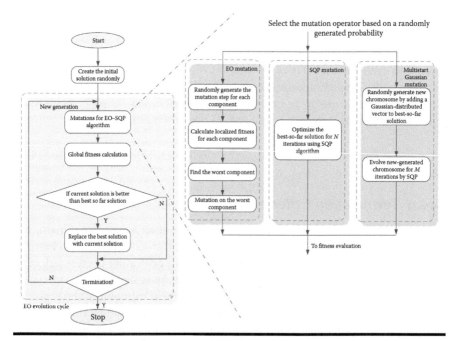

Figure 5.8 Flowchart of the EO–SQP algorithm. (From Chen, P. and Lu, Y. Z. Memetic algorithms based real-time optimization for nonlinear model predictive control. *International Conference on System Science and Engineering*, Macau, China, pp. 119–124. © 2011, IEEE.)

The workflow of the proposed EO–SQP algorithm can be described below:

1. Create the initial solution randomly.
2. For the first iteration of EO, calculate the object fitness function, set $S_{best} = S$.
3. Decide what kind of mutation operators should be imposed to the current chromosome S based on randomly generated probability parameter p_m, if $P_m \leq P_{m\_basic}$, go to (i); else if $P_{m\_basic} < P_m \leq P_{m\_SQP}$, go to (ii); else if $P_m > P_{m\_SQP}$, go to (iii).
   a. Perform the standard EO mutation on the best-so-far solution S.
      i. Change the value of each component in S and get a set of new solutions S'_k $k \in [1, 2, ..., n]$.
      ii. Sequentially evaluate the localized fitness λ_k specified in Equation 5.28 for every S'_k and rank them according to their fitness values.
      iii. Choose the best solution S' from the new solution set $[S']$.
   b. Perform the SQP mutation on the best-so-far chromosome S.
      i. The best-so-far solution S is updated for N iterations by

      $$S' = S + a_k \cdot d_k \qquad (5.29)$$

      where a_k, d_k are the optimum search length and the search direction described in Equations 5.23 and 5.25, respectively.
   c. Perform the multistart Gaussian mutation on the best-so-far chromosome S.
      i. Generate a new chromosome S'_0 by adding a Gaussian distribution random vector with n dimensions to the best-so-far chromosome S.

      $$S'_0 = S + Scale * N(0,1) \qquad (5.30)$$

      where n is the length of chromosome, $Scale$ is the mutation step size.
      ii. Update S'_0 by SQP algorithm in Equation 5.29 for M iterations and get S'.
4. Calculate the global object fitness. If $F(S') < F(S_{best})$, set $S_{best} = S'$.
5. If the termination criteria aren't satisfied, go to Step (3), else go to the next step.
6. Return S_{best}.

5.4.8 Experimental Tests on Benchmark Functions

In this section, 12 widely used NLP benchmark problems are introduced, which have been already extensively discussed in the literature (Runarsson and Yao, 2000; Mezura-Montes and Carlos, 2005; Suganthan et al., 2005; Zhu and Ali, 2009). They represent classes of constrained, unconstrained, unimodal, and multimodal test functions summarized in Table 5.4. It is obvious that the benchmark problems include different types of objective functions (e.g., linear, nonlinear, quadratic,

Table 5.4 Main Characteristics of 12 Benchmark Problems

Benchmark Function		Dimension	Type of the Objective Function	Type and the Number of Constraints		
				LI	*NE*	*NI*
Unconstrained	Michalewicz	10	Continuous, multimodal function	N/A	N/A	N/A
	Schwefel	30	Continuous, unimodal function	N/A	N/A	N/A
	Griewank	30	Continuous, multimodal function	N/A	N/A	N/A
	Rastrigin	30	Continuous, multimodal function	N/A	N/A	N/A
	Ackley	30	Continuous, multimodal function	N/A	N/A	N/A
	Rosenbrock	30	Continuous, unimodal function	N/A	N/A	N/A
Constrained	g04	5	Quadratic	0	0	6
	g05	4	Nonlinear	2	3	0
	g07	10	Quadratic	3	0	5
	g09	7	Nonlinear	0	0	4
	g10	8	Linear	3	0	3
	g12	3	Quadratic	0	0	729

unimodal, and multimodal) and constraints (e.g., linear inequalities [LIs], nonlinear equalities [NEs], and nonlinear inequalities [NIs]). These benchmark functions make it possible to study the proposed EO–SQP algorithm in comparison with other state-of-the-art methods and some well-known results published recently.

5.4.8.1 Unconstrained Problems

First, the performance of the proposed algorithm is tested on six well-known unconstrained problems: the Michalewicz, Schwefel, Griewank, Rastrigin, Ackley, and Rosenbrock functions, with the dimension listed in Table 5.4 and detailed description in Table 5.5. For close comparison of solution accuracy, the runtime, success rate, best, average, worst, and standard deviation values obtained from total 10 independent runs of proposed EO–SQP, standard GA, standard PSO,

Table 5.5 Formulation of Unconstrained Benchmark Problems

Function	Function Expression	Search Space	Global Minimum
Michalewicz	$f_1(X) = -\sum_{i=1}^{n} \sin(x_i)$ $\sin^{2m}\left(\dfrac{i - x_i^2}{\pi}\right), m = 10$	$(0, \pi)^n$	−9.66
Schwefel	$f_2(X) = -\sum_{i=1}^{n} x_i\left(\sin\left(\sqrt{|x_i|}\right)\right)$	$(-500, 500)^n$	−12569.5
Griewank	$f_3(X)$ $= \dfrac{1}{4000}\sum_{i=1}^{n} x_i^2 - \prod_{i=1}^{n}\cos\left(\dfrac{x_i}{\sqrt{i}}\right) + 1$	$(-600, 600)^n$	0
Rastrigin	$f_4(X) = \sum_{i=1}^{n}\left[x_i^2 - 10\cos(2\pi\,x_i) + 10\right]$	$(-5.12, 5.12)^n$	0
Ackley	$f_5(X) = 20 + e - 20e^{\left[-0.2\sqrt{\frac{1}{n}\sum_{i=1}^{n} x_i^2}\right]}$ $- e^{\frac{1}{n}\sum_{i=1}^{n}\cos(2\pi x_i)}$	$(-32.768, 32.768)^n$	0
Rosenbrock	$f_6(X) = \sum_{i=1}^{n-1}\left[100(x_{i+1} - x_i^2)^2 + (x_i - 1)^2\right]$	$(-30, 30)^n$	0

Source: Adapted from Chen, P. et al., *International Journal of Computational Intelligence Systems* 3: 622–631, 2010.

and standard SQP on the six benchmark functions are presented in Table 5.6. The best results among the four approaches are shown in bold. "Success" represents the success rate (percentage of success to discover the global minimum), and "runtime" is the average runtime when the algorithm stops according to the termination criteria defined in Section 5.4.6. In our experiments, the population size of GA and PSO are set to 100 and 50, respectively. The four algorithms are implemented in MATLAB and the experiments are carried out on a P4 E5200 (2.5)-GHz machine with 2-GB RAM under WINXP platform, and the source codes of GA and PSO can be obtained from the MATLAB website.

As shown in Table 5.6, the EO–SQP algorithm proposed in this study is able to find the global optima consistently with a success rate of 100% for all six unconstrained benchmark functions, while GA, PSO, and SQP have a very low success rate for most benchmark problems (Michalewicz, Schwefel, Rastrigin, and Rosenbrock functions). Moreover, EO–SQP is quite an efficient method; the computational time is significantly reduced in comparison with GA and PSO. Although

Table 5.6 Comparison Results for Six Benchmark-Unconstrained Functions

Problems ([a]Optimal Value)	Algorithm	Run Time (s)	Success (%)	Worst	Mean	Best	Standard Deviation
Michalewicz (−9.66)	EO–SQP	25.6757	**100**	**−9.66**	**−9.66**	**−9.66**	**1.39e − 07**
	GA	179.2503	0	−8.7817	−9.2020	−9.4489	0.1989
	PSO	93.7919	0	−8.6083	−9.0733	−9.4796	0.2264
	SQP	0.1497	0	−3.0995	−4.7615	−6.4559	1.0656
Schwefel (−12569.5)	EO–SQP	0.8257	**100**	**−1.2569e + 04**	**−1.2569e + 04**	**−1.2569e + 04**	**6.2674e − 08**
	GA	1233.5	0	−1.014e + 04	−1.0719e + 04	−1.1227e + 04	321.5311
	PSO	218.7393	0	−7.150e + 03	−8.8250e + 03	−1.0037e + 04	832.7256
	SQP	0.1046	0	−5.5547e + 03	−6.8011e + 03	−7.7497e + 03	720.3097
Griewank (0)	EO–SQP	0.0962	**100**	**0**	**0**	**0**	**0**
	GA	64.1398	80	0.1346	0.0172	7.8931e − 04	0.0422
	PSO	102.9561	0	5.1443	0.7026	0.0074	1.5786
	SQP	0.0167	**100**	**0**	**0**	**0**	**0**

(Continued)

Table 5.6 (Continued) Comparison Results for Six Benchmark-Unconstrained Functions

Problems (ᵃOptimal Value)	Algorithm	Run Time (s)	Success (%)	Worst	Mean	Best	Standard Deviation
Rastrigin (0)	EO–SQP	13.2099	100	0	0	0	0
	GA	137.1764	30	2.9849	0.9999	$7.7225e-004$	1.1942
	PSO	102.3654	0	109.8554	70.4078	44.4531	22.6459
	SQP	0.7289	0	267.6421	188.2451	92.5310	62.4574
Ackley (0)	EO–SQP	3.8563	100	0	0	0	0
	GA	16.8079	90	1.5017	0.1511	$9.3102e-04$	0.4619
	PSO	103.9701	0	11.7419	7.0320	3.8700	2.7755
	SQP	0.0404	0	19.8725	19.5720	19.1787	0.2086
Rosenbrock (0)	EO–SQP	2.6241	100	$4.3065e-04$	$4.3200e-05$	$1.5547e-08$	$1.3614e-04$
	GA	1195.6	20	22.1819	3.5821	$9.9942e-04$	6.6396
	PSO	94.3400	0	853.3601	150.7588	27.1222	242.7105
	SQP	1.6760	50	3.9866	1.9933	$1.1961e-007$	2.1011

the deterministic SQP is the fastest method among the four, it is easily trapped in local minima as shown in simulation results (Michalewicz, Schwefel, Rastrigin, Ackley, and Rosenbrock functions). The proposed EO–SQP method can successfully prevent solutions from falling into the deep local minimal, reduce evolution process significantly with efficiency, and converge to the global optimum or its close vicinity.

5.4.8.2 Constrained Problems

In this section, we selected six (*g*04, *g*05, *g*07, *g*09, *g*10, and *g*12) out of 13 published benchmark functions (Runarsson and Yao, 2000; Mezura-Montes and Carlos, 2005; Zhu and Ali, 2009) as constrained test problems, since the characteristics of those functions contain the "difficulties" in having global optimization problems by using an EA. The formulation of the six benchmark functions are listed below:

g04:

$$\text{Min } f(x) = 5.3578547 x_3^2 + 0.8356891 x_1 x_5 + 37.293239 x_1 - 40792.141$$

$$s.t. \quad g_1(x) = 85.334407 + 0.0056858 x_2 x_5 + 0.0006262 x_1 x_4 - 0.0022053 x_3 x_5 - 92 \le 0$$

$$g_2(x) = -85.334407 - 0.0056858 x_2 x_5 - 0.0006262 x_1 x_4 + 0.0022053 x_3 x_5 \le 0$$

$$g_3(x) = 80.51249 + 0.0071317 x_2 x_5 + 0.0029955 x_1 x_2 + 0.0021813 x_3^2 - 110 \le 0$$

$$g_4(x) = -80.51249 - 0.0071317 x_2 x_5 - 0.0029955 x_1 x_2 - 0.0021813 x_3^2 + 90 \le 0$$

$$g_5(x) = 9.300961 + 0.0047026 x_3 x_5 + 0.0012547 x_1 x_3 + 0.0019085 x_3 x_4 - 25 \le 0$$

$$g_6(x) = -9.300961 - 0.0047026 x_3 x_5 - 0.0012547 x_1 x_3 - 0.0019085 x_3 x_4 + 20 \le 0$$

where $78 \le x_1 \le 102$, $33 \le x_2 \le 45$ and $27 \le x_i \le 45$ ($i = 3, 4, 5$). The optimum solution is $x^* = (78, 33, 29.995256025682, 45, 36.775812905788)$ where $f(x^*) = -30665.539$. Two constraints are active (g_1 and g_6).

g05:

$$\text{Min } f(x) = 3x_1 + 0.000001 x_1^3 + 2x_2 + (0.000002/3) x_2^3$$

$$s.t. \quad g_1(x) = -x_4 + x_3 - 0.55 \le 0$$

$$g_2(x) = -x_3 + x_4 - 0.55 \le 0$$

$$h_3(x) = 1000 \sin(-x_3 - 0.25) + 1000 \sin(-x_4 - 0.25) + 894.8 - x_1 = 0$$

$$h_4(x) = 1000 \sin(x_3 - 0.25) + 1000 \sin(x_3 - x_4 - 0.25) + 894.8 - x_2 = 0$$

$$h_5(x) = 1000 \sin(x_4 - 0.25) + 1000 \sin(x_4 - x_3 - 0.25) + 1294.8 = 0$$

where $0 \le x_1 \le 1200$, $0 \le x_2 \le 1200$ $-0.55 \le x_3 \le 0.55$, and $-0.55 \le x_4 \le 0.55$. The best known solution $x^* = (679.9453, 1026.067, 0.1188764, -0.3962336)$ where $f(x^*) = 5126.4981$.

g07:

$$\text{Min } f(x) = x_1^2 + x_2^2 + x_1 x_2 - 14x_1 - 16x_2 + (x_3 - 10)^2 + 4(x_4 - 5)^2 + (x_5 - 3)^2$$

$$+ 2(x_6 - 1)^2 + 5x_7^2 + 7(x_8 - 11)^2 + 2(x_9 - 10)^2 + (x_{10} - 7)^2 + 45$$

s.t. $\quad g_1(x) = -105 + 4x_1 + 5x_2 - 3x_7 + 9x_8 \le 0$

$\qquad g_2(x) = 10x_1 - 8x_2 - 17x_7 + 2x_8 \le 0$

$\qquad g_3(x) = -8x_1 + 2x_2 + 5x_9 - 2x_{10} - 12 \le 0$

$\qquad g_4(x) = 3(x_1 - 2)^2 + 4(x_2 - 3)^2 + 2x_3^2 - 7x_4 - 120 \le 0$

$\qquad g_5(x) = 5x_1^2 + 8x_2 + (x_3 - 6)^2 - 2x_4 - 40 \le 0$

$\qquad g_6(x) = x_1^2 + 2(x_2 - 2)^2 - 2x_1 x_2 + 14x_5 - 6x_6 \le 0$

$\qquad g_7(x) = 0.5(x_1 - 8)^2 + 2(x_2 - 4)^2 + 3x_5^2 - x_6 - 30 \le 0$

$\qquad g_8(x) = -3x_1 + 6x_2 + 12(x_9 - 8)^2 - 7x_{10} \le 0$

where $10 \le x_i \le 10$ ($i = 1, \ldots, 10$). The optimum solution is $x^* = (2.171996, 2.363683, 8.773926, 5.095984, 0.9906548, 1.430574, 1.321644, 9.828726, 8.280092, 8.375927)$ where $f(x^*) = 24.3062091$. Six constraints are active (g_1, g_2, g_3, g_4, g_5, and g_6).

g09:

$$\text{Min } f(x) = (x_1 - 10)^2 + 5(x_2 - 12)^2 + x_3^4 + 3(x_4 - 11) + 10x_5^6$$

$$+ 7x_6^2 + x_7^4 - 4x_6 x_7 - 10x_6 - 8x_7$$

s.t. $\quad g_1(x) = -127 + 2x_1^2 + 3x_2^4 + x_3 + 4x_4^2 + 5x_5 \le 0$

$\qquad g_2(x) = -282 + 7x_1^2 + 3x_2 + 10x_3^2 + x_4 - x_5 \le 0$

$\qquad g_3(x) = -196 + 23x_1 + x_2^2 + 6x_6^2 - 8x_7 \le 0$

$\qquad g_4(x) = 4x_1^2 + x_2^2 - 3x_1 x_2 + 2x_3^2 + 5x_6 - 11x_7 \le 0$

where for $10 \le x_i \le 10$ ($i = 1, \ldots, 7$). The optimum solution is $x^* = (2.330499, 1.951372, 0.4775414, 4.365726, 0.6244870, 1.038131, 1.594227)$ where $f(x^*) = 680.6300573$. Two constraints are active (g_1 and g_4).

g10:

Min $f(x) = x_1 + x_2 + x_3$

s.t. $g_1(x) = -1 + 0.0025(x_4 + x_6) \leq 0$

$g_2(x) = -1 + 0.0025(x_5 + x_7 - x_4) \leq 0$

$g_3(x) = -1 + 0.01(x_8 - x_5) \leq 0$

$g_4(x) = -x_1 x_6 + 833.33252 x_4 + 100 x_1 - 83333.333 \leq 0$

$g_5(x) = -x_2 x_7 + 1250 x_5 + x_2 x_4 - 1250 x_4 \leq 0$

$g_6(x) = -x_3 x_8 + 1250000 + x_3 x_5 - 2500 x_5 \leq 0$

where $100 \leq x_1 \leq 10{,}000$, $1000 \leq x_i \leq 10{,}000$ ($i = 2$, 3), and $10 \leq x_i \leq 1000$ ($i = 4, \ldots, 8$). The optimum solution is $x^* = (579.19, 1360.13, 5109.92, 182.0174, 295.5985, 217.9799, 286.40, 395.5979)$ where $f(x^*) = 7049.248$. Three constraints are active (g_1, g_2, and g_3).

g12:

Max $f(x) = (100 - (x_1 - 5)^2 - (x_2 - 5)^2 - (x_3 - 5)^2)/100$

s.t. $g(x) = (x_1 - p)^2 + (x_2 - q)^2 + (x_3 - r)^2 - 0.0625 \leq 0$

where $0 \leq x_i \leq 10$ ($i = 1$, 2, 3) and p, q, $r = 1$, 2, $\ldots$, 9. The feasible region of the search space consists of 9^3 disjointed spheres. A point (x_1, x_2, x_3) is feasible if and only if there exist p, q, r such that the above inequality holds. The optimum is located at $x^* = (5, 5, 5)$ where $f(x^*) = 1$. The solution lies within the feasible region.

Note that test function g12 is a maximization problem, and the others are minimization problems. In this study, the maximization problem is transformed into a minimization problem using $-F$. Table 5.7 shows the performance comparisons among SR (Runarsson and Yao, 2000), SMES (Mezura-Montes and Carlos, 2005), auxiliary function method (AFM) (Zhu and Ali, 2009), and the hybrid EO–SQP. The best results among the four approaches are shown in bold.

Among these four methods (see Table 5.7), the EO–SQP appears to be more promising. It provided three better "best" results (g05, g07, and g10) among six functions, and two similar "best" results (g04 and g12). Moreover, the EO–SQP provided better "mean" results for three problems (g05, g07, and g10), and similar "mean" results in other two (g04 and g12). Finally, the EO–SQP obtained better "worst" results in two problems (g05 and g07), and it reached similar "worst" solutions in the other two problems (g04 and g12). The proposed EO–SQP can produce better, if not optimal, solutions for most of the six benchmark problems

Table 5.7 Comparison Results for Six Benchmark Functions with Constraints

Function and Optimum	Statistical Features	Approaches for Constrained Optimization			
		SR	SMES	AFM	EO–SQP
g04	Best	−30665.539	−30665.539	−30665.50	−30665.539
	Mean	−30665.539	−30665.539	−30665.32	−30665.539
−30665.539	Worst	−30665.539	−30665.539	−30665.23	−30665.539
	Standard deviation	2.0e − 05	0	0.063547	4.4238e − 07
g05	Best	a5126.497	5126.599	5126.5	**5126.498**
	Mean	5128.881	5174.492	5126.65	**5126.498**
5126.498	Worst	5142.472	5304.167	5126.96	**5126.498**
	Standard deviation	3.5	5.006e + 01	0.145896	**7.4260e − 13**
g07	Best	24.307	24.327	24.30694	**24.306**
	Mean	24.374	24.475	24.30789	**24.306**
24.306	Worst	24.642	24.843	24.30863	**24.306**
	Standard deviation	6.6e − 02	1.32e − 01	4.9999e − 04	**3.0169e − 14**
g09	Best	**680.630**	680.632	680.6376	680.6387
	Mean	680.656	**680.643**	680.67833	680.8047
680.630	Worst	680.763	680.719	**680.6980**	680.9844
	Standard deviation	3.4e − 02	**1.55e − 02**	0.016262	0.1145
g10	Best	7054.316	7051.903	7049.333	**7049.248**
	Mean	7559.192	7253.047	7049.545	**7049.312**
7049.248	Worst	8835.655	7638.366	**7049.603**	7049.891
	Standard deviation	5.3e + 02	1.3602e + 02	**0.071513**	0.2034

(Continued)

Table 5.7 (*Continued*) **Comparison Results for Six Benchmark Functions with Constraints**

Function and Optimum	Statistical Features	Approaches for Constrained Optimization			
		SR	SMES	AFM	EO–SQP
g12	Best	1	1	1	1
	Mean	1	1	0.999988	1
1	Worst	1	1	0.999935	1
	Standard deviation	0	0	1.7e – 05	0

[a] The best result of problem g05 by SR is even better than the optimal solution of 5126.498. This is the consequence of transforming equality constraints into inequality constraints by a relaxed parameter ε (Runarsson and Yao, 2000).

with the exception of test function g09. With respect to test function g09, although the EO–SQP fails to provide superior results, the performance of the four methods are actually very close. Generally, constrained optimization problems with equality constraints are very difficult to solve. It should be noted that for the three test functions with equality constraints (g05, g07, and g10), the EO–SQP can provide better performance than the other three methods, the optimum solutions are found by the EO–SQP for all the three problems with equality constraints; while the SR, SMES, and AFM fail to find the global optimums. This is due to the hybrid mechanism that the EO–SQP can benefit from the strong capability of SQP to deal with constraints during the EO evolution.

5.4.9 Dynamics Analysis of the Hybrid EO–SQP

The convergence and dynamics during the optimization of EO and its derivatives remain up to now challenging open problems. In this section, we use a typical optimization run of the Ackley function ($F^* = 0$) as an example to analyze evolution dynamics of the proposed EO–SQP and show the mechanism strength of the proposed algorithm.

Figure 5.9 shows the search dynamics of the basic EO and the hybrid EO–SQP, which demonstrates the fitness evolution of the current solution in a typical run on the Ackley function. The figure shows that both the basic EO and hybrid EO–SQP descend sufficiently fast to a near-optimal solution with enough fluctuations to escape from local optima and explore new regions of configuration space. It can be seen quite clearly that with the help of SQP, the hybrid EO–SQP can execute a deeper search in comparison with the basic EO. The search efficiency of basic EO can be improved significantly by incorporating the LS method SQP into evolution, when the EO–SQP converges to the close vicinity of the global optimum, the SQP

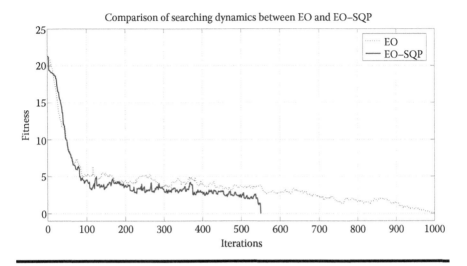

Figure 5.9 Comparison of the current solution fitness generation by generation between EO and EO–SQP on Ackley function. (From Chen, P. and Lu, Y. Z. Memetic algorithms based real-time optimization for nonlinear model predictive control. *International Conference on System Science and Engineering,* **Macau, China, pp. 119–124. © 2011, IEEE.)**

mutation will help to find the global optimum point in just a few runs, as shown in Figure 5.9.

Evolutions of best solution fitness as a function of time for the EO–SQP, GA, PSO, and SQP on the Ackley function are also shown in Figure 5.10. The convergence rate of the proposed EO–SQP algorithm is a little slower than the GA and PSO at the early stage, due to the better solution diversity of population-based methods (GA and PSO); however, when approaching a near region of the global optimum, the EO–SQP keeps a high convergence rate and reaches the global minimum very fast due to the efficiency of gradient-based SQP LS. On the other hand, the conventional SQP converges to a local minimal far from the global optimum with high efficiency and cannot escape from it due to the weakness of GS.

As a general remark on the comparisons above, the EO–SQP shows better performance with respect to state-of-the-art approaches in terms of quality, robustness, and efficiency of search. The results show that the proposed EO–SQP finds optimal or near-optimal solutions quickly, and has more statistical soundness and faster convergence rate than the compared algorithms. It should be noted that the factors contributing to the performance of the proposed EO–SQP method are the global-search capability of EO and the capability of the gradient-based SQP method to search the local optimum efficiently with high accuracy and to deal with various constraints.

In this section, a novel MA-based hybrid EO–SQP algorithm is proposed for global optimization of NLP problems, which are typically quite difficult to solve

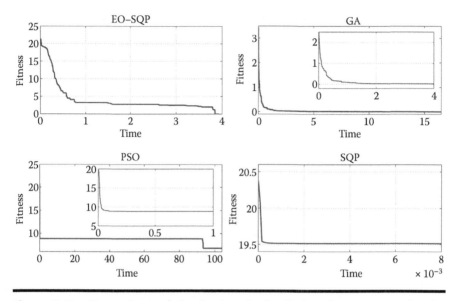

Figure 5.10 Comparison of the best-evolved solution fitness generation by generation between EO–SQP, GA, PSO, and SQP on Ackley function.

exactly. Traditional deterministic methods are more vulnerable to get trapped in the local minima; while most CI-based optimization methods with global-search capability tend to suffer from high computation cost. Therefore, under the framework of MAs, the general-purpose heuristic EO and deterministic LS method SQP are combined together in order to develop a robust and fast optimization technique with global-search capability and mechanism to deal with constraints. The hybrid method avoids the possibility of local minimum by providing the GS method with the exploration ability of EO. Those advantages have been clearly demonstrated by the comparison with some other state-of-the-art approaches over 12 widely used benchmark functions.

5.5 EO–PSO Integration

5.5.1 Introduction

The PSO algorithm is a recent addition to the list of global-search methods. This derivative-free method is particularly suited to continuous variable problems and has received increasing attention in the optimization community. PSO was originally developed by Kennedy and Eberhart (1995) and inspired by the paradigm of birds flocking. PSO consists of a swarm of particles and each particle flies through the multidimensional search space with a velocity, which is constantly updated by the particle's previous best performance and by the previous best performance of the particle's neighbors. PSO can be easily implemented and is computationally inexpensive in terms of both memory requirements and CPU speed (Kennedy and Eberhart, 1995). However, even though PSO is a good and fast search algorithm, it has premature convergence, especially in complex multi-peak-search problems. This means that it does not "know how" to sacrifice short-term fitness to gain longer-term fitness. The likelihood of this occurring depends on the shape of the fitness landscape: certain problems may provide an easy ascent toward a global optimum; others may make it easier for the function to find the local optima. So far, there have been many researchers devoted to this field to deal with this problem (Shelokar et al., 2007; Jin et al., 2008; Chen and Zhao, 2009).

To avoid premature convergence of PSO, an idea of combining PSO with EO was addressed by Chen et al. (2010b). Such a hybrid approach expects to enjoy the merits of PSO with those of EO. In other words, PSO contributes to the hybrid approach in a way to ensure that the search converges faster, while EO makes the search jump out of local optima due to its strong LS ability. Chen et al. (2010b) developed a novel hybrid optimization method, called the hybrid PSO–EO algorithm, to solve those complex unimodal/multimodal functions which may be difficult for the standard PSOs. The performance of PSO–EO was testified on six unimodal/multimodal benchmark functions and provided comparisons with the PSO–GA-based hybrid algorithm (PGHA) (Shi et al., 2005), standard PSO, standard GA, and PEO (Chen et al., 2006). Experimental results indicate that PSO–EO has better performance and strong capability of escaping from local optima. Hence, the hybrid PSO–EO algorithm may be a good alternative to deal with complex numerical optimization problems.

5.5.2 Particle Swarm Optimization

PSO is a population-based optimization tool, where the system is initialized with a population of random particles and the algorithm searches for optima by updating generations (Kennedy and Eberhart, 1995). Suppose that the search space is D-dimensional. The position of the ith particle can be represented by a D-dimensional vector $X_i = (x_{i1}, x_{i2}, ..., x_{iD})$ and the velocity of this particle is $V_i = (v_{i1}, v_{i2}, ..., v_{iD})$. PSO is a population-based optimization tool, where the system is initialized with a population of random particles and the algorithm searches for optima by updating generations. The best previously visited position of the ith particle is represented by $P_i = (p_{i1}, p_{i2}, ..., p_{iD})$ and the global best position of the swarm found so far is denoted by $P_g = (p_{g1}, p_{g2}, ..., p_{gD})$. The fitness of each particle can be evaluated through putting its position into a designated objective function. The particle's velocity and its new position are updated as follows (Chen et al., 2010b):

$$v_{id}^{t+1} = w^t v_{id}^t + c_1 r_1^t (p_{id}^t - x_{id}^t) + c_1 r_2^t (p_{gd}^t - x_{id}^t) \qquad (5.31)$$

$$x_{id}^{t+1} = x_{id}^t + v_{id}^{t+1} \qquad (5.32)$$

where $d \in \{1, 2, ..., D\}$, $i \in \{1, 2, ..., N\}$, N is the population size, the superscript t denotes the iteration number, w the inertia weight, r_1 and r_2 are two random values in the range [0, 1], c_1 and c_2 are the cognitive and social scaling parameters which are positive constants.

5.5.3 PSO–EO Algorithm

Note that PSO has great global-search ability, while EO has strong LS capability. In Chen et al. (2010b), a novel hybrid PSO–EO algorithm was proposed, which combines the merits of PSO and EO. This hybrid approach makes full use of the exploration ability of PSO and the exploitation ability of EO. Consequently, through introducing EO to PSO, the PSO–EO algorithm may overcome the limitation of PSO and have capability of escaping from local optima. However, if EO is introduced to PSO at each iteration, the computational cost will increase sharply. And at the same time, the fast convergence ability of PSO may be weakened. In order to perfectly integrate PSO with EO, EO is introduced to PSO at *INV*-iteration intervals (here a parameter *INV* is used to represent the frequency of introducing EO to PSO). For instance, *INV* = 10 means that EO is introduced to PSO every 10 iterations. Therefore, the hybrid PSO–EO approach is able to keep fast convergence in most of the time with the help of PSO, and capable of escaping from a local optimum with the aid of EO. The value of parameter *INV* is predefined by the user. Usually, according to Chen et al. (2010b), the value of *INV* can be set to 50–100 when the test function is unimodal or multimodal with relatively less local optima. As a consequence, PSO will play a key role in this case. When the test function is multimodal with many local optima, the value of *INV* can be set to 1–50 and thus EO can help PSO to jump out of local optima.

For a minimization problem with D dimensions, the PSO–EO algorithm works as shown in Figure 5.11 and its flowchart is illustrated in Figure 5.12. In the main procedure of the PSO–EO algorithm, the fitness of each particle is evaluated through putting its position into the objective function. However, in the EO procedure, in order to find out the worst component, each component of a solution should be assigned a fitness value. The fitness of each component of a solution for an unconstrained minimization problem is defined as follows. For the ith particle, the fitness λ_{ik} of the kth component is defined as the mutation cost, that is, $OBJ(X_{ik}) - OBJ(P_g)$, where X_{ik} is the new position of the ith particle obtained by performing mutation only on the kth component and leaving all other components fixed, $OBJ(X_{ik})$ is the objective value of X_{ik}, and $OBJ(P_g)$ is the objective value of the best position in the swarm found so far. The EO procedure is described in Figure 5.13.

5.5.4 Mutation Operator

Since there is merely a mutation operator in EO, the mutation plays a key role in EO search. Chen et al. (2010b) adopted the hybrid Gaussian–Cauchy mutation

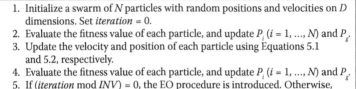

1. Initialize a swarm of N particles with random positions and velocities on D dimensions. Set *iteration* = 0.
2. Evaluate the fitness value of each particle, and update P_i (i = 1, ..., N) and P_g.
3. Update the velocity and position of each particle using Equations 5.1 and 5.2, respectively.
4. Evaluate the fitness value of each particle, and update P_i (i = 1, ..., N) and P_g.
5. If (*iteration* mod *INV*) = 0, the EO procedure is introduced. Otherwise, continue the next step.
6. If the terminal condition is satisfied, go to the next step; otherwise, set *iteration* = *iteration* + 1, and go to Step (3).
7. Output the optimal solution and the optimal objective function value.

Figure 5.11 Pseudocode of PSO–EO algorithm. (Reprinted from *International Journal of Computational Intelligence Systems*, 3, Chen, P. et al., Extremal optimization combined with LM gradient search for MLP network learning, 622–631, Copyright 2010, with permission from Elsevier.)

(*GC* mutation for short) presented by Chen and Lu (2008). This mutation method mixes Gaussian mutation and Cauchy mutation. The mechanisms of Gaussian and Cauchy mutation operations have been studied by Yao et al. (1999). They pointed out that Cauchy mutation is better at coarse-grained search while Gaussian mutation is better at fine-grained search. In the hybrid *GC* mutation, the Cauchy mutation is first used. It means that the large step size will be taken first at each mutation.

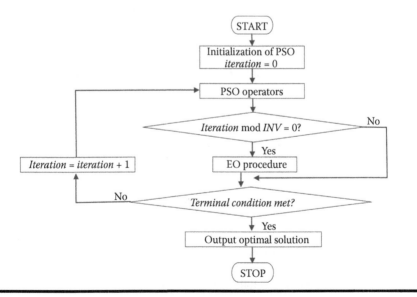

Figure 5.12 Flowchart of PSO–EO algorithm. (Reprinted from *International Journal of Computational Intelligence Systems*, 3, Chen, P. et al., Extremal optimization combined with LM gradient search for MLP network learning, 622–631, Copyright 2010, with permission from Elsevier.)

1. Set the index of the current particle $i = 1$.
2. For the position $X_i = (x_{i1}, x_{i2}, ..., x_{iD})$ of the ith particle
   a. Perform mutation on each component of X_i one by one, while keeping the other component fixed. Then D new position X_{ik} ($k = 1, ..., D$) can be obtained;
   b. Evaluate the fitness $\lambda_{ik} = OBJ(X_{ik}) - OBJ(P_g)$ of each component x_{ik}, $k \in \{1, ..., D\}$;
   c. Compare all the components according to their fitness values and find out the worst adapted component x_{iw}, and then x_{iw} is the new position corresponding to x_{iw}, $w \in \{1, ..., D\}$;
   d. If $OBJ(X_{iw}) < OBJ(X_i)$, then set $X_i = X_{iw}$ and $OBJ(X_i) = OBJ(X_{iw})$, and continue the next step. Otherwise, X_i keeps unchanged and go to Step 3;
   e. Update P_i and P_g.
3. If it equals to the population size N, return the results. Otherwise, set $i = i + 1$ and go to Step 2.

Figure 5.13   Pseudocode of EO procedure in PSO-EO algorithm. (Reprinted from *International Journal of Computational Intelligence Systems*, **3, Chen, P. et al., Extremal optimization combined with LM gradient search for MLP network learning, 622–631, Copyright 2010, with permission from Elsevier.)**

If the new generated variable after mutation goes beyond the range of the variable, the Cauchy mutation will be used repeatedly for some times (parameter TC is used to represent the times of Cauchy mutation), until the new generated offspring falls into the range. Otherwise, the Gaussian mutation will be carried out repeatedly for another set of times (parameter TG is used to represent the times of the Gaussian mutation), until the offspring satisfies the requirement. That is, the step size will become smaller than before. If the new generated variable after mutation still goes beyond the range of the variable, then the upper or lower bound of the decision variable will be chosen as the new generated variable. Thus, the hybrid GC mutation combines the advantages of coarse-grained search and fine-grained search. Unlike some switching algorithms which have to decide when to switch between different mutations during search, the hybrid GC mutation does not need to make such decisions. The Gaussian mutation performs with the following representation (Chen et al., 2010b):

$$x_k' = x_k + N_k(0,1) \tag{5.33}$$

where x_k and x_k' denote the kth decision variables before mutation and after mutation, respectively, $N_k(0, 1)$ denotes the Gaussian random number with mean zero and standard deviation one and is generated anew for kth decision variable. The Cauchy mutation performs as follows (Chen et al., 2010b):

$$x_k' = x_k + \delta_k \tag{5.34}$$

where δ_k denotes the Cauchy random variable with the scale parameter equal to one and is generated anew for the kth decision variable.

In the hybrid *GC* mutation, the values of parameters *TC* and *TG* are set by the user beforehand. The value of *TC* decides the coarse-grained searching time, while the value of *TG* has an effect on the fine-grained searching time. Therefore, both values of the two parameters cannot be large because it will prolong the search process and hence increase the computational overhead. According to the literature (Chen and Lu, 2008), the moderate values of *TC* and *TG* can be set to 2–4.

5.5.5 Computational Complexity

Let us now look at the complexity of an iteration of the entire algorithm. The basic operations being performed and their worst-case complexities are as follows:

1. The PSO procedure is $O(2N + ND)$, where D is the number of decision variables, N is the number of the particles in the swarm.
2. The EO procedure is $O(2ND)$.

As a consequence, the overall complexity of the hybrid PSO–EO algorithm is $O(2N + 3ND)$, while the overall complexity of the standard PSO algorithm is $O(2N + ND)$. As can be seen, the computational complexity of the hybrid PSO–EO algorithm is mainly affected by the problem dimension D and the population size N. The larger problem dimension D or population size N is, the higher computational complexity of hybrid PSO–EO algorithm becomes. It can be also observed that the computational complexity of the hybrid PSO–EO is a little higher than that of the standard PSO, when EO being introduced to PSO at *INV*-iteration intervals (here $INV > 0$). However, this hybrid approach has great capability of escaping from local optima, and thus can avoid the premature convergence which is one drawback of the standard PSO. If the test function is easy to solve, we can set *INV* to be zero, and thus the computational complexity of the PSO–EO will be the same as that of the standard PSO. On the other hand, if the test function is hard to deal with, *INV* can be set to a large number, and hence the complexity of the PSO–EO will be higher than that of the standard PSO. From the above analysis, it can be concluded that, compared with the standard PSO, the hybrid PSO–EO obtains stronger capability of jumping out of local optima at the expense of a little higher computational complexity.

5.5.6 Experimental Results

In order to demonstrate the performance of the hybrid PSO–EO, Chen et al. (2010b) used six well-known benchmark functions shown in Table 5.5. All the functions are to be minimized. For these functions, there are many local optima and/or saddles in their solution spaces. The amount of local optima and saddles increases with increasing complexity of the functions, that is, with increasing dimension.

To testify the efficiency and effectiveness of the PSO–EO, in Chen et al. (2010b), the experimental results of PSO-EO were compared with those of the PGHA, standard PSO, standard GA, and PEO. Note that all the algorithms were run on the same hardware (i.e., Intel Pentium M with 900-MHz CPU and 256M memory) and software (i.e., JAVA) platform. Each algorithm was run independently for 20 trials.

Table 5.8 shows the settings of problem dimension, maximum generation, population size, initialization range of each algorithm, and the value of parameter *INV for each test function*.

For the hybrid PSO–EO, the cognitive and social scaling parameters, that is, c_1 and c_2, were both set to 2, the inertia weight w varied from 0.9 to 0.4 linearly with the iterations, the upper and lower bounds for velocity on each dimension, that is, v_{min} and v_{max}, were set to be the upper and lower bounds of each dimension, that is, $(v_{min}, v_{max}) = (x_{min}, x_{max})$. The parameters TC and TG in the hybrid GC mutation were both set to 3. From the values of parameter INV shown in Table 5.8, we can see that EO was introduced to PSO more frequently on functions f_1, f_2, and f_6 than other three functions due to the complexity of problems. For the standard PSO, all the parameters, that is, c_1, c_2, w, and (v_{min}, v_{max}), were set to the same as those used in the hybrid PSO–EO. For the standard GA, elitism mechanism, roulette wheel selection mechanism, single-point uniform crossover with the rate of 0.3, nonuniform mutation with the rate of 0.05 and the system parameter of 0.1 were used. For the PGHA, all the parameters were set to the same as those used in the standard PSO and standard GA. For the PEO, the hybrid GC mutation with both parameters TC and TG equal to 3 were adopted.

In order to compare the different algorithms, a fair time measure must be selected. The number of iterations cannot be used as a time measure, as these algorithms do different amounts of work in their inner loops. It is also noticed that each component of a solution in the EO procedure has to be evaluated at each

Table 5.8 Parameter Settings for Six Test Functions

Function	Dimension	Maximum Generation	Population Size	Initialization Range	INV
f_1	10	20,000	10	$(0, \pi)^n$	20
f_2	30	20,000	30	$(-500, 500)^n$	1
f_3	30	20,000	30	$(-600, 600)^n$	100
f_4	30	20,000	10	$(-5.12, 5.12)^n$	100
f_5	30	10,000	30	$(-32.768, 32.768)^n$	100
f_6	30	10,000	30	$(-30, 30)^n$	1

Source: Reprinted from *International Journal of Computational Intelligence Systems*, 3, Chen, P. et al., Extremal optimization combined with LM gradient search for MLP network learning, 622–631, Copyright 2010, with permission from Elsevier.

iteration, and thus the calculation of the number of function evaluations will bring some trouble. Therefore, the number of function evaluations is not very suitable as a time measure. Chen et al. (2010b) adopted the average runtime of 20 runs when the near-optimal solution is found or otherwise when the maximum generation is reached as a time measure.

After 20 trials of running each algorithm for each test function, the simulation results were obtained and shown in Tables 5.9 through 5.14. The bold numerals in Tables 5.9 through 5.14 mean that they are the best optima among all the optima. Denote F as the result found by the algorithms and F^* as the optimum value of the functions. The simulation is considered successful, or in other words, the near-optimal solution is found, if F satisfies that $|(F^* - F)/F^*| < 1E - 3$ (for the case $F^* \neq 0$) or $|F^* - F| < 1E - 3$ (for the case $F^* = 0$). In these tables, "*success*"

Table 5.9 Comparison Results for Michalewicz Function f_1 ($F_1^* = -9.66$)

Algorithm	Runtime (s)	Success (%)	Worst	Mean	Best	Standard Deviation
PSO–EO	**1.9**	**100**	**−9.66**	**−9.66**	**−9.66**	**2.15E − 3**
PGHA	3.3	**100**	−9.65	−9.65	**−9.66**	3.00E − 3
PSO	35.6	5	−9.06	−9.52	**−9.66**	0.17
GA	23.5	60	−9.50	−9.62	**−9.66**	0.06
PEO	58.9	0	−9.50	−9.55	−9.61	0.03

Source: Reprinted from *International Journal of Computational Intelligence Systems,* 3, Chen, P. et al., Extremal optimization combined with LM gradient search for MLP network learning, 622–631, Copyright 2010, with permission from Elsevier.

Table 5.10 Comparison Results for Schwefel Function f_2 ($F_2^* = -12569.5$)

Algorithm	Runtime (s)	Success (%)	Worst	Mean	Best	Standard Deviation
PSO–EO	**63.8**	**100**	**−12562.6**	**−12568.0**	**−12569.5**	**2.01**
PGHA	77.8	0	−10575.8	−10895.5	−11069.5	193.9
PSO	67.7	0	−9577.7	−10139.3	−11026.2	625.7
GA	70.1	0	−8404.5	−8846.0	−9549.3	481.0
PEO	180.5	0	−11977.3	−12083.3	−12214.2	90.3

Source: Reprinted from *International Journal of Computational Intelligence Systems,* 3, Chen, P. et al., Extremal optimization combined with LM gradient search for MLP network learning, 622–631, Copyright 2010, with permission from Elsevier.

Table 5.11   Comparison Results for Griewank Function f_3 ($F_3^* = 0$)

Algorithm	Runtime (s)	Success (%)	Worst	Mean	Best	Standard Deviation
PSO–EO	0.23	100	0.0	0.0	0.0	0.0
PGHA	0.12	100	0.0	0.0	0.0	0.0
PSO	0.30	100	0.0	0.0	0.0	0.0
GA	19.1	0	0.101	0.074	0.039	0.21E – 03
PEO	12.0	100	0.0	0.0	0.0	0.0

Source: Reprinted from *International Journal of Computational Intelligence Systems*, 3, Chen, P. et al., Extremal optimization combined with LM gradient search for MLP network learning, 622–631, Copyright 2010, with permission from Elsevier.

Table 5.12   Comparison Results for Rastrigin Function f_4 ($F_4^* = 0$)

Algorithm	Runtime (s)	Success (%)	Worst	Mean	Best	Standard Deviation
PSO–EO	0.61	100	0.0	0.0	0.0	0.0
PGHA	0.11	100	0.0	0.0	0.0	0.0
PSO	0.53	100	0.0	0.0	0.0	0.0
GA	39.2	0	0.046	0.014	9.93E – 4	0.014
PEO	74.7	0	2.47	2.14	1.85	0.25

Source: Reprinted from *International Journal of Computational Intelligence Systems*, 3, Chen, P. et al., Extremal optimization combined with LM gradient search for MLP network learning, 622–631, Copyright 2010, with permission from Elsevier.

Table 5.13   Comparison Results for Rastrigin Function f_5 ($F_5^* = 0$)

Algorithm	Runtime (s)	Success (%)	Worst	Mean	Best	Standard Deviation
PSO–EO	0.26	100	−8.88E – 16	−8.88E – 16	−8.88E – 16	0.0
PGHA	0.75	100	−8.88E – 16	−8.88E – 16	−8.88E – 16	0.0
PSO	0.23	100	−8.88E – 16	−8.88E – 16	−8.88E – 16	0.0
GA	11.6	0	0.094	0.054	0.03	0.02
PEO	77.8	0	0.12	0.11	0.09	8.4E – 3

Source: Reprinted from *International Journal of Computational Intelligence Systems*, 3, Chen, P. et al., Extremal optimization combined with LM gradient search for MLP network learning, 622–631, Copyright 2010, with permission from Elsevier.

Table 5.14 Comparison Results for Rosenbrock Function f_6 ($F_6^* = 0$)

Algorithm	Runtime (s)	Success (%)	Worst	Mean	Best	Standard Deviation
PSO–EO	117.1	100	9.99E – 4	9.88E – 4	9.54E – 4	2.39E – 5
PGHA	237.1	0	22.8	22.2	21.2	0.65
PSO	202.5	0	26.8	26.0	25.4	0.59
GA	206.8	0	39.7	33.1	30.1	3.95
PEO	199.8	0	9.63	9.42	9.30	0.13

Source: Reprinted from *International Journal of Computational Intelligence Systems*, 3, Chen, P. et al., Extremal optimization combined with LM gradient search for MLP network learning, 622–631, Copyright 2010, with permission from Elsevier.

represents the success rate, and *"runtime"* is the average runtime of 20 runs when the near-optimal solution is found or otherwise when the maximum generation is reached. The worst, mean, best, and standard deviation of solutions found by the five algorithms are also listed in these tables. Note that the standard deviation of solutions indicates the stability of the algorithms, and the successful rate represents the robustness of the algorithms.

The Michalewicz function (Molga and Smutnicki, 2005) is a highly multimodal test function (with $n!$ local optima). The parameter m defines the "steepness" of the valleys or edges. Larger m leads to a more difficult search. For a very large m, the function behaves like a needle in the haystack (the function values for points in the space outside the narrow peaks give very little information on the location of the global optimum) (Molga and Smutnicki, 2005). As can be seen from Table 5.9, PSO–EO significantly outperformed PSO, GA, and PEO in terms of solution quality, convergence speed, and success rate. PSO–EO had the same success rate as PGHA, but converged faster and found more accurate solution than PGHA. It is interesting to notice that PSO–EO converged to the global optimum more than 10 times faster than PSO, GA, and PEO on this function.

With regard to the Schwefel function (Molga and Smutnicki, 2005), its surface is composed of a great number of peaks and valleys. The function has a second best minimum far from the global minimum where many search algorithms are trapped. Moreover, the global minimum is near the bounds of the domain. The search algorithms are potentially prone to convergence in the wrong direction in optimization of this function (Molga and Smutnicki, 2005). Consequently, this function is very hard to solve for many state-of-the-art optimization algorithms. Nevertheless, PSO–EO showed its great search ability and the prominent capability of escaping from local optima when dealing with this function. From Table 5.10, it is clear that PSO–EO was the winner which was capable of converging to the global optimum with a 100% success rate, whereas the other four algorithms were not able to find the global optimum at all. It is important to point out that, in order to help the PSO to jump out of local optima, the EO procedure was introduced to PSO in each iteration. Thus, the exploitation search played a critical role in optimization of the Schwefel function (Chen et al., 2010b).

The Griewank function is a highly multimodal function with many local minima distributed regularly (Srinivasan and Seow, 2006). The difficult part about finding the optimal solution to this function is that an optimization algorithm easily can be trapped in a local optimum on its way toward the global optimum. From Table 5.11, we can see that the PSO–EO, PGHA, PSO, and PEO could find the global optimum with a 100% success rate. It can be observed that the PSO–EO converged to the global optimum almost as quickly as the PGHA and PSO, and faster than the PEO.

Tables 5.12 and 5.13 show the simulation results of each algorithm on the Rastrigin and Ackley functions respectively. Both these functions are highly multimodal and their patterns are similar to that observed with the Griewank function. As can be seen from Tables 5.12 and 5.13, only the PSO–EO, PGHA, and PSO were capable of finding the global optimum on the two functions with a 100% success rate, and the PSO–EO performed nearly as well as the PSO with respect to convergence speed. The PGHA converged a little faster than the PSO–EO on the Rastrigin function, but slower than the PSO–EO on the Ackley function.

Note that the EO procedure was introduced to the PSO every 100 iterations in optimization of the Griewank, Rastrigin, and Ackley functions. In other words, PSO played a key role in optimization of these functions, while EO brought the diversity to the solutions generated by the PSO at intervals. In this way, the hybrid PSO–EO algorithm is able to preserve the fast convergence ability of PSO. This offers the explanation as to almost the same convergence speed of PSO–EO in comparison with PSO on the Griewank, Rastrigin, and Ackley functions.

The Rosenbrock function is a unimodal optimization problem. Its global optimum is inside a long, narrow, parabolic shaped flat valley, popularly known as Rosenbrock's valley (Molga and Smutnicki, 2005). It is trivial to find the valley, but it is a difficult task to achieve convergence to the global optimum. The Rosenbrock function proves to be hard to solve for all the algorithms, as can be observed from Table 5.14. On this function, only the PSO–EO could find the global optimum with a 100% success rate. Due to the difficulty in finding the global optimum of this function, the EO procedure was introduced to PSO at each iteration.

In a general analysis of the simulation results shown in Tables 5.9 through 5.14, the following conclusions can be drawn:

1. On the convergence speed, the PSO–EO was the fastest algorithm in comparison with the PGHA, standard PSO, standard GA, and PEO for the Michalewicz, Schwefel, and Rosenbrock functions. The PSO–EO converged to the global optimum almost as quickly as the PGHA and PSO for the other three functions, that is, Griewank, Rastrigin, and Ackley. Thus, the hybrid

PSO–EO can be considered a very fast algorithm when solving numerical optimization problems, especially complex multimodal functions.

2. With respect to the stability of algorithms, the PSO–EO showed the best stability as compared to the other four algorithms. The standard deviations of solutions found by the PSO–EO were the smallest among all the algorithms on the Michalewicz, Schwefel, and Rosenbrock functions, and as good as those found by the PGHA and PSO on the other three functions. Overall, the PSO–EO is a highly stable algorithm that is capable of obtaining reasonable consistent results.

3. In terms of the robustness of algorithms, the PSO–EO significantly outperformed the other four algorithms. The PSO–EO was capable of finding the global optimum with a 100% success rate on all the test functions, especially on the Schwefel and Rosenbrock functions, which are very difficult to solve for many state-of-the-art optimization algorithms. Hence, the PSO–EO can be regarded as a highly robust optimization algorithm.

4. The PSO–EO showed its strong capability of escaping from local optima when dealing with those highly multimodal functions, for example, the Michalewicz and Schwefel functions. The PSO–EO also exhibited its great search ability when handling those unimodal functions which are hard to solve for many optimization problems, for example, the Rosenbrock function.

5. In this study, the dimensions of all the test functions are 30, except for the Michalewicz function with the dimension of 10. It is observed that the PSO–EO performed well in optimization of those functions with high dimension. This suggests that the PSO–EO is very suitable for functions with high dimension.

6. The value of parameter *INV* determines the frequency of introducing the EO procedure to PSO. As a consequence, the hybrid PSO–EO algorithm elegantly combines the exploration ability of PSO with the exploitation ability of EO via tuning the parameter *INV*. This offers the explanation as to the better performance of the PSO–EO algorithm in comparison with the other four algorithms.

From the above summary, it can be concluded that the PSO–EO algorithm possesses superior performance in accuracy, convergence speed, stability, and robustness, as compared to the PGHA, standard PSO, standard GA, and PEO. As a result, the PSO–EO algorithm is a perfectly good performer in optimization of those complex high-dimensional functions.

5.6 EO–ABC Integration

The ABC algorithm is a novel swarm intelligent algorithm inspired by the foraging behavior of the honeybee. It was first introduced by Karaboga (2005). After that, ABC was applied to solving the binding numerical optimization problems by Karaboga and Basturk (2008), and satisfactory results were achieved. Since the

ABC algorithm has many advantages, such as being simple in concept, easy to implement, and having fewer control parameters, it has attracted the attention of many researchers and has been used in solving many real-world optimization problems (Potschka, 2010; Yan et al., 2011, 2013; Rangel, 2012).

However, the standard ABC algorithm also has its limitations, such as premature convergence, slow convergence speed at the later stage of evolution, and low convergence accuracy. In order to overcome the limitations of ABC, inspired by the PSO–EO algorithm proposed by Chen et al. (2010b), an idea of combining ABC with EO was addressed in Chen et al. (2014a). Chen et al. (2014a) developed a hybrid optimization method, called the ABC–EO algorithm, which makes full use of the global-search ability of ABC and the LS ability of EO. The performance of the ABC–EO algorithm was testified on six unimodal/multimodal benchmark functions and furthermore the ABC–EO algorithm was compared with other five state-of-the-art optimization algorithms, that is, the standard ABC, PSO–EO (Chen et al., 2010b), standard PSO, PEO (Chen et al., 2006), and standard GA. The experimental results indicate that the ABC–EO algorithm may be a good alternative for complex numerical optimization problems.

5.6.1 Artificial Bee Colony

The ABC algorithm is a recently proposed optimization algorithm that simulates the foraging behavior of a bee colony. In the ABC algorithm, the search space corresponds to a food source that the artificial bees can exploit. The position of a food source represents a possible solution to the optimization problem. The nectar amount of a food source represents the fitness of the associated solution. There are three kinds of bees in a bee colony: employed bees, onlooker bees, and scout bees. Half of the colony comprises employed bees and the other half includes the onlooker bees (Karaboga, 2005).

Artificial colony search activities can be summarized as follows (Karaboga and Basturk, 2008): Initially, the ABC generates a randomly distributed initial population of $SN/2$ solutions (i.e., food source positions), where SN denotes the size of population. Each solution $X_i(i = 1,2, \ldots, SN/2)$ is a D-dimensional vector. Here, D is the number of optimization parameters. After initialization, the population of solutions is subject to repeated cycles of the search processes of the employed bees, the onlooker bees, and the scout bees. Employed bees exploit the specific food sources they have explored before and give the quality information about the food sources to the onlooker bees waiting outside the hive. Onlooker bees receive information about the food sources and choose a food source to exploit depending on the quality information. The more nectar the food source contains, the larger probability the onlooker bees choose it. In the ABC algorithm, one of the employed bees is selected and classified as the scout bee. The classification is controlled by a control parameter called "limit." If a solution representing a food source is not improved by a predetermined number of trials, then that food source is abandoned

1. Initialize the food source positions
2. Evaluate the nectar amount (i.e., fitness) of each food source
3. Cycle = 1
4. Repeat (if the termination conditions are not met)
5. Employed bees phase
6. Calculate probabilities for onlooker bees
7. Onlooker bees phase
8. Scout bees phase
9. Memorize the best solution found so far
10. Cycle = cycle + 1
11. Until cycle = maximum cycle number

Figure 5.14 Pseudocode of ABC algorithm. (Reprinted from *Applied Soft Computing*, 11, Karaboga, D. and Akay, B., A modified artificial bee colony (ABC) algorithm for constrained optimization, 3021–3031, Copyright 2011, with permission from Elsevier.)

by its employed bee and the employed bee associated with that food source becomes a scout. Here "trial" is used to record the nonimprovement number of the solution X_i, used for the abandonment. Finally, scout bees search the whole environment randomly. Note that each food source is exploited by only one employed bee. That is, the number of the employed bees or the onlooker bees is equal to the number of food sources.

The pseudocode of the standard ABC algorithm is described in Figure 5.14.

In order to produce a candidate food position X_i' from the old one X_i in memory, the ABC uses the following expression (Karaboga and Akay, 2011):

$$X_{i,j}' = X_{i,j} + \varphi_{i,j}(X_{i,j} - X_{k,j}) \tag{5.35}$$

where $k = \{1, 2, \ldots, SN/2\}$ and $j = \{1, 2, \ldots, D\}$ are randomly chosen indexes; k has to be different from i; D is the number of variables (problem dimension); and φ_{ij} is a random number between [−1 and 1].

The pseudocode of the employed bees phase of the ABC algorithm is as shown in Figure 5.15.

An artificial onlooker bee chooses a food source depending on the probability value (denoted as P), which is associated with that food source. P is calculated by the following expression (Karaboga and Akay, 2009):

$$P = \frac{\text{fit}_i}{\displaystyle\sum_{n=1}^{SN/2} \text{fit}_n} \tag{5.36}$$

where fit_i is the fitness value of the solution X_i, which is proportional to the nectar amount of the food source in the position X_i, and $SN/2$ is the number of food sources, which is equal to the number of employed bees or onlooker bees.

```
1. For i = 1 to SN/2 do
2. For j = 1 to D do
3. Produce a new food source X′ᵢ for the employed
   bee of the food source Xᵢ using Equation 5.35
4. End for
5. Evaluate the fitness of X′ᵢ
6. Apply the selection process between X′ᵢ and Xᵢ
   based on greedy selection
7. If the solution X′ᵢ does not improve, let
   trial = trial + 1, otherwise trial = 0
8. End for
```

Figure 5.15 Pseudocode of employed bees phase. (Reprinted from *Applied Soft Computing*, 11, Karaboga, D. and Akay, B., A modified artificial bee colony (ABC) algorithm for constrained optimization, 3021–3031, Copyright 2011, with permission from Elsevier.)

The pseudocode of the onlooker bees phase of the ABC algorithm is as shown in Figure 5.16.

The positions of the new food sources found by the scout bees will be produced by the following expression (Karaboga and Akay, 2009):

$$X'_{i,j} = X_{\min,j} + \text{rand}(0,1)(X_{\max,j} - X_{\min,j}) \tag{5.37}$$

where i is the index of the employed bees whose "trial" value reaches the "limit" value first, $j = 1, 2, \ldots, D$, $X_{\min}$ and $X_{\max}$ are the lower bound and the upper bound of each solution, respectively, and rand(0,1) is a random number between [0 and 1].

```
1. t = 0, i = 1
2. Repeat (if the termination conditions are not met)
3. If random < P then (note that P is calculated by Eq. (5.36))
4. t = t + 1
5. For j = 1 to D do
6. Produce a new food source X′ᵢ for the onlooker bee of the
   food source Xᵢ by using E
7. End for
8. Apply the selection process between X′ᵢ and Xᵢ based on
   greedy selection;
9. If the solution Xᵢ does not improve, then let
   trial = trial + 1, otherwise let trial = 0
10. End if
11. i = i + 1
12. i = i mod (SN/2 + 1)
13. Until t = SN/2.
```

Figure 5.16 Pseudocode of onlooker bees phase. (Reprinted from *Applied Soft Computing*, 11, Karaboga, D. and Akay, B., A modified artificial bee colony (ABC) algorithm for constrained optimization, 3021–3031, Copyright 2011, with permission from Elsevier.)

> 1. If max(*trial*) > limit then
> 2. Replace X_i with a new randomly produced solution X_i' by Equation 5.37
> 3. End if

Figure 5.17   Pseudocode of scout bees phase. (Reprinted from *Applied Soft Computing*, 11, Karaboga, D. and Akay, B., A modified artificial bee colony (ABC) algorithm for constrained optimization, 3021–3031, Copyright 2011, with permission from Elsevier.)

The pseudocode of the scout bees phase of the ABC algorithm is worked as in Figure 5.17.

The fitness of the ABC algorithm is proportional to the nectar amount of that food source. The fitness is determined by the following expressions (Karaboga and Akay, 2011):

$$fitness_i = 1/(1+f_i) \quad \text{if } f_i \geq 0 \tag{5.38}$$

$$fitness_i = 1 + abs(f_i) \quad \text{if } f_i < 0 \tag{5.39}$$

where f_i is the cost value of the solution X_i and $abs(f_i)$ is the absolute value of f_i.

5.6.2 ABC–EO Algorithm

Inspired by the PSO–EO algorithm (Chen et al., 2010b), which makes full use of the exploration ability of PSO and the exploitation ability of EO, Chen et al. (2014a) presented a novel hybrid ABC–EO algorithm which combines the merits of ABC and EO. Like PSO, ABC is also a population-based global-search method. Therefore, ABC can be combined with EO in the same way as the PSO–EO algorithm. In other words, when the global optimum found by ABC algorithm is unchanged for several iterations, which indicates that ABC has got trapped into local optima, EO can be used to help ABC to escape from local optima. As a result, through introducing EO to ABC, the hybrid ABC–EO algorithm may compensate for the weakness of ABC with EO and have the capability to jump out of local optima. In order to decrease the computational cost, Chen et al. (2014a) adopted the same way as the PSO–EO to integrate ABC with EO. That is, EO is introduced to ABC when the global optimal solution (i.e., X_{best}) is unchanged continuously for *INV* iterations. As a consequence, the hybrid ABC–EO approach is capable of converging quickly in most of the time with the aid of ABC, and able to escape from a local optimum under the help of EO. The value of parameter *INV* is predefined by the user according to the complexity of problems.

To improve the efficiency and accuracy of the standard ABC, Chen et al. (2014a) presented two improved versions of ABC–EO. One is the combination

1. Initialize the food source positions and set iteration = 0.
2. Evaluate the nectar amount (i.e., fitness) of food sources, and the search way of employed bees is changed according to Equation 5.35 (for ABC–EO algorithm) or Equation 5.40 (for IABC–EO algorithm).
3. If the global optimal solution X_{best} is unchanged for *INV* iterations, then the EO procedure is introduced to change the positions of food sources. Otherwise, continue the next step.
4. If the terminal condition is satisfied, go to the next step; otherwise, set *iteration = iteration* + 1, and go to Step (2).
5. Output the optimal solution and the optimal objective function value.

Figure 5.18 Pseudocode of algorithms ABC–EO and IABC–EO. (From Chen, M. R. et al., Handling multiple objectives with integration of particle swarm optimization and extremal optimization. *Proceedings of the Eighth International Conference on Intelligent Systems and Knowledge Engineering (ISKE 2013)***, Vol. 277, pp. 287–297. © 2014, IEEE.)**

of standard ABC and EO, and the other is the combination of the improved ABC (Gao et al., 2012) (IABC for short) and EO, in which its search way of employed bees is changed as follows (Gao et al., 2012):

$$X'_{i,j} = X_{best,j} + \varphi_{i,j}(X_{best,j} - X_{k,j}) \tag{5.40}$$

Chen et al. (2014a) called them ABC–EO and IABC–EO, respectively. The pseudocode of ABC–EO and IABC–EO for a minimization problem with D dimensions is described in Figure 5.18.

In the main procedure of the ABC–EO algorithm, the fitness of each individual is evaluated by Equations 5.38 and 5.39. However, in the EO procedure, in order to find out the worst component, each component of a solution should be assigned a fitness value. Chen et al. (2014a) defined the fitness of each component of a solution for an unconstrained minimization problem as follows. For the ith position of food sources, the fitness $\lambda_{i,k}$ of the kth component is defined as the mutation cost, that is, $OBJ(X'_{i,k}) - OBJ(X_{best})$, where $X'_{i,k}$ is the new position of the ith position obtained by performing mutation only on the kth component and leaving all other components fixed, $OBJ(X'_{i,k})$ is the objective value of $X'_{i,k}$, and $OBJ(X_{best})$ is the objective value of the best position in the bee colony found so far. The EO procedure is described in Figure 5.19.

5.6.3 Mutation Operator

Since there is merely a mutation operator in EO, the mutation plays a key role in the EO search. In Chen et al. (2014a), the hybrid *GC* mutation is adopted for ABC–EO algorithms, which combines the coarse search and grained search perfectly. For more details, interested readers are referred to Section 5.5.4.

1. For each position $X_i = (X_{i,1}, X_{i,2}, ..., X_{i,D})$ of the food source, $i = 1, ..., SN/2$. Perform mutation on each component of X_i one by one, while keeping other components fixed. Then D new positions $X'_{i,k} (k = 1, ..., D)$ can be obtained;
2. Evaluate the fitness $\lambda_{ik} = OBJ(X'_{i,k}) - OBJ(X_{best})$ of each component $X_{i,k}, k \in \{1, ..., D\}$;
3. Compare all the components according to their fitness values and find out the worst adapted component $X_{i,w}$, and then $X'_{i,w}$ is the new position corresponding to $X_{i,w}, w \in \{1, ..., D\}$;
4. If $OBJ(X'_{i,w}) < OBJ(X_i)$, then set $X_i = X'_{i,w}$ and $OBJ(X_i) = OBJ(X'_{i,w})$, and update X_{best} using X_i; Otherwise X_i keeps unchanged.

Figure 5.19 Pseudocode of EO procedure in ABC–EO and IABC–EO. (From Chen, M. R. et al., Handling multiple objectives with integration of particle swarm optimization and extremal optimization. *Proceedings of the Eighth International Conference on Intelligent Systems and Knowledge Engineering (ISKE 2013),* **Vol. 277, pp. 287–297. © 2014, IEEE.)**

5.6.4 Differences between ABC–EO and Other Hybrid Algorithms

Note that Azadehgan et al. (2011) have also proposed a hybrid algorithm combining ABC with EO, called EABC. In the EABC algorithm, EO was used to determine how to choose the neighbor of employed bees or onlooker bees, that is, X_k in Equation 5.35. While in ABC–EO, EO is introduced to update the positions of food sources when the global optimal position is unchanged for several iterations. The EABC in the literature (Azadehgan et al., 2011) was applied to solving three numerical optimization problems. However, they did not explain the mechanism of the proposed algorithm in detail and the experimental results were unconvincing (Azadehgan et al., 2011). Readers may refer to the literature (Azadehgan et al., 2011) for more details.

5.6.5 Experimental Results

In order to demonstrate the performance of ABC–EO, Chen et al. (2014a) used six well-known benchmark functions shown in Table 5.5. All the functions are to be minimized. The experimental results of ABC–EO and IABC–EO are compared with five state-of-the-art algorithms, that is, the standard ABC, PSO–EO, standard PSO, PEO, and GA.

In Chen et al. (2014a), all the algorithms were run on the same hardware and software platform. Each algorithm was run independently for 50 trials. *INV* in ABC–EO and IABC–EO is set to 100 for each test function. Table 5.15 shows the settings of problem dimension, maximum generation, population size, and initialization range of each algorithm.

Table 5.15 Parameter Settings

Function	Dimension	Maximum Generation	Population Size	Initialization Range
Michalewicz	10	20,000	10	$(0, \pi)$
Schwefel	30	20,000	30	$(-500, 500)$
Griewank	30	20,000	30	$(-600, 600)$
Rastrigin	30	20,000	10	$(-5.12, 5.12)$
Ackley	30	10,000	30	$(-32.768, 32.768)$
Rosenbrock	30	100,000	30	$(-30, 30)$

Source: From Chen, M. R. et al., Handling multiple objectives with integration of particle swarm optimization and extremal optimization. *Proceedings of the Eighth International Conference on Intelligent Systems and Knowledge Engineering (ISKE 2013)*, Vol. 277, pp. 287–297. © 2014, IEEE.

After 50 trials of running each algorithm for each test function, the simulation results were obtained and shown in Tables 5.16 through 5.21. The bold numerals in Tables 5.16 through 5.21 mean that they are the best optima among all the optima. Denote F as the result found by the algorithms and F^* as the optimum value of the functions ($F_1^* = -9.66$, $F_2^* = -12569.5$, $F_3^* = F_4^* = F_5^* = F_6^* = 0$). The simulation is

Table 5.16 Comparison Results for Michalewicz Function f_1 ($F_1^* = -9.66$)

Algorithm	Runtime (s)	Success (%)	Best	Mean	Worst	Standard Deviation
IABC–EO	0.27	**100**	**−9.66**	**−9.66**	**−9.66**	**8.12E − 5**
ABC–EO	**0.258**	**100**	**−9.66**	**−9.66**	**−9.66**	8.75E − 5
ABC	0.272	**100**	**−9.66**	**−9.66**	**−9.66**	8.75E − 5
PSO–EO	0.563	**100**	**−9.66**	**−9.66**	**−9.66**	9.02E − 5
PSO	0.71	0	−9.65	−9.34	−8.44	0.28
PEO	8.131	0	−9.61	−9.55	−9.49	0.029
GA	0.567	56	**−9.66**	−9.63	−9.46	0.037

Source: From Chen, M. R. et al., Handling multiple objectives with integration of particle swarm optimization and extremal optimization. *Proceedings of the Eighth International Conference on Intelligent Systems and Knowledge Engineering (ISKE 2013)*, Vol. 277, pp. 287–297. © 2014, IEEE.

Table 5.17 Comparison Results for Schwefel Function f_2 ($F_2^* = -12569.5$)

Algorithm	Runtime (s)	Success (%)	Best	Mean	Worst	Standard Deviation
IABC–EO	**0.084**	100	−12569.5	−12569.5	−12569.4	**0.03**
ABC–EO	0.172	100	−12569.5	−12569.4	−12569.4	0.04
ABC	0.163	100	−12569.5	−12569.4	−12569.4	**0.03**
PSO–EO	28.38	78	−12569.5	−12543.4	−12451	49.55
PSO	1.508	0	−11532.1	−9382.2	−7599.5	933.74
PEO	45.193	2	−12561.7	−12254.3	−12095.7	115.84
GA	1.685	0	−9845.2	−8690.9	−7693.5	489.03

Source: From Chen, M. R. et al., Handling multiple objectives with integration of particle swarm optimization and extremal optimization. *Proceedings of the Eighth International Conference on Intelligent Systems and Knowledge Engineering (ISKE 2013)*, Vol. 277, pp. 287–297. © 2014, IEEE.

Table 5.18 Comparison Results for Griewank Function f_3 ($F_3^* = 0$)

Algorithm	Runtime (s)	Success (%)	Best	Mean	Worst	Standard Deviation
IABC–EO	0.029	100	1.95E − 6	7.82E − 6	9.96E − 6	1.98E − 6
ABC–EO	0.045	100	1.22E − 6	7.67E − 6	1.00E − 5	2.35E − 6
ABC	0.043	100	1.91E − 6	7.63E − 6	9.96E − 6	2.07E − 6
PSO–EO	0.018	100	0	0	0	0
PSO	0.013	100	0	0	0	0
PEO	3.288	100	8.30E − 4	9.44E − 4	9.99E − 4	4.76E − 5
GA	2.189	0	0.0055	0.061	0.1949	0.036

Source: From Chen, M. R. et al., Handling multiple objectives with integration of particle swarm optimization and extremal optimization. Proceedings of the Eighth International Conference on Intelligent Systems and Knowledge Engineering (ISKE 2013), Vol. 277, pp. 287–297. © 2014, IEEE.

considered successful, or in other words, the near-optimal solution is found, if F satisfies that $|(F^* − F)/F^*| < 1E − 3$ (for the case $F^* \neq 0$) or $|F^*−F| < 1E − 3$ (for the case $F^* = 0$). In these tables, "success" represents the success rate, and "runtime" is the average runtime of 50 runs when the near-optimal solution is found or otherwise when the maximum generation is reached. The worst, mean, best, and standard deviation of solutions found by all the algorithms are also listed in these tables.

Table 5.19   Comparison Results for Rastrigin Function f_4 ($F_4^* = 0$)

Algorithm	Runtime (s)	Success (%)	Best	Mean	Worst	Standard Deviation
IABC–EO	0.314	100	2.94E − 7	5.62E − 6	9.98E − 6	3.02E − 6
ABC–EO	0.163	98	3.90E − 8	1.96E − 3	0.098	0.014
ABC	0.179	100	1.62E − 7	7.27E − 6	1.27E − 4	1.77E − 5
PSO–EO	0.018	100	0	0	0	0
PSO	0.011	100	0	0	0	0
PEO	16.04	0	1.544	2.242	2.727	0.273
GA	0.596	8	8.33E − 4	0.02	0.533	0.075

Source: From Chen, M. R. et al., Handling multiple objectives with integration of particle swarm optimization and extremal optimization. *Proceedings of the Eighth International Conference on Intelligent Systems and Knowledge Engineering (ISKE 2013)*, Vol. 277, pp. 287–297. © 2014, IEEE.

Table 5.20   Comparison Results for Ackley Function f_5 ($F_5^* = 0$)

Algorithm	Runtime (s)	Success (%)	Best	Mean	Worst	Standard Deviation
IABC–EO	0.083	100	2.80E − 7	7.63E − 6	9.94E − 6	2.22E − 6
ABC–EO	0.091	100	1.27E − 6	8.24E − 6	9.98E − 6	1.71E − 6
ABC	0.092	100	3.29E − 6	7.79E − 6	9.97E − 6	1.95E − 6
PSO–EO	0.016	100	−8.88E − 16	−8.88E − 16	−8.88E − 16	9.96E − 32
PSO	0.015	100	−8.88E − 16	−8.88E − 16	−8.88E − 16	9.96E − 32
PEO	0.083	100	2.80E − 7	7.63E − 6	9.94E − 6	2.22E − 6
GA	0.091	100	1.27E − 6	8.24E − 6	9.98E − 6	1.71E − 6

Source: From Chen, M. R. et al., Handling multiple objectives with integration of particle swarm optimization and extremal optimization. *Proceedings of the Eighth International Conference on Intelligent Systems and Knowledge Engineering (ISKE 2013)*, Vol. 277, pp. 287–297. © 2014, IEEE.

The Michalewicz function is a highly multimodal function. As can be seen from Table 5.16, IABC–EO, ABC–EO, ABC, and PSO–EO could find the global optimum with a 100% success rate, GA could find the global optimum with a 56% success rate, but the PSO and PEO algorithms could not find the global optimum.

It can be observed that the ABC–EO has the fastest convergence speed, and the IABC–EO converged to the global optimum almost as quickly as the ABC, and faster than the PSO–EO, PSO, PEO, and GA. The IABC–EO also had a good performance in terms of stability.

With regard to the Schwefel function, it has a second best minimum far from the global minimum where many search algorithms may get trapped. From Table 5.17, it is clear that the IABC–EO was the winner which was capable of converging to the global optimum with the 100% success rate, the fastest convergence speed, and the lowest standard deviation. The ABC–EO and ABC were better than the PSO–EO, PSO, PEO, and GA in terms of success rate, convergence speed, and solution accuracy. It is interesting to notice that the IABC–EO and ABC–EO converged to the global optimum more than 100 times faster than the PSO–EO and PEO.

Tables 5.18 and 5.20 show the simulation results of each algorithm on the Griewank and Ackley functions respectively. Both the functions are highly multimodal. As can be seen from Table 5.18, all algorithms could find the optimal solution with a 100% success rate, except for GA. But the IABC–EO, ABC–EO, and ABC were a little worse than the PSO–EO and PSO with respect to solution accuracy. From Table 5.20, we can see that the IABC–EO and ABC–EO could find the optimum with a 100% success rate in a short time. At the same time, Table 5.20 indicates that the ABC–EO and IABC–EO were not better than the PSO and PSO–EO, but better than the PEO and GA with respect to convergence speed and solution accuracy.

From Table 5.19, which shows the simulation results of each algorithm on the Rastrigin function, we can see that the PSO and PSO–EO were the best performers in all aspects, and the IABC–EO and ABC–EO could find the optimal solution in a short time with a higher success rate and solution accuracy, almost as good as the ABC.

The last test function, Rosenbrock, is a unimodal function, but for a lot of optimization algorithms, it is very difficult to converge to the global optimal point. As can be seen from Table 5.21, the IABC–EO was capable of finding the optimal solution quickly with a 96% success rate. Moreover, the IABC–EO converged to the global optimum more than 60 times faster than the PSO–EO, although the PSO–EO could find the optimum with a 100% success rate. The IABC–EO significantly outperformed other algorithms, except for the PSO–EO, in terms of solution quality, convergence speed, and success rate.

From the simulation results, it can be concluded that the ABC–EO and IABC–EO possess good or superior performance in solution accuracy, convergence speed, and success rate, as compared to the standard ABC, PSO–EO, standard PSO, PEO, and standard GA. As a result, the ABC–EO and IABC–EO can be considered as perfectly good performers in optimization of complex high-dimensional functions.

Table 5.21 Comparison Results for Rosenbrock Function f_6 ($F_6^* = 0$)

Algorithm	Runtime (s)	Success (%)	Best	Mean	Worst	Standard Deviation
IABC–EO	**3.88**	96	**2.09E − 6**	7.43E − 4	0.026	3.79E − 3
ABC–EO	7.75	2	8.03E − 4	0.012	0.045	0.011
ABC	7.64	4	6.41E − 4	0.014	0.057	0.013
PSO–EO	247.7	**100**	9.31E − 6	**8.17E − 5**	**2.73E − 4**	**5.93E − 5**
PSO	8.46	2	2.25E − 4	24.89	27.39	4.79
PEO	253.9	0	5.425	7.497	8.761	0.788
GA	10.214	0	20.970	29.542	51.995	6.385

Source: From Chen, M. R. et al., Handling multiple objectives with integration of particle swarm optimization and extremal optimization. *Proceedings of the Eighth International Conference on Intelligent Systems and Knowledge Engineering (ISKE 2013)*, Vol. 277, pp. 287–297. © 2014, IEEE.

5.7 EO–GA Integration

GAs belong to the larger class of EAs, which mimic the process of natural selection, such as crossover, mutation, and selection. In a GA, a population of candidate solutions to an optimization problem is evolved toward better solutions through iterations. GAs have several merits, such as easy implementation, fewer adjustable parameters, and parallel computation. GAs have been widely applied in many fields, such as bioinformatics, phylogenetics, computational science, engineering, economics, chemistry, manufacturing, mathematics, physics, and pharmacometrics. However, there is also premature convergence of GAs. When dealing with many optimization problems, GAs may tend to converge to local optima rather than to the global optimum of the problem. This problem may be alleviated by increasing the mutation rate, adopting a different fitness function, or by using selection techniques that keep a diverse population of solutions (Taherdangkoo et al., 2012). A possible technique to maintain diversity would be mutation operation, which simply replaces part of the population with randomly generated individuals, while most of the population is too similar to each other. Diversity is of importance in GAs because crossing over a homogeneous population does not generate new solutions. Hence, increasing the diversity of the population of solutions may prevent GAs from getting trapped in local optima. Note that in EO, there is only mutation operation, which can maintain the diversity of solutions. So EO may be introduced to GAs to increase the diversity of solutions, and thus the hybrid algorithm, which is the combination of GA and EO, may not easily go into local optima.

EO can be integrated with GA through the following three frameworks:

1. *GA–EO-I*: As in PSO–EO and ABC–EO, the EO procedure can be introduced to GA at *INV*-iteration intervals (here we use a parameter *INV* to represent the frequency of introducing EO to GA). The value of parameter *INV* is predefined by the user. For more information about the EO procedure and the setting of parameter *INV*, readers may refer to Section 5.5.3. The flowchart of this framework is shown in Figure 5.20.
2. *GA–EO-II*: While GA cannot find better solutions for *num* iterations (here the value of *num* is predefined by users according to the complexity of problem), the EO procedure can be introduced to GA to help GA to go out of local optima. Figure 5.21 is the flowchart of this framework.
3. *GA–EO-III*: Due to only a mutation operation in EO, the EO procedure can be used as a mutation operation of GA. After the selection and crossover operations, each individual in the population will perform mutation (here the mutation operation can be replaced with the EO procedure) according to the mutation rate. Figure 5.22 is the flowchart of this configuration.

The above three frameworks have their own merits. GA–EO-I may be manipulated more easily for users. However, EO will be introduced to GA when GA does not get trapped in local optima. This will increase the computational cost. As far as GA–EO-II is concerned, EO will be adopted by GA when GA has gone into local optima. Therefore, the computational cost of GA–EO-II will be lower than that

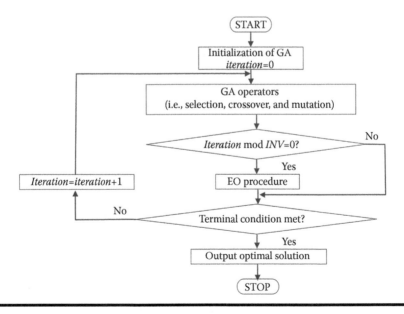

Figure 5.20   Flowchart of GA–EO-I.

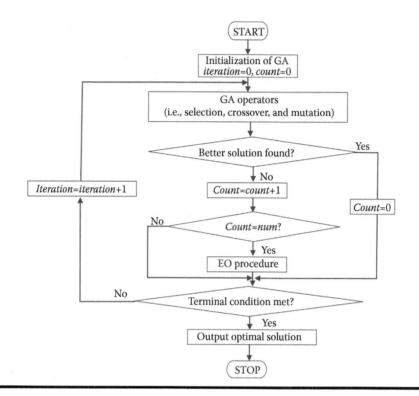

Figure 5.21   Flowchart of GA–EO-II.

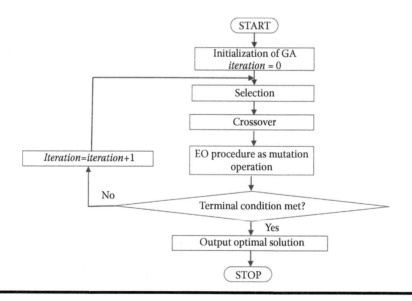

Figure 5.22   Flowchart of GA–EO-III.

of GA–EO-I. But the setting of parameter *num* will depend on the experience of users. GA–EO-III proposes a new idea that the EO procedure can play the role of mutation. This operation will bring more diversity for the population of solutions. But at the same time, the computational complexity of the GA–EO-III will rise. As a result, how to make the balance of effectiveness and efficiency of an algorithm will be our vital objective in future studies.

5.8 Summary

Regarding the disadvantages of current evolutionary computation in theoretical foundation and LS effectiveness, the interrelations between coevolution, MAs, and extremal dynamics are studied. This chapter first introduces two hybrid EO algorithms under the umbrella of MAs, called "EO–LM" and "EO–SQP," which combine the global-search capability of EO and the LS efficiency of the deterministic methods. The proposed methods are employed for the numerical optimization and NN learning. The search dynamics of the proposed methods are studied in detail. The effectiveness and the efficiency of the proposed methods are proven by the comparison with traditional methods through a number of benchmark systems. Second, this chapter introduces the hybrid PSO–EO, which makes full use of the exploration ability of PSO and the exploitation ability of EO, and thus can overcome the limitation of PSO and has the capability of escaping from local optima. Furthermore, another hybrid algorithm called ABC–EO is introduced in this chapter, which combines the merits of ABC and EO. Experimental results show that PSO–EO and ABC–EO possess good performance in terms of solution accuracy, convergence speed, and success rate. As a consequence, PSO–EO and ABC–EO can be considered as perfectly good performers in the optimization of complex high-dimensional functions. Finally, to avoid premature convergence of GA, an idea of combining GA with EO is proposed in this chapter and three frameworks through which EO can be integrated with GA are also addressed.

Chapter 6

Multiobjective Optimization with Extremal Dynamics

6.1 Introduction

Multiobjective optimization (MOO), or so-called multicriteria optimization, Pareto optimization, involves a problem solution with more than one objective function to be optimized simultaneously. MOO has been in high demand and widely applied in a variety of real-world areas, including sciences, engineering, economics, logistics, etc. The solutions and/or decisions taken for MOO problems need to consider the trade-off between two or more conflicting criteria. In general, MOO may provide a set of solutions; the decision maker will then select one that meets the desired requirement.

The operations research (OR) community has developed several mathematical programming techniques to solve multiobjective optimization problems (MOPs) since the 1950s (Miettinen, 1999). However, mathematical programming techniques have some limitations when dealing with MOPs (Coello, 2006). For example, many of them may not work when the Pareto front is concave or disconnected. Some require differentiability of the objective functions and the constraints. In addition, most of them only generate a single solution from each run. During the past two decades, a considerable amount of MOEAs have been presented to solve MOPs (Sarker et al., 2002; Beausoleil, 2006; Coello, 2006; Elaoud et al., 2007; Hanne, 2007). EAs seem particularly suitable to solve MOPs, because they can deal with a set of possible solutions (or so-called population) simultaneously. This allows us to find several members of the Pareto-optimal set in a single run of the algorithm (Coello, 2006). There is also no requirement for differentiability of the objective functions and the constraints. Moreover, EAs are susceptible to the shape of the Pareto front and can easily deal with discontinuous or concave Pareto fronts.

So far there have been some papers studying MOO using extremal dynamics. Ahmed and Elettreby (2004) introduced a random version of the BS model. They also generalized the single objective BS model to a multiobjective one by the weighted sum aggregation method. The method is easy to implement but its most serious drawback is that it cannot generate proper members of the Pareto-optimal set when the Pareto front is concave regardless of the weights used (Das and Dennis, 1997). Galski et al. (2004) applied the generalized extremal optimization (GEO) algorithm (De Sousa et al., 2004) to design a spacecraft thermal control system. The design procedure was tackled as an MOP and they also resorted to the weighted sum aggregation method to solve the problem. In order to extend GEO to solve MOPs effectively, Galski et al. (2005) further presented a revised multiobjective version of the GEO algorithm, called M-GEO. The M-GEO algorithm does not use the weighted sum aggregation method. Instead, the Pareto dominance concept was introduced to M-GEO in order to find out the approximate Pareto front, and at the same time the approximate Pareto front was stored and updated with each run. Since the fitness assignment in the M-GEO is not based on the Pareto dominance strategy, M-GEO belongs to the non-Pareto approach (Galski et al., 2005). M-GEO was successfully applied to the inverse design of a remote sensing satellite constellation (Galski et al., 2005).

In order to extend EO to solve MOPs in an efficient way, Chen and Lu (2008) developed a novel elitist single-parent single-offspring MOO method, called multiobjective extremal optimization (MOEO). Being different from the aforementioned methods, MOEO introduces the Pareto dominance strategy to EO for the first time. It is interesting to note that, similar to the PAES (Knowles and Corne, 1999), MOEO is also a single-parent single-offspring MOO algorithm. MOEO has been validated using five benchmark functions reported in the specialized literature and compared with four competitive MOEAs: the Nondominated Sorting Genetic Algorithm-II (NSGA-II) (Deb et al., 2002), the Pareto Archived Evolution Strategy (PAES) (Knowles and Corne, 1999), the Strength Pareto Evolutionary Algorithm (SPEA) (Zitzler and Thiele, 1998), and the Strength Pareto Evolutionary Algorithm2 (SPEA2) (Zitzler et al., 2001). The simulation results demonstrate that MOEO is highly competitive with the state-of-the-art MOEAs. Hence, MOEO may be a good alternative to solve numerical MOPs. So far, MOEO has been extended to solve more real-world problems, such as multiobjective mechanical components design problems (M. R. Chen et al., 2007), multiobjective portfolio optimization problems (Chen et al., 2009), and multiobjective 0/1 knapsack problems (MOKP) (P. Chen et al., 2010).

6.2 Problem Statement and Definition

Without loss of generality, the MOPs are mathematically defined as follows (Coello, 2005):

Find X which minimizes

$$F(X) = [f_1(X), f_2(X), \ldots, f_k(X)] \qquad (6.1)$$

subject to

$$g_i(X) \le 0, \quad i = 1, 2, \ldots, m$$

$$h_j(X) = 0, \quad j = 1, 2, \ldots, p$$

where $X = (x_1, x_2, \ldots, x_n)^T$ is a vector of decision variables, each decision variable is bounded by lower and upper limits $l_l \le x_l \le u_l$, $l = 1, \ldots, n$, k is the number of objectives, m is the number of inequality constraints, and p is the number of equality constraints. The following four concepts are of importance (Fonseca and Fleming, 1995).

Definition 6.1 Pareto dominance

A solution u is said to *dominate* another solution v (denoted by $u \succ v$) if and only if: $\forall i \in \{1, \ldots, M\} : f_i(u) \le f_i(v) \wedge (\exists k \in \{1, \ldots, M\} : f_k(u) < f_k(v))$.

Definition 6.2 Pareto optimal

A solution u is said to be Pareto optimal (or nondominated) if and only if: $\neg \exists v \in X : v \succ u$.

Definition 6.3 Pareto-optimal set

The Pareto-optimal set P_S is defined as the set of all Pareto-optimal solutions: $P_S = \{u \mid \neg \exists v \in X : v \succ u\}$.

Definition 6.4 Pareto-optimal front

The Pareto-optimal front P_F is defined as the set of all objective functions values corresponding to the solutions in P_S: $P_F = \{f(x) = (f_1(x), \ldots, f_M(x)) \mid x \in P_S\}$.

6.3 Solutions to Multiobjective Optimization

The potential of EAs for solving MOPs was hinted at in the late 1960s by Ronsenberg in his PhD thesis (Rosenberg, 1967). The first actual implementation of a MOEA was Schaffer's Vector Evaluated Genetic Algorithm (VEGA) (Schaffer, 1985), which was introduced in the mid-1980s. Since then, a wide variety of algorithms have been proposed in the literature (Coello, 2005).

MOEAs can be roughly divided into the following types (Coello, 2005):

1. Aggregating functions
2. Population-based non-Pareto approaches
3. Pareto-based approaches

We will briefly discuss each of them in the following subsections.

6.3.1 Aggregating Functions

Perhaps the most straightforward approach to handling multiple objectives with any technique is to combine all the objectives into a single one using either an addition, multiplication, or any other combination of arithmetical operations. These techniques are known as "aggregating functions." An example of this approach is a linear sum of weights of the form:

$$\text{minimize} \sum_{i=1}^{k} w_i f_i(x) \tag{6.2}$$

where $w_i \geq 0$ are the weighting coefficients representing the relative importance of the k objective functions. It is usually assumed that

$$\sum_{i=1}^{k} w_i = 1 \tag{6.3}$$

This approach does not require any changes to the basic mechanism of an EA and it is therefore very simple, easy to implement, and efficient. This approach can work well in simple MOPs with few objective functions and convex search spaces. One obvious problem of this approach is that it may be difficult to generate a set of weights that properly scales the objectives when little is known about the problem. However, its most serious drawback is that it cannot generate proper members of the Pareto-optimal set when the Pareto front is concave regardless of the weights used (Coello, 2005).

6.3.2 Population-Based Non-Pareto Approaches

In these techniques, the population of an EA (EA) is used to diversify the search, but the concept of Pareto dominance is not directly incorporated into the selection process. The classical example of this sort of approach is the Vector Evaluated Genetic Algorithm (VEGA) proposed by Schaffer (1985). VEGA basically consists of a simple GA with a modified selection mechanism. At each generation, a number of subpopulations are generated by performing proportional selection according to each objective function in turn. Thus, for a problem with k objectives, k subpopulations of size M/k each are generated (assuming a total population size of M). These subpopulations are then shuffled together to obtain a new population of size M, on which GA applies the crossover and mutation operators. VEGA has several problems, of which the most serious is that its selection scheme is opposed to the concept of Pareto dominance. If, for example, there is an individual that encodes a good compromise solution for all the objectives, but it is not the best in any of them, it will be discarded. Note however, that such an individual should really be preserved because it encodes a Pareto-optimal solution. So the fact that Pareto dominance is not directly incorporated into the selection process of the algorithm remains its main disadvantage.

6.3.3 Pareto-Based Approaches

Pareto-based approaches can be historically studied as covering two generations. The first generation is characterized by the use of fitness sharing and niching combined with Pareto ranking. The most representative algorithms from the first generation are the following: Nondominated Sorting Genetic Algorithm (NSGA) (Srinivas and Deb, 1994), Niched-Pareto Genetic Algorithm (NPGA) (Horn et al., 1994), and Multi-objective Genetic Algorithm (MOGA) (Fonseca and Fleming, 1993). The second generation of MOEAs was born with the introduction of the notion of elitism. In the context of MOO, elitism usually refers to the use of an external population to preserve the nondominated individuals. Note that elitism can also be introduced through the use of a $(\mu + \lambda)$-selection in which parents compete with their children, and those which are nondominated are selected for the following generation. In a study by Zitzler et al. (2000), it was clearly shown that elitism helps in achieving better convergence in MOEAs. Among the existing elitist MOEAs, NSGA-II (Deb et al., 2002), SPEA (Zitzler and Thiele, 1998), and SPEA2 (Zitzler et al., 2001), Pareto-archived PAES (Knowles and Corne, 1999) enjoyed more attention.

1. *Nondominated Sorting Genetic Algorithm II (NSGA-II):* This algorithm is proposed by Deb et al. (2002). NSGA-II is a revised version of the NSGA proposed by Srinivas and Deb (1994). The original NSGA is based on several layers of classifications of the individuals as suggested by Goldberg (1989). Before selection, all nondominated individuals are classified into one

category. This group of classified individuals is then ignored and another layer of nondominated individuals is considered. The process continues until all individuals in the population are classified. Since individuals in the first front have the maximum fitness value, they always get more of a chance of surviving than the remainder of the population. The NSGA-II is more efficient in terms of computational complexity than the original NSGA. To keep diversity, NSGA-II uses a crowded comparison operator without specifying any additional parameters, while the original NSGA used fitness sharing. Besides, NSGA-II adopts $(\mu + \lambda)$-selection as its elitist mechanism.

2. *Strength Pareto Evolutionary Algorithm (SPEA) and Strength Pareto Evolutionary Algorithm 2 (SPEA2):* Zitzler and Thiele (1998) suggested SPEA. They suggested maintaining an external population at every generation storing all nondominated solutions discovered so far. This external population participates in all genetic operations. Presented by Zitzler et al. (2001), SPEA2 is an improvement over SPEA. In contrast to SPEA, SPEA2 uses a fine-grained fitness assignment strategy that incorporates density information. Furthermore, the archive size in SPEA2 is fixed, that is, whenever the number of nondominated individuals is less than the predefined archive size, the archive is filled up by dominated individuals. In addition, the clustering technique, which is used in SPEA as the archive truncation method, has been replaced by an alternative truncation method which has similar features but does not lose boundary points. Finally, another difference to SPEA is that only members of the archive participate in the mating selection process.

3. *Pareto Archived Evolution Strategy (PAES):* Knowles and Corne (1999) suggested a simple MOEA using an $(1 + 1)$ evolution strategy (a single parent generates a single offspring). PAES is purely based on the mutation operator to search for new individuals. An archive is used to store the nondominated solutions found in the evolutionary process. Such a historical archive is the elitist mechanism adopted in PAES. An interesting aspect of this algorithm is the adaptive grid method used as the archive truncation method. They also presented $(1 + \lambda)$ and $(\mu + \lambda)$ variations of the basic approach. The former is identical to PAES $(1 + 1)$ except that λ offspring are generated from the current solution. The $(\mu + \lambda)$ version maintains a population of size μ from which λ copies are made. Then the fittest μ from the $\mu + \lambda$ solutions replaces the current population.

6.4 EO for Numerical MOPs

In order to further extend EO to solve MOPs, Chen and Lu (2008) proposed a novel elitist MOO algorithm, the so-called MOEO, through introducing Pareto dominance strategy to EO for the first time. Similar to EO, MOEO performs only one operation, that is, mutation, on each decision variable of the current solution.

It is well known that there exist two fundamental goals in MOO design: one is to minimize the distance of the generated solutions to the Pareto-optimal set, the other is to maximize the diversity of the achieved Pareto set approximation. MOEO mainly consists of three components: fitness assignment, diversity preservation, and external archive. A good fitness assignment is beneficial for guiding the search toward the Pareto-optimal set. In order to increase the diversity of the nondominated solutions, the diversity-preserving mechanism is introduced to MOEO. For the sake of preventing nondominated solutions from being lost, MOEO also adopts an external archive to store the nondominated solutions found in the evolutionary process.

6.4.1 MOEO Algorithm

For a numerical MOP, MOEO algorithm works as shown in Figure 6.1. The flowchart of MOEO is given in Figure 6.2. In the following, fitness assignment, diversity preservation, external archive, and mutation operator of MOEO will be addressed in more detail.

6.4.1.1 Fitness Assignment

In MOEO, the dominance ranking method (Fonseca and Fleming, 1993) is used to evaluate the special fitness, that is, the fitness value of one solution equals to the number of other solutions by which it is dominated, to determine the fitness value for each solution. Therefore, the nondominated solutions are ranked as zero, while

1. Randomly generate an initial solution $S = (x_1, x_2, ..., x_n)$. Set the external archive empty. Set *iteration* = 0.
2. Generate n offspring of the current solution S by performing mutation on each decision variable one by one.
3. Perform dominance ranking on the n offspring and then obtain their rank numbers, i.e., $r_j \in [0, n-1], j \in \{1, ..., n\}$.
4. Assign the fitness $\lambda_j = r_j$ for each variable, $x_j, j \in \{1, ..., n\}$.
5. If there is only one variable with fitness value of zero, the variable will be considered the worst component; otherwise, the diversity preservation mechanism is invoked. Assuming that the worst component is x_w with fitness $\lambda_w = 0, w \in \{1, ..., n\}$.
6. Perform mutation only on x_w while keeping other variables unchanged, then get a new solution s_w.
7. Accept $S = S_w$ *unconditionally*.
8. Apply UpdateArchive (S, archive) to update the external archive (see Fig. 6.4)).
9. If the iterations reach the predefined maximum number of the generations, go to Step 10; otherwise, set *iteration* = *iteration* + 1, and go to Step 2.
10. Output the external archive as the Pareto-optimal set.

Figure 6.1 Pseudo-code of MOEO algorithm for numerical MOPs. (Reprinted from *European Journal of Operational Research*, 3 (188), Chen, M. R. and Lu, Y. Z., A novel elitist multi-objective optimization algorithm: Multi-objective extremal optimization, 637–651, Copyright 2008, with permission from Elsevier.)

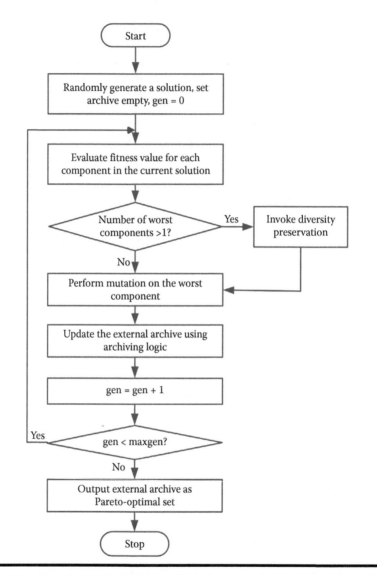

Figure 6.2 Flowchart of MOEO algorithm for numerical MOPs. (Adapted from Chen, M. R. and Lu, Y. Z., *European Journal of Operational Research* 3 (188): 637–651, 2008.)

the worst possible ranking is the number of all the solutions minus one. It is important to note that there exists only one individual in the search process of MOEO, and each decision variable in the individual is considered a species. In order to find out the worst species via fitness assignment, MOEO generates a population of new individuals (or so-called offspring) by performing mutation on the decision variables of the current individual one by one. Then the dominance ranking is carried

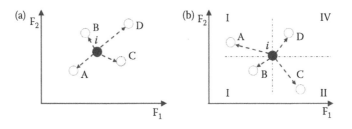

Figure 6.3 (a) Dominance ranking in MOEO and (b) diversity preservation in MOEO. (Adapted from Chen, M. R. and Lu, Y. Z., *European Journal of Operational Research* **3 (188): 637–651, 2008.)**

out in the newly generated individuals. In MOEO, the species corresponding to the nondominated offspring (i.e., its fitness value equals to zero) is considered the weakest one. Then the new individual in the next generation will be generated by mutating the worst species, while keeping the other species unchanged.

To be clearer, Chen and Lu (2008) illustrated the process of dominance ranking in MOEO using Figure 6.3a. Assume that this is a two-objective optimization problem and both objectives are to be minimized. The locations of the current individual in the objective space (marked with black solid circle) will change to new one in the next generation via mutating its weakest species. For example, given an individual $S_i = (x_1, x_2, x_3, x_4)$, of which the location in the objective space is denoted by the circle i, we can identify the weakest species by mutating the four variables one by one and then performing dominance ranking on the newly generated offspring. First, a new individual $S_{iA} = (x_1', x_2, x_3, x_4)$ can be obtained by mutating x_1 to x_1', keeping the other variables unchanged. Similarly, the other three offspring, that is, $S_{iB} = (x_1, x_2', x_3, x_4)$, $S_{iC} = (x_1, x_2, x_3', x_4)$, and $S_{iD} = (x_1, x_2, x_3, x_4')$, are generated. The four white dashed circles (i.e., A, B, C, D) in Figure 6.3a stand for the locations of four offspring (i.e., S_{iA}, S_{iB}, S_{iC}, S_{iD}) in the objective space, respectively. The next location of the current individual S_i in the objective space depends on the dominance ranking number of the four newly generated offspring. It can be seen from Figure 6.3a that for the circle i, circle A is nondominated by the other three dashed circles (i.e., B, C, and D). The rank numbers of A, B, C, and D are 0, 1, 1, and 3, respectively. Hence, the species x_1 corresponding to A is considered the weakest species and the individual S_i will change to S_{iA} in the next generation.

If there exists more than one worst adapted species, that is, at least two species with the same species fitness value of zero, then the following diversity-preserving mechanism will be invoked.

6.4.1.2 Diversity Preservation

The goal of introducing the diversity preservation mechanism is to maintain a good spread of solutions in the obtained set of solutions. In MOEO, a new approach is proposed to keep good diversity of nondominated solutions. It is worth pointing out that MOEO does not require any user-defined parameter for maintaining diversity among population members.

It can be seen from Figure 6.3b that the locations of the nondominated offspring (marked with white dash circle) relative to that of the parent i (indicated by black circle) can be categorized into four cases, that is, regions I, II, III, and IV. The offspring (circle B) residing in the region II dominates its parent, those (circles A and C) in the regions I and III do not dominate the parent and vice versa, and the one (circle D) in the region IV is dominated by the parent. It should be pointed out that those species lying in region IV will increase the searching time to approach the Pareto-optimal set. Thus, the species residing in region IV will not be chosen as the worst species unless there exists no nondominated offspring lying in the other three regions. In addition, for the purpose of keeping a good spread of nondominated solutions, MOEO not only regards the nondominated offspring lying in region I but also those in regions II and III as the candidates of the new individual. The simplest method is that we can randomly choose one nondominated species in the above three regions as the weakest one.

6.4.1.3 External Archive

The main objective of the external archive is to keep a historical record of the nondominated individuals found along the search process. Note that the archive size is fixed in MOEO. This external archive provides the elitist mechanism for MOEO. The archive in MOEO consists of two main components as follows:

1. *Archiving logic:* The function of the archiving logic is to decide whether the current solution should be added to the archive or not. The archiving logic works as follows. The newly generated solution S is compared with the archive to check if it dominates any member of the archive. If yes, the dominated solutions are eliminated from the archive and S is added to the archive. If S does not dominate any member of the archive, the archive does not need to be updated and the iteration continues. However, if S and any member of the archive do not dominate each other, there exist three cases:
   a. If the archive is not full, S is added to the archive.
   b. If the archive is full and S resides in the most crowded region in the parameter space among the members of the archive, the archive does not need to be updated.
   c. If the archive is full and S does not reside in the most crowded region of the archive, the member in the most crowded region of the archive is replaced by S.

   Based on the above archiving logic, the pseudo-code of function UpdateArchive (S, archive) is shown in Figure 6.4.

```
function UpdateArchive (S, archive)
begin function
    if the current solution S is dominated by at least one member of the archive, then
        the archive does not need to be updated;
    else if some members of the archive are dominated by S, then
        remove all the dominated members from the archive and add S to the archive;
    end if
    else
        if the archive is not full, then
            add S to the archive;
        else
            if S resides in the most crowded region of the archive, then
                the archive does not need to be updated;
            else
                replace the member in the most crowded region of the archive by S;
            end if
        end if
    end if
end function
```

Figure 6.4 **Pseudo-code of function UpdateArchive (S, archive). (Adapted from Chen, M. R. and Lu, Y. Z., *European Journal of Operational Research* 3 (188): 637–651, 2008.)**

2. *Crowding-distance metric:* In MOEO, the crowding-distance metric proposed by Deb et al. (2002) is adopted to judge whether the current solution resides in the most crowded region of the archive. For more details about crowding-distance metric, the interested readers are referred to Deb et al. (2002).

6.4.1.4 Mutation Operation

Since there is merely a mutation operation in MOEO, the mutation plays a key role in the MOEO search that generates new solutions through adding or removing genes at the current solutions, that is, chromosomes. In MOEO, a new mutation method based on mixing Gaussian mutation and Cauchy mutation is used as the mutation operator.

The mechanisms of Gaussian and Cauchy mutation operations have been studied by Yao et al. (1999). They pointed out that Cauchy mutation can get to the global optimum faster than Gaussian mutation due to its higher probability of making longer jumps. However, they also indicated that Cauchy mutation spends less time in exploiting the local neighborhood and thus has a weaker fine-tuning ability than Gaussian mutation from small to mid-range regions. Therefore, Cauchy mutation is better at coarse-grained search while Gaussian mutation is better at fine-grained search. At the same time, they pointed out that Cauchy mutation performs better when the current search point is far away from the global optimum, while Gaussian mutation is better at finding a local optimum in a good region.

Thus, it would be ideal if Cauchy mutation is used when search points are far away from the global optimum and Gaussian mutation is adopted when search points are in the neighborhood of the global optimum. Unfortunately, the global optimum is usually unknown in practice, making the ideal switch from Cauchy to Gaussian mutation very difficult.

A new method based on mixing (rather than switching) different mutation operators has been proposed by Yao et al. (1999). The idea is to mix different search biases of Cauchy and Gaussian mutations. The method generates two offspring from the parents, one by Cauchy mutation and the other by Gaussian. The better one is then chosen as the offspring. Inspired by the above idea, Chen and Lu (2008) presented a new mutation method-"hybrid *GC* mutation," which is based on mixing Gaussian mutation and Cauchy mutation. It must be indicated that, unlike the method in Yao et al. (1999), the hybrid *GC* mutation does not compare the anticipated outcomes between Gaussian mutation and Cauchy mutation due to the characteristics of EO. In the hybrid *GC* mutation, Cauchy mutation is first adopted. It means that the large step size will be taken first at each mutation. If the new generated variable after mutation goes beyond the intervals of the decision variables, Cauchy mutation will be used repeatedly for some times (*TC*), which is defined by the user, until the offspring satisfies the requirements. Otherwise, Gaussian mutation will be carried out repeatedly for some times (*TG*), which is also defined by the user, until the offspring satisfies the requirements. That is, the step size will become smaller than before. If the new generated variable after mutation still goes beyond the intervals of the decision variables, then the upper limit or lower limit of the decision variables is chosen as the new generated variable. Thus, the hybrid *GC* mutation combines the advantages of coarse-grained search and fine-grained search. The above analyses show that the hybrid *GC* mutation is very simple yet effective. Unlike some switching algorithms which have to decide when to switch between different mutations during search, the hybrid *GC* mutation does not need to make such decisions.

6.4.2 Unconstrained Numerical MOPs with MOEO

To test the efficiency of MOEO, Chen and Lu (2008) applied MOEO to solving five traditional unconstrained numerical MOPs (see Table 6.1), and compared the performance of MOEO with NSGA-II, PAES, SPEA, and SPEA2 obtained in Deb et al. (2002) under the same conditions.

6.4.2.1 Performance Metrics

Chen and Lu (2008) used two performance metrics proposed by Deb et al. (2002) to assess the performance of MOEO. The first metric Υ measures the extent of convergence to a known set of Pareto-optimal solutions. Deb et al. (2002) pointed out that even when all solutions converge to the Pareto-optimal front, the convergence

Table 6.1 Test Problems

Problem	n	Variable Bounds	Objective Functions	Optimal Solutions	Comments
ZDT1	30	[0, 1]	$f_1(X) = x_1$ $f_2(X) = g(X)\left[1 - \sqrt{x_1/g(X)}\right]$ $g(X) = 1 + 9\left(\sum_{i=2}^{n} x_i\right)\Big/(n-1)$	$x_1 \in [0, 1]$, $x_i = 0$, $i = 2,\ldots, n$	Convex
ZDT2	30	[0, 1]	$f_1(X) = x_1$ $f_2(X) = g(X)[1 - (x_1/g(X))^2]$ $g(X) = 1 + 9\left(\sum_{i=2}^{n} x_i\right)\Big/(n-1)$	$x_1 \in [0, 1]$, $x_i = 0$, $i = 2,\ldots, n$	Nonconvex
ZDT3	30	[0, 1]	$f_1(X) = x_1$ $f_2(X) = g(X)\left[1 - \sqrt{x_1/g(X)} - (x_1/g(X))\sin(10\pi x_1)\right]$ $g(X) = 1 + 9\left(\sum_{i=2}^{n} x_i\right)\Big/(n-1)$	$x_1 \in [0, 1]$, $x_i = 0$, $i = 2,\ldots, n$	Convex, disconnected

(Continued)

Table 6.1 (*Continued*) Test Problems

Problem	n	Variable Bounds	Objective Functions	Optimal Solutions	Comments
ZDT4	10	$x_1 \in [0, 1]$ $x_2 \in [-5, 5]$, $i = 2, \ldots, n$	$f_1(X) = x_1$ $f_2(X) = g(X)\left[1 - \sqrt{x_1/g(X)}\right]$ $g(X) = 1 + 10(n-1) + \sum_{i=2}^{n} [x_i^2 - 10\cos(4\pi x_i)]$	$x_1 \in [0, 1]$ $x_i = 0,$ $i = 2, \ldots, n$	Nonconvex
ZDT6	10	$[0, 1]$	$f_1(X) = 1 - \exp(-4x_1)\sin^6(6\pi x_1)$ $f_2(X) = g(X)[1 - (f_1(X)/g(X))^2]$ $g(X) = 1 + 9\left[\left(\sum_{i=2}^{n} x_i\right)\Big/(n-1)\right]^{0.25}$	$x_1 \in [0, 1]$ $x_i = 0,$ $i = 2, \ldots, n$	Nonconvex, nonuniformly spaced

Source: Reprinted from *European Journal of Operational Research*, 3 (188), Chen, M. R. and Lu, Y. Z., A novel elitist multi-objective optimization algorithm: Multi-objective extremal optimization, 637–651, Copyright 2008, with permission from Elsevier.

n is the number of decision variables.

metric does not have a value of zero. The metric will yield zero only when each obtained solution lies exactly on each of the chosen solutions. The second metric Δ measures the extent of spread of the obtained nondominated solutions. It is desirable to get a set of solutions that spans the entire Pareto-optimal region. The second metric Δ can be calculated as follows:

$$\Delta = \frac{d_f + d_l + \sum_{i=1}^{N-1} |d_i - \bar{d}|}{d_f + d_l + (N-1)\bar{d}} \tag{6.4}$$

where d_f and d_l are the Euclidean distances between the extreme solutions and the boundary solutions of the obtained nondominated set, d_i is the Euclidean distance between consecutive solutions in the obtained nondominated set of solutions, and $\bar{d}$ is the average of all distances d_i ($i = 1, 2, \ldots, N-1$), assuming that there are N solutions on the best nondominated front. Note that a good distribution would make all distances d_i equal to $\bar{d}$ and would make $d_f = d_l = 0$ (with existence of extreme solutions in the nondominated set). Consequently, for the most widely and uniformly spreadout set of nondominated solutions, Δ would be zero.

6.4.2.2 Experimental Settings

For MOEO, the algorithm is encoded in the floating point representation and uses an archive of size 100. The maximum number of generations for ZDT1, ZDT2, and ZDT3 is 830 (FFE = maximum generation × number of decision variables = 830 × 30 = 24,900 ≈ 25,000) and for ZDT4 and ZDT6 is 2500 (FFE = 2500 × 10 = 25,000). The parameters in the hybrid *GC* mutation, that is, *TC* and *TG*, are both set to 3.

MOEO was compared with NSGA-II, PAES, SPEA, and SPEA2 under the same conditions: 10 independent runs and a maximum of 25,000 fitness function evaluations (FFE for short) (Chen and Lu, 2008).

6.4.2.3 Experimental Results and Discussion

Table 6.2 shows the mean and variance of the convergence metric Υ obtained using five algorithms, that is, MOEO, NSGA-II (real-coded), PAES, SPEA, and SPEA2. The bold numeral in Table 6.2 means it is the best optimum among all the optima. It can be observed from Table 6.2 that MOEO is able to converge better than any other algorithm on three problems (ZDT1, ZDT4, and ZDT6). For problem ZDT2, MOEO performs better than NSGA-II and PAES, but a little worse than SPEA and SPEA2 in terms of convergence. For problem ZDT3, MOEO outperforms NSGA-II, PAES, and SPEA with respect to the convergence to the Pareto-optimal front, where SPEA2 performs the best. In all cases with MOEO, the variance of the convergence metric in 10 runs is very small.

Table 6.2 Mean (First Rows) and Variance (Second Rows) of the Convergence Metric Υ

Algorithm	ZDT1	ZDT2	ZDT3	ZDT4	ZDT6
MOEO	**0.001277**	0.001355	0.004385	**0.008145**	**0.000630**
	0.000697	0.000897	0.00191	0.004011	3.26E–05
NSGA-II (real-coded)	0.033482	0.072391	0.114500	0.513053	0.296564
	0.004750	0.031689	0.007940	0.118460	0.013135
PAES	0.082085	0.126276	0.023872	0.854816	0.085469
	0.008679	0.036877	0.00001	0.527238	0.006664
SPEA	0.001799	0.001339	0.047517	7.340299	0.221138
	0.000001	0	0.000047	6.572516	0.000449
SPEA2	0.001448	**0.000743**	**0.003716**	0.028492	0.011643
	0.000317	8.33E–05	0.000586	0.047482	0.002397

Source: Reprinted from *European Journal of Operational Research*, 3 (188), Chen, M. R. and Lu, Y. Z., A novel elitist multi-objective optimization algorithm: Multi-objective extremal optimization, 637–651, Copyright 2008, with permission from Elsevier.

Table 6.3 shows the mean and variance of the diversity metric Δ obtained using all the algorithms. The bold numeral in Table 6.3 means it is the best optimum among all the optima. As can be seen from Table 6.3, MOEO is capable of finding a better spread of solutions than any other algorithm on all the problems except ZDT3. This indicates that MOEO has the ability to find a well-distributed set of nondominated solutions than many other state-of-the-art MOEAs. In all cases with MOEO, the variance of the diversity metric in 10 runs is also small.

It is interesting to note that MOEO can perform well with respect to convergence and diversity of solutions in problem ZDT4, where there exist 21^9 different local Pareto-optimal fronts in the search space, of which only one corresponds to the global Pareto-optimal front. This indicates that MOEO is capable of escaping from the local Pareto-optimal fronts and approaching the global nondominated front. Furthermore, MOEO is suitable to deal with those problems with nonuniformly spaced Pareto front, for example, problem ZDT6.

For illustration, 1 of 10 runs of MOEO on three test problems (ZDT1, ZDT2, and ZDT3) is shown in Figures 6.5 through 6.7, respectively. The figures show all nondominated solutions obtained after 25,000 FFE with MOEO. From Figures 6.5 through 6.7, we can see that MOEO is able to converge to the true Pareto-optimal front on all the three problems. Moreover, MOEO can find a well-distributed set

Table 6.3   Mean (First Rows) and Variance (Second Rows) of the Diversity Metric Δ

Algorithm	ZDT1	ZDT2	ZDT3	ZDT4	ZDT6
MOEO	**0.32714**	**0.285062**	0.965236	**0.275664**	**0.225468**
	0.065343	0.056978	0.046958	0.183704	0.033884
NSGA-II (real-coded)	0.390307	0.430776	0.738540	0.702612	0.668025
	0.001876	0.004721	0.019706	0.064648	0.009923
PAES	1.229794	1.165942	0.789920	0.870458	1.153052
	0.004839	0.007682	0.001653	0.101399	0.003916
SPEA	0.784525	0.755148	0.672938	0.798463	0.849389
	0.004440	0.004521	0.003587	0.014616	0.002713
SPEA2	0.472254	0.473808	**0.606826**	0.705629	0.670549
	0.097072	0.0939	0.191406	0.266162	0.077009

Source: Reprinted from *European Journal of Operational Research*, 3 (188), Chen, M. R. and Lu, Y. Z., A novel elitist multi-objective optimization algorithm: Multi-objective extremal optimization, 637–651, Copyright 2008, with permission from Elsevier.

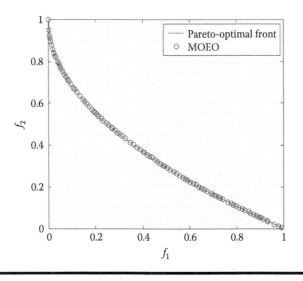

Figure 6.5 Nondominated solutions with MOEO on ZDT1. (Adapted from Chen, M. R. and Lu, Y. Z., *European Journal of Operational Research* 3 (188): 637–651, 2008.)

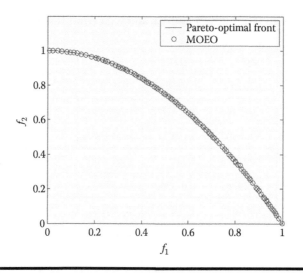

Figure 6.6 Nondominated solutions with MOEO on ZDT2. (Adapted from Chen, M. R. and Lu, Y. Z., *European Journal of Operational Research* 3 (188): 637–651, 2008.)

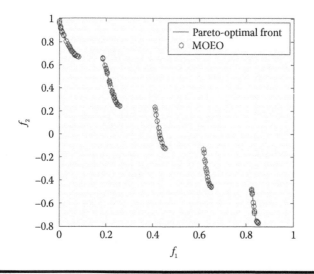

Figure 6.7 Nondominated solutions with MOEO on ZDT3. (Reprinted from *European Journal of Operational Research*, 3 (188), Chen, M. R. and Lu, Y. Z., A novel elitist multi-objective optimization algorithm: Multi-objective extremal optimization, 637–651, Copyright 2008, with permission from Elsevier.)

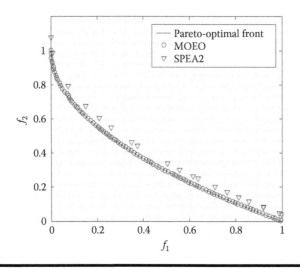

Figure 6.8 **MOEO converges to the true Pareto-optimal front and finds a better spread of solutions than SPEA2 on ZDT4. (Reprinted from *European Journal of Operational Research*, 3 (188), Chen, M. R. and Lu, Y. Z., A novel elitist multi-objective optimization algorithm: Multi-objective extremal optimization, 637–651, Copyright 2008, with permission from Elsevier.)**

of nondominated solutions on ZD1 and ZDT2. But the diversity of the nondominated solutions found by MOEO on ZDT3 is not very good.

Figure 6.8 shows 1 of 10 runs with MOEO and SPEA2 on ZDT4. As can be observed from Figure 6.8, MOEO is able to find a better convergence and spread of solutions than SPEA2 on ZDT4. In the literature (Deb et al., 2002), it can be observed that NSGA-II can find better convergence and spread of solutions than PAES on ZDT4, but both of them cannot converge to the true Pareto-optimal front. For more details, readers can refer to Deb et al. (2002). It is interesting to notice that MOEO can converge to the true Pareto-optimal front, which indicates that MOEO outperforms NSGA-II and PAES on ZDT4 in terms of solution convergence.

Figure 6.9 shows 1 of 10 runs with MOEO and SPEA2 on ZDT6. From Figure 6.9, it can be seen that MOEO is able to find a better convergence and spread of solutions than SPEA2 on ZDT6. It is worth noting that MOEO is capable of converging to the true Pareto-optimal front and finding a well-distributed set of nondominated solutions that cover the whole front on ZDT6.

In order to compare the running times of MOEO with those of the other three algorithms, that is, NSGA-II, SPEA2, and PAES, Chen and Lu (2008) also showed the mean and variance of running times of each algorithm in 10 runs in Table 6.4. The bold numeral in Table 6.4 means it is the best optimum among all the optima. The FFE for all the algorithms was 25,000.

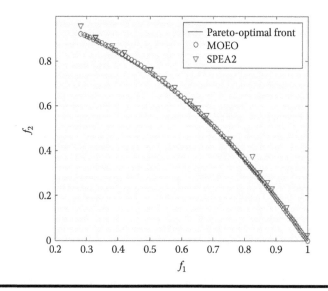

Figure 6.9 MOEO can converge to the true Pareto-optimal front and finds a better spread of solutions than SPEA2 on ZDT6. (Reprinted from *European Journal of Operational Research*, 3 (188), Chen, M. R. and Lu, Y. Z., A novel elitist multi-objective optimization algorithm: Multi-objective extremal optimization, 637–651, Copyright 2008, with permission from Elsevier.)

Table 6.4 Mean (First Rows) and Variance (Second Rows) of the Running Times (in Milliseconds)

Algorithm	ZDT1	ZDT2	ZDT3	ZDT4	ZDT6
MOEO	**427.93**	**436.20**	**330.81**	**1971.27**	2741.30
	35.56	61.17	39.66	466.79	100.26
NSGA-II (real-coded)	2,906.90	2,889.83	3,154.90	2,475.57	2,746.61
	159.51	287.70	489.04	166.31	401.19
SPEA2	2,782.37	2,802.36	2,884.46	2,303.01	**2,687.53**
	303.95	167.87	474.73	267.51	302.44
PAES	12,051.63	12,088.77	8,890.13	2,661.80	25,716.33
	1,044.91	240.52	357.17	139.36	1,685.56

Source: Reprinted from *European Journal of Operational Research*, 3 (188), Chen, M. R. and Lu, Y. Z., A novel elitist multi-objective optimization algorithm: Multi-objective extremal optimization, 637–651, Copyright 2008, with permission from Elsevier.

From Table 6.4, we can observe that MOEO is the fastest algorithm in comparison with the other three algorithms on problems ZDT1, ZDT2, ZDT3, and ZDT4. It is important to notice that MOEO is about 6 times faster than NSGA-II and SPEA2, even approximately 30 times as fast as PAES on problems ZDT1, ZDT2, and ZDT3. For problem ZDT6, MOEO runs almost as fast as NSGA-II and SPEA2, nearly 10 times as fast as PAES. It is worthwhile noting that although MOEO and PAES are both single-parent single-child MOO algorithms, MOEO runs much faster than PAES when the same archive truncation method is adopted. The results in Table 6.4 are remarkable if NSGA-II and SPEA2 are normally considered "very fast" algorithms. Therefore, MOEA may be considered a very fast approach.

6.4.2.4 Conclusions

From the above analysis, it can be seen that MOEO has the following advantages:

1. Similar to MOEAs, MOEO is not susceptible to the shape of Pareto front.
2. Only one operator, that is, mutation operator, exists in MOEO, which makes MOEO simple and convenient to be implemented.
3. The hybrid *GC* mutation operator suggested in MOEO combines the advantages of coarse-grained search and fine-grained search.
4. The historical external archive provides the elitist mechanism for MOEO.
5. MOEO provides good performance in both aspects of convergence and distribution of solutions.
6. MOEO is capable of handling those problems with multiple local Pareto-optimal fronts or nonuniformly spaced Pareto-optimal front.
7. Compared with three competitive MOEAs in terms of running times on five test problems, MOEO tested very fast.

6.4.3 Constrained Numerical MOPs with MOEO

In order to further extend MOEO to solve the constrained MOPs, Chen et al. (2007) presented a constrained MOEO algorithm, through introducing a constraint handling strategy to MOEO.

The constrained MOEO (Chen et al., 2007) adopts the constrained-domination principle proposed by Deb et al. (2002) for solving the constrained MOPs. This principle can be incorporated into the dominance comparison between any two solutions. A solution i is said to constrain-dominate a solution j, if any of the following conditions is true (Deb et al., 2002):

1. Solution i is feasible and solution j is not.
2. Solutions i and j are both infeasible, but solution i has a smaller overall constraint violation.
3. Solutions i and j are feasible and solution i dominates solution j.

Table 6.5 Constrained Test Problems

Problem	n	Variable Bounds	Objective Functions	Constraints
CONSTR	2	$x_1 \in [0.1, 1.0]$ $x_2 \in [0, 5]$	$f_1(X) = x_1$ $f_2(X) = (1 + x_2)/x_1$	$g_1(x) = x_2 + 9x_1 \geq 6$ $g_2(x) = -x_2 + 9x_1 \geq 1$
SRN	2	$x_i \in [-20, 20]$ $i = 1, 2$	$f_1(X) = (x_1 - 2)^2$ $\quad + (x_2 - 1)^2 + 2$ $f_2(X) = 9x_1 - (x_2 - 1)^2$	$g_1(x) = x_1^2 + x_2^2 \leq 225$ $g_2(x) = x_1 - 3x_2 \leq -10$
TNK	2	$x_i \in [0, \pi]$ $i = 1, 2$	$f_1(X) = x_1$ $f_2(X) = x_2$	$g_1(x) = -x_1^2 - x_2^2 + 1 +$ $\quad 0.1\cos(16\arctan(x_1/x_2)) \leq 0$ $g_2(x) = (x_1 - 0.5)^2$ $\quad + (x_2 - 0.5)^2 \leq 0.5$

Source: Reproduced from Chen, M. R. et al. *Journal of Zhejiang University, Science A* 8 (12): 1905–1911, 2007. With permission.

n is the number of decision variables.

The effect of using this constrained-domination principle is that any feasible solution has a better nondomination rank than any infeasible solution. All feasible solutions are ranked according to their nondomination level based on the objective function values. However, among two infeasible solutions, the solution with a smaller constraint violation has a better rank. Moreover, this modification in the nondomination principle does not change the computational complexity of MOEO. The rest of the MOEO procedure as described earlier can be used as usual.

In the constrained MOEO, four constrained numerical MOPs (see Table 6.5) are chosen as the test problems. In the first problem, a part of the unconstrained Pareto-optimal region is not feasible. Thus, the resulting constrained Pareto-optimal region is a concatenation of the first constraint boundary and some part of the unconstrained Pareto-optimal region. In the second problem SRN, the constrained Pareto-optimal set is a subset of the unconstrained Pareto-optimal set. The third problem TNK has a discontinuous Pareto-optimal region, falling entirely on the first constraint boundary.

6.4.3.1 Performance Metrics

The constrained MOEO (Chen et al., 2007) used the following two performance metrics to assess its performance.

1. *Front spread (FS)* (Bosman and Thierens, 2003): It indicates the size of the objective space covered by an approximate set. A larger *FS* metric is preferable. The *FS* metric for an approximation set S is defined to be the maximum Euclidean distance inside the smallest m-dimensional bounding-box that contains S. Here, m is the number of objectives. This distance can be computed using the maximum distance among the solutions in S in each dimension separately.

$$FS(S) = \sqrt{\sum_{i=0}^{m-1} \max_{(z^0, z^1) \in S \times S} \left\{ \left(f_i(z^0) - f_i(z^1) \right)^2 \right\}} \tag{6.5}$$

2. *Coverage of two sets* (Zitzler and Thiele, 1999): Assume, without loss of generality, a minimization problem and consider two decision vectors $a, b \in X$. Then a is said to cover b (written as $a \prec b$) if a dominates b (also written as $a \prec b$)) or a equals to b. Let X', $X'' \subseteq X$ be two sets of decision vectors. The function C maps the ordered pair (X', X'') to the interval $[0, 1]$.

$$C(X', X'') = \frac{\left| \left\{ a'' \in X''; \exists a' \in X' : a' \preceq a'' \right\} \right|}{|X''|} \tag{6.6}$$

Here, $C(X', X'') = 1$ means that all points in X'' are covered by points in X'. The opposite, $C(X', X'') = 0$ represents the situation when none of points in X'' are covered by the set X'. Note that both $C(X', X'')$ and $C(X'', X')$ have to be considered, since $C(X', X'')$ is not necessarily equal to $C(X'', X')$ (e.g., if X' dominates X'', then $C(X', X'') = 1$ and $C(X'', X') = 0$).

The first metric FS can be used to measure the spread of an approximate front, while the second metric C can be used to show that the outcomes of one algorithm dominate the outcomes of another algorithm.

6.4.3.2 Experimental Settings

In the literature (Chen et al., 2007), the simulation results of the constrained MOEO were compared with those of NSGA-II, SPEA2, and PAES under the same FFE. Note that for MOEO, FFE = maximum generation × problem dimension, for PAES, FFE = maximum generation, and for NSGA-II and SPEA2, FFE = maximum generation × population size. All the algorithms were run for a maximum of 50,000 FFE for problem TNK, 20,000 FFE for problem SRN, and 40,000 FFE for CONSTR. For MOEO, it was encoded in the floating point representation and an archive of size 100 was used. The parameters in the hybrid *GC* mutation, that is, *TC* and *TG*, were both set to 3. For NSGA-II (real-coded), the population size was 100. For SPEA2, a population of size 80 and an external population of size 20 were adopted, so that the overall population size became 100. For PAES, a depth value *d* equal to 4 and the archive size of 100 were used. The crossover probability of $p_c = 0.9$ and a mutation probability of $p_m = 0.05$ for real-coded GAs or $p_m = 1/l$ for binary-coded GAs (where l is the string length) were used. Additionally, all the algorithms were executed 50 times independently on each problem.

6.4.3.3 Experimental Results and Discussion

Table 6.6 shows the mean and standard deviation of the *FS* metric obtained using four algorithms, that is, MOEO, NSGA-II, SPEA2, and PAES. The bold numeral in Table 6.6 means it is the best optimum among all the optima. As can be seen from Table 6.6, MOEO is capable of finding a wider spread of solutions than any other algorithms on the problems CONSTR, TNK. This indicates that MOEO has the ability to find a wider distributed set of nondominated solutions than many other state-of-the-art MOEAs. In all cases with MOEO, the standard deviation of *FS* metric in 50 runs is also small.

The direct comparison of the different algorithms based on the *C* metric is shown in Table 6.7. It can be observed from Table 6.7 that MOEO is able to perform better than NSGA-II on the problems SRN concerning the *C* metric. It is interesting to note that MOEO outperforms SPEA2 on all the problems in terms of the *C* metric. It is also clear from Table 6.7 that MOEO significantly outperforms PAES on two problems, that is, SRN, TNK, with respect to the *C* metric.

For illustration, Chen and Lu (2008) also showed 1 of 50 runs of all the algorithms on each problem in Figures 6.10 through 6.12, respectively. The figures show the trade-off fronts obtained by MOEO, NSGA-II, SPEA2, and PAES. The insets in all the figures show the parts which may be unclear in the main plots. From Figures 6.10 through 6.12, we can see that MOEO is able to converge nearly as well as or better than the other three algorithms on all the problems. Moreover, MOEO can find a well-distributed set of nondominated solutions on all the problems.

It is interesting to note that MOEO can perform well in both aspects of convergence and diversity of solutions on the problem TNK. The problem TNK is slightly harder—the Pareto front is discontinuous and, as shown by Deb et al. (2002), some

Table 6.6 Comparisons of *FS* Metric

Algorithm	CONSTR		SRN		TNK	
	Mean	St. Dev	Mean	St. Dev	Mean	St. Dev
MOEO	**7.87**	0.12	343.98	2.33	**9.43**	0.08
NSGA-II	7.36	0.69	343.22	3.29	9.37	0.18
SPEA2	7.74	0.27	**344.47**	2.63	9.42	0.13
PAES	2.19	2.05	336.03	49.29	5.01	3.04

Source: Reproduced from Chen, M. R. et al. *Journal of Zhejiang University, Science A* 8 (12): 1905–1911, 2007. With permission.

Table 6.7 Comparisons of C Metric

Coverage of Two Sets	CONSTR		SRN		TNK	
	Mean	St. Dev	Mean	St. Dev	Mean	St. Dev
C(NSGA-II, MOEO)	0.30	0.04	0.06	0.02	0.35	0.05
C(MOEO, NSGA-II)	0.09	0.03	0.09	0.03	0.16	0.04
C(SPEA2, MOEO)	0.03	0.02	0.01	0.01	0.06	0.02
C(MOEO, SPEA2)	0.39	0.08	0.46	0.13	0.38	0.10
C(PAES, MOEO)	0.25	0.03	0.03	0.02	0.01	0.01
C(MOEO, PAES)	0.01	0.02	0.26	0.20	0.45	0.24

Source: Reproduced from Chen, M. R. et al. *Journal of Zhejiang University, Science A* 8 (12): 1905–1911, 2007. With permission.

algorithms have difficulty finding the entire central continuous region. Figure 6.12 shows the trade-off fronts obtained by all the algorithms on this problem. It is obvious that MOEO performs significantly better than SPEA2 and PAES in terms of the convergence and diversity of solutions. Note that MOEO and NSGA-II nearly converge to the same trade-off front. However, MOEO is able to find a much more diverse set of nondominated solutions than NSGA-II.

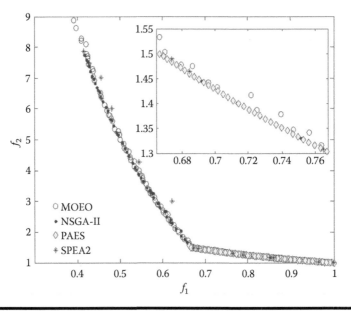

Figure 6.10 Approximate Pareto fronts for CONSTR. (Reproduced from Chen, M. R. et al. *Journal of Zhejiang University, Science A* 8 (12): 1905–1911, 2007. With permission.)

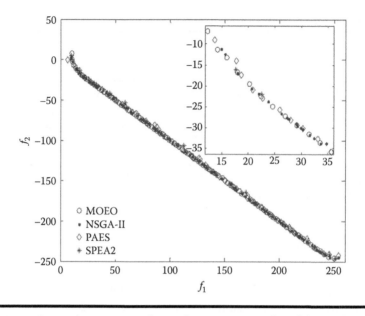

Figure 6.11 Approximate Pareto fronts for SRN. (Reproduced from Chen, M. R. et al. *Journal of Zhejiang University, Science A* 8 (12): 1905–1911, 2007. With permission.)

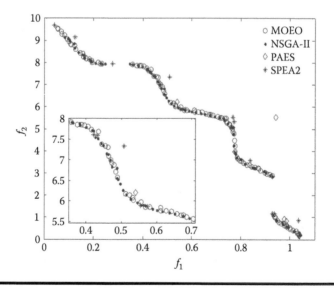

Figure 6.12 Approximate Pareto fronts for TNK. (Reproduced from Chen, M. R. et al. *Journal of Zhejiang University, Science A* 8 (12): 1905–1911, 2007. With permission.)

6.4.3.4 Conclusions

In this section, we introduce the constrained MOEO algorithm in application to solving constrained MOPs. The simulation results indicate that MOEO is highly competitive with three state-of-the-art MOEAs, that is, NSGA-II, SPEA2, and PAES in terms of convergence and diversity of solutions. Thus, MOEO may be a good alternative to deal with constrained MOPs.

6.5 Multiobjective 0/1 Knapsack Problem with MOEO

Multiobjective 0/1 knapsack problems (MOKP) is a maximization problem. It is a generalization of the 0-1 simple knapsack problem, and is a well-known member of the NP-hard class of problems. The problem is formulated to maximize the function:

$$\text{maximize} \quad f = (f_1(X), f_2(X), \ldots, f_n(X))$$

$$\text{subject to} \quad \sum_{j=1}^{m} w_{i,j} \cdot x_j \leq C_i, \quad i = 1, 2, \ldots, n \tag{6.7}$$

$$X = (x_1, x_2, \ldots x_m) \in \{0,1\}^m$$

where $f_i(X) = \sum_{j=1}^{m} p_{i,j} \cdot x_j$, $i = 1, 2, \ldots, n$. A feasible solution is represented by a vector $X = (x_1, x_2, \ldots, x_m)$ of binary decision variable x_j, such that $x_j = 1$ if item j is included in the solution and 0 otherwise. Also, $p_{i,j}$ and $w_{i,j}$ denote profit and weight of item j according to knapsack (objective) i. C_i is the capacity of knapsack i.

6.5.1 Extended MOEO for MOKP

Based on the above numerical MOEO, P. Chen et al. (2010) further extended MOEO to deal with MOKP. The extended MOEO algorithm works as shown in Figure 6.13.

Note that fitness assignment, diversity preservation, and external archive in the extended MOEO are the same as those in the numerical MOEO. Different from the numerical MOEO, the extended MOEO adopts the following mutation operation and repair strategy.

6.5.1.1 Mutation Operation

The extended MOEO adopts the following mutation operation: assuming that the kth component (i.e., x_k) is the current component, perform flipping on the

1. Generate an initial feasible solution $X = (x_1, x_2, ..., x_m) \in \{0, 1\}^m$ using the greedy algorithm. Add X to the external archive. Set *iteration* = 0.
2. Generate m offspring of the current solution X by performing mutation on each component one by one, while keeping other components unchanged.
3. Apply the repair strategy to repair m offspring.
4. Perform dominance ranking on the m offspring and then obtain their rank numbers, i.e., $r_j \in [0, m - 1], j \in \{1, ..., m\}$.
5. Assign the fitness $\lambda_j = r_j$ for each component $x_j, j \in \{1, ..., m\}$.
6. If there is only one component with fitness value of zero, the component will be considered the worst one. Otherwise, if there are two or more components with fitness value of zero, the diversity preservation mechanism is invoked to decide the worst component. Assuming that the kth component is the worst one, and $\bar{x}_k$ is the corresponding offspring generated by performing mutation only on the kth component, $k \in \{1, ..., m\}$.
7. Accept $\bar{x} = \bar{x}_k$ *unconditionally*.
8. Apply the archiving logic to updating the external archive. Otherwise, go to Step 9.
9. If the iterations reach the predefined maximum number of the generations, go to Step 10; otherwise, set *iteration* = *iteration* + 1, and go to Step 2.
10. Output the external archive as the Pareto-optimal set.

Figure 6.13 Pseudo-code of extended MOEO for MOKP. (Chen, M. R. et al. A novel multi-objective optimization algorithm for 0/1 multi-objective knapsack problems. *Proceedings of the 4th IEEE Conference on Industrial Electronics and Applications (ICIEA 2010)*, pp. 1511–1516. © 2010 IEEE.)

current component, that is, set $x_k = 0$ when its value equals to 1; otherwise, set $x_k = 1$. At the same time, randomly choose another component with different value from that of the current component, and perform flipping on it. For instance, if the current solution $X = 1010$, then four offspring individuals will be generated through the above mutation, that is, $X_1 = 0110$, $X_2 = 1100$, $X_3 = 1001$, and $X_4 = 0011$.

6.5.1.2 Repair Strategy

In order to ensure the new offspring to be feasible solution, MOEO uses the repair strategy to repair the new generated X. The pseudo-code of repair strategy is shown in Figure 6.14. When the knapsack i is overfilled, the repair strategy will enter the remove phase. The order in which the items is removed is determined by the maximum (*profit · capacity*)/*weight* ratio per item. The item with the minimum value ratio will be first removed from the knapsack. In this phase, the repair strategy step by step removes items from the solution coded by X until knapsack i is not overfilled. On the other hand, when knapsack i is unfilled, the repair strategy will go into the add phase. The order in which the items are added is also determined by the maximum (*profit · capacity*)/*weight* ratio per item. In this phase, the item with the maximum value ratio outside knapsack i will be added to this knapsack one by one until the knapsack is not unfilled.

```
Procedure Repair (x̄)
begin
    knapsack-overfilled = false;
    knapsack-unfilled = false;
    x̄′ = x̄, i.e., (x′₁,…, x′ₘ): = (x₁,…, xₘ);
```

sort all the items in the increasing order of the value ratios $\left(\text{i.e., } \max_{i=1}^{n} \left\{ \dfrac{C_i \cdot p_{ij}}{w_{ij}} \right\} \right)$;

```
    /*Remove phase */
```

if $\sum_{j=1}^{m} w_{ij} \cdot x′_j > C_i, \quad i \in \{1,…,n\}$

```
            knapsack-overfilled = true;
            while (knapsack-overfilled)
            {j := select the item with the minimum value
                ratio from the knapsack;
            remove the selected item from the knapsack, i.e., x′_j := 0;
```

if $\sum_{j=1}^{m} w_{ij} \cdot x′_j \le C_i, \quad i = 1,…,n$

```
                knapsack-overfilled = false;
            end if }
        end if
    /* Add phase */
```

if $\sum_{j=1}^{m} w_{ij} \cdot x′_j \le C_i, \quad i = 1,…,n$

```
            knapsack-unfilled = true;
            while (knapsack-unfilled)
            {j := select the item with the maximum value ratio outside the knapsack;
            add the selected item to the knapsack, i.e., x′_j := 1;
```

if $\sum_{j=1}^{m} w_{ij} \cdot x′_j > C_i, \quad i = 1,…,n$

```
                knapsack-unfilled = false and x′_j := 0;
                end if }
        end if
end
```

Figure 6.14   Pseudo-code of repair strategy in the extended MOEO. (Chen, M. R. et al. A novel multi-objective optimization algorithm for 0/1 multi-objective knapsack problems. *Proceedings of the 4th IEEE Conference on Industrial Electronics and Applications (ICIEA 2010)*, **pp. 1511–1516. © 2010 IEEE.)**

6.5.2 *Experimental Settings*

In the literature (P. Chen et al., 2010), the extended MOEO was validated by three benchmark problems with $n = \{100, 250, 500\}$ items and $m = 2$ objectives downloaded from http://www.tik.ee.ethz.ch/~zitzler/testdata.html, for which we know the true Pareto-optimal fronts only in case of two objectives $m = 2$. The three instances are named 2-100, 2-250, and 2-500. The simulation results of MOEO were compared with those of NSGA, SPEA, and NPGA which come from the literature (Zitzler and Thiele, 1999). In addition, to conduct a fair comparison, all algorithms were performed under the same FFE. The parameters of NSGA, SPEA, and NPGA were set as follows: the population sizes of instances 2-100, 2-250, 2-500 are 100, 150, 200, respectively. The maximum iterations for the three algorithms are 500. Thus, the maximum iterations for MOEO are 500, 300, 200 for instances 2-100, 2-250, 2-500, respectively. For MOEO, it was encoded in the floating point representation and an archive of size 100 was used. All algorithms were run 50 times independently on each instance. Two performance metrics, that is, *FS* and coverage of two sets (*C* metric) were used to assess the performance of the extended MOEO.

6.5.3 Experimental Results and Discussion

The direct comparison of the different algorithms based on the metric *C* is shown in Figure 6.15. It is clear that MOEO significantly outperformed NSGA and NPGA

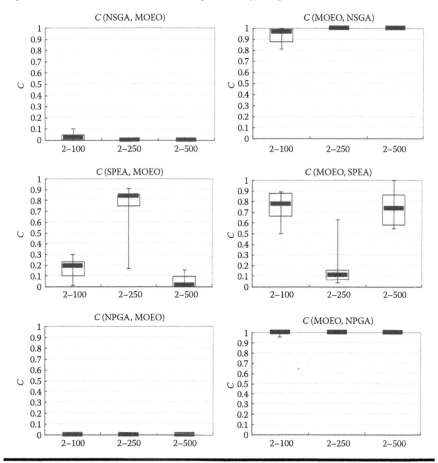

Figure 6.15 Comparisons of metric C. (Chen, M. R. et al. A novel multi-objective optimization algorithm for 0/1 multi-objective knapsack problems. *Proceedings of the 4th IEEE Conference on Industrial Electronics and Applications (ICIEA 2010)***, pp. 1511–1516. © 2010 IEEE.)**

Table 6.8 Comparisons of *FS* Metric

Algorithms	2–100		2–250		2–500	
	Mean	*St. Dev.*	*Mean*	*St. Dev.*	*Mean*	*St. Dev.*
MOEO	**973.2**	107.8	1267.8	125.3	1435.4	224.7
NSGA	793.1	97.0	1262.6	181.4	**1573.7**	274.9
SPEA	844.4	129.2	**1369.9**	181.2	1532.6	201.7
NPGA	782.4	99.2	1282.3	113.5	1333.6	139.5

Source: Chen, M. R. et al. 2010b. A novel multi-objective optimization algorithm for 0/1 multi-objective knapsack problems. *Proceedings of the 4th IEEE Conference on Industrial Electronics and Applications (ICIEA 2010),* pp. 1511–1516. © 2010 IEEE.

on all the instances in terms of the metric *C*. It can be also observed from Figure 6.15 that MOEO performed better than SPEA except the instance 2-250 concerning the metric *C*. This illustrates that MOEO is capable of performing nearly as well as or better than other three algorithms with respect to the convergence of solutions.

Table 6.8 shows the mean and standard deviation of the *FS* metric obtained using four algorithms, that is, MOEO, NSGA, SPEA, and NPGA. The bold numeral in Table 6.8 means it is the best optimum among all the optima. As can be seen from Table 6.8, MOEO was capable of finding a wider spread of solutions than the other three algorithms on the instance 2-100. MOEO performed better than NSGA on instance 2-250, and better than NPGA on instance 2-500. This indicates that the extended MOEO has the ability to find a wide distributed set of nondominated solutions.

For illustration, we also show 1 of 50 runs of all the algorithms on each instance in Figure 6.16a–c, respectively. Figure 6.16 shows the approximate Pareto fronts obtained by MOEO, NSGA, SPEA, and NPGA. From Figure 6.16, we can see that MOEO was able to converge nearly as well as or better than the other three algorithms on all the instances. Moreover, MOEO could find a well-distributed set of nondominated solutions on all the instances.

6.5.4 Conclusions

The extended MOEO for MOKP was validated using three benchmark problems. The simulation results demonstrated that the extended MOEO is highly competitive with three state-of-the-art MOEAs, that is, the NSGA, the SPEA, and the NPGA. As a result, the extended MOEO may be a good alternative to solve MOKP.

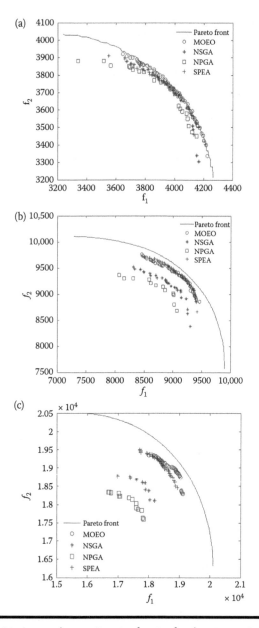

Figure 6.16 (a)–(c) Approximate Pareto fronts for instances 2-100, 2-250, and 2-500. (Chen, M. R. et al. A novel multi-objective optimization algorithm for 0/1 multi-objective knapsack problems. *Proceedings of the 4th IEEE Conference on Industrial Electronics and Applications (ICIEA 2010)*, pp. 1511–1516. © 2010 IEEE.)

6.6 Mechanical Components Design with MOEO

6.6.1 Introduction

Most engineering design problems involve multiple and often conflicting objectives. In principle, the presence of conflicting objectives results in a number of optimal solutions, commonly known as Pareto-optimal solutions. Since no one Pareto-optimal solution can be said to be better than another without further considerations, it is desired to find as many such Pareto-optimal solutions as possible.

In order to further extend MOEO to solve mechanical components design problems with multiple objectives, in the literature (Chen et al., 2007), the MOEO algorithm for mechanical components design problems was presented as shown in Figure 6.17.

In order to extend MOEO to handle the constrained engineering MOPs, the constrained-domination principle (Chen et al., 2007) was incorporated into the dominance ranking method, which is used as the fitness assignment of MOEO. That is, the fitness value of one solution equals the number of other solutions by which it is constrained-dominated. Therefore, the nondominated solutions are ranked as zero, while the worst possible ranking is the number of all the solutions minus one.

Note that fitness assignment, diversity preservation, external archive, and mutation operation used in the extended MOEO for the mechanical components design problems are the same as those in the numerical MOEO (Chen and Lu, 2008).

1. Randomly generate an initial feasible solution $S = (x_1, x_2, ..., x_n)$ and add S to the external archive. Set *iteration* = 0.
2. Generate n offspring of the current solution S by performing mutation on each component one by one, while keeping other components unchanged.
3. Perform dominance ranking on the n offspring and then obtain their rank numbers, i.e., $r_j \in [0, n-1], j \in \{1, ..., n\}$.
4. Assign the fitness $\lambda_j = r_j$ for each component $x_j, j \in \{1, ..., n\}$.
5. If there is only one component with fitness value of zero, the component will be considered the worst one. Otherwise, if there are two or more components with fitness value of zero, the diversity preservation mechanism is invoked to decide the worst component. Assuming that the worst component is x_w, and S_w is the corresponding offspring generated by performing mutation only on $x_w, w \in \{1, ..., n\}$.
6. Accept $S = S_w$ *unconditionally.*
7. If S is a feasible solution, then apply the archiving logic to updating the external archive. Otherwise, go to Step 8.
8. If the iterations reach the predefined maximum number of the generations, go to Step 9; otherwise, set *iteration = iteration + 1*, and go to Step 2.
9. Output the external archive as the Pareto-optimal set.

Figure 6.17 Pseudo-code of MOEO algorithm for mechanical components design. (Reproduced from Chen, M. R. et al. *Journal of Zhejiang University, Science A* 8(12): 1905–1911, 2007. With permission.)

6.6.2 Experimental Settings

In the literature (M. R. Chen et al., 2007), MOEO was used to solve three benchmark mechanical components design problems as follows:

6.6.2.1 Two-Bar Truss Design (Two Bar for Short)

A truss has to carry a certain load without elastic failure (Deb et al., 2000). Thus, in addition to the objective of designing the truss for minimum volume (which is equivalent to designing for minimum cost of fabrication), there are additional objectives of minimizing stresses in each of the two members AC and BC. A two-objective optimization problem for three variables y (vertical distance between B and C in m), x_1 (length of AC in m), and x_2 (length of BC in m) is constructed as follows (Deb et al., 2000):

$$\text{minimize} \quad f_1(x) = x_1\sqrt{16 + y^2} + x_2\sqrt{1 + y^2}$$

$$\text{minimize} \quad f_2(x) = \max(\sigma_{AC}, \sigma_{BC})$$

$$\text{subject to} \quad \max(\sigma_{AC}, \sigma_{BC}) \le 10^5$$

$$1 \le y \le 3 \quad \text{and} \quad x \ge 0$$

(6.8)

The stresses are calculated as follows:

$$\sigma_{AC} = \frac{20\sqrt{16 + y^2}}{yx_1}, \quad \sigma_{BC} = \frac{80\sqrt{1 + y^2}}{yx_2}$$

(6.9)

In M. R. Chen et al. (2007), MOEO is applied with $0 \le x_i \le 0.01$.

6.6.2.2 Welded Beam Design (Welded Beam for Short)

A beam needs to be welded on another beam and must carry a certain load F (Deb et al., 2000). The overhang portion of the beam has a length of 14 in. and $F = 6000$ lb force is applied at the end of the beam. The objective of the design is to minimize the cost of fabrication and minimize the end deflection. In the following, the two-objective optimization problems are formulated (Deb et al., 2000):

$$\text{minimize} \quad f_1(x) = 1.10471h^2l + 0.04811tb(14.0 + l)$$

$$\text{minimize} \quad f_2(x) = \delta(x),$$

$$\text{subject to} \quad g_1(x) = 13,600 - \tau(x) \ge 0,$$

$$g_2(x) = 30,000 - \sigma(x) \geq 0,$$

$$g_3(x) = b - h \geq 0,$$
$$\qquad (6.10)$$

$$g_4(x) = P_c(x) - 6000 \geq 0$$

The deflection term $\delta(x)$ is given as follows:

$$\delta(x) = \frac{2.1952}{t^3 b} \qquad (6.11)$$

The stress and buckling terms are given as follows:

$$\tau(x) = \sqrt{(\tau')^2 + (\tau'')^2 + (l\tau'\tau'')}\Big/\sqrt{0.25(l^2 + (h+t)^2)},$$

$$\tau' = \frac{6000}{\sqrt{2}hl},$$

$$\tau'' = \frac{6000(14 + 0.5l)\sqrt{0.25(l^2 + (h+t)^2)}}{2\{0.707hl(l^2/12 + 0.25(h+t)^2)\}}, \qquad (6.12)$$

$$\sigma(x) = \frac{5,04,000}{t^2 b},$$

$$P_C = 64,746.022(1 - 0.0282346t)tb^3$$

The variables are initialized in the following range: $0.125 \leq h, b \leq 5$ and $0.1 \leq l, t \leq 10$.

6.6.2.3 Machine Tool Spindle Design (Spindle for Short)

This is a two-objective constrained optimization problem (Coello, 1996). The objective of the problem is to minimize the volume of the spindle ($f(x_1)$) and minimize the static displacement under the force F ($f_2(x)$). The formulation of the problem is defined below (Coello, 1996):

$$\text{minimize } f(x_1) = \frac{\pi}{4}\left[a\left(d_a^2 - d_0^2\right) + l\left(d_b^2 - d_0^2\right)\right]$$

$$\text{minimize } f(x_2) = \frac{Fa^3}{3EI_a}\left(1 + \frac{l}{a}\frac{I_a}{I_b}\right) + \frac{F}{c_a}\left[\left(1 + \frac{a}{l}\right)^2 + \frac{c_a a^2}{c_b l^2}\right] \qquad (6.13)$$

$$\text{subject to } g_1(x) = l - l_g \le 0$$

$$g_2(x) = l_k - l \le 0$$

$$g_3(x) = d_{a1} - d_a \le 0$$

$$g_4(x) = d_a - d_{a2} \le 0$$

$$g_5(x) = d_{b1} - d_b \le 0$$

$$g_6(x) = d_b - d_{b2} \le 0$$

$$g_7(x) = d_{om} - d_o \le 0$$

$$g_8(x) = p_1 d_o - d_b \le 0$$

$$g_9(x) = p_2 d_b - d_a \le 0$$

$$g_{10}(x) = \left| \Delta_a + (\Delta_a - \Delta_b) \frac{a}{l} \right| - \Delta \le 0$$

where there are two continuous variables (l and d_o) and two discrete variables (d_a and d_b), I_a and I_b are the moments of inertia:

$$I_a = 0.049 \left(d_a^4 - d_o^4 \right), \quad I_b = 0.049 \left(d_b^4 - d_o^4 \right) \tag{6.14}$$

and c_a and c_b are bearing stiffnesses:

$$c_a = 35,400 |\delta_{ra}|^{1/9} d_a^{10/9}, \quad c_b = 35,400 |\delta_{rb}|^{1/9} d_b^{10/9} \tag{6.15}$$

where δ_{ra} and δ_{rb} are the preloads of the bearings, $g_8(x)$ and $g_9(x)$ are the designer's proportion requirements, $g_{10}(x)$ is the maximal radial run out of the spindle nose Δ, and Δ_a and Δ_b are the radial run outs of the front and the end bearings.

For this example, it is assumed that d_a must be chosen from the set $X_3 = \{80, 85, 90, 95\}$ and d_b must be chosen from the set $X_4 = \{75, 80, 85, 90\}$. Additionally, the following constant parameters are assumed:

$d_{om} = 25.00$ mm, $\quad d_{a1} = 80.00$ mm, $\quad d_{a2} = 95.00$ mm, $\quad d_{b1} = 75.00$ mm,

$d_{b2} = 90.00$ mm, $\quad p_1 = 1.25, \quad p_2 = 1.05, \quad l_k = 150.00$ mm, $\quad l_g = 200.00$ mm,

$a = 80.00$ mm, $\quad E = 210,000.0$ N/mm^2, $\quad F = 10,000$ N, $\quad \Delta_a = 0.0054$ mm,

$\Delta_b = -0.0054$ mm, $\quad \Delta = 0.01$ mm, $\quad \delta_{ra} = -0.001$ mm, $\quad \delta_{rb} = -0.001$ mm

6.6.3 Experimental Results and Discussion

In the literature (Chen et al., 2007), the simulation results of MOEO were compared with those of NSGA-II, SPEA2, and PAES. To conduct a fair comparison, all algorithms were performed under the same FFE. Note that for MOEO, FFE = maximum generation × problem dimension, for PAES, FFE = maximum generation, and for NSGA-II and SPEA2, FFE = maximum generation × population size. All the algorithms were run for a maximum of 30,000 FFE for problem Two Bar and 40,000 FFE for the rest two problems. For MOEO, it was encoded in the floating point representation and an archive of size 100 was used. The parameters in the hybrid *GC* mutation, that is, *TC* and *TG*, were both set to 3. For NSGA-II (real-coded), the population size was 100. For SPEA2, a population of size 80 and an external population of size 20 were used, so that the overall population size became 100. For PAES, a depth value *d* equal to 4 and the archive size of 100 were adopted. The crossover probability of $p_c = 0.9$ and a mutation probability of $p_m = 0.05$ for real-coded GAs or $p_m = 1/l$ for binary-coded GAs (where *l* is the string length) were used. Additionally, all the algorithms were executed 50 times independently on each problem. In the literature (Chen et al., 2007), two performance metrics, that is, *FS* and coverage of two sets (C metric), were used to assess the performance of all the algorithms.

Table 6.9 shows the mean and standard deviation of the *FS* metric obtained using four algorithms, that is, MOEO, NSGA-II, SPEA2, and PAES. The bold numeral in Table 6.9 means it is the best optimum among all the optima. As can be seen from Table 6.9, MOEO is capable of finding a wider spread of solutions than any other algorithms on the problems Welded Beam and Spindle. This indicates that MOEO has the ability to find a wider distributed set of nondominated solutions than many other state-of-the-art MOEAs. In all cases with MOEO, the standard deviation of *FS* metric in 50 runs is also small except for the problem Two Bar.

The direct comparison of the different algorithms based on the *C* metric is shown in Table 6.10. It can be observed from Table 6.10 that MOEO is able to

Table 6.9 Comparisons of *FS* Metric

Algorithm	Two Bar		Welded Beam		Spindle	
	Mean	*St. Dev*	*Mean*	*St. Dev*	*Mean*	*St. Dev*
MOEO	8.47E4	4.2E3	**37.45**	3.22	**1.143E6**	1.2E4
NSGA-II	**9.15E4**	1.0E2	33.62	1.89	1.129E6	2.6E4
SPEA2	**9.15E4**	8.4E1	33.81	1.52	1.136E6	2.5E4
PAES	8.80E4	3.7E3	31.35	4.90	1.073E6	7.7E4

Source: Reproduced from Chen, M. R. et al. *Journal of Zhejiang University, Science A* 8 (12): 1905–1911, 2007. With permission.

Table 6.10 Comparisons of C Metric

Coverage of Two	Sets Two Bar		Welded Beam		Spindle	
	Mean	St. Dev	Mean	St. Dev	Mean	St. Dev
C(NSGA-II, MOEO)	0.38	0.04	0.52	0.11	0.01	0.01
C(MOEO, NSGA-II)	0.15	0.04	0.09	0.09	0.03	0.03
C(SPEA2, MOEO)	0.06	0.03	0.09	0.04	0.004	0.01
C(MOEO, SPEA2)	0.58	0.10	0.32	0.17	0.29	0.10
C(PAES, MOEO)	0.15	0.07	0.38	0.18	0.002	0.005
C(MOEO, PAES)	0.53	0.15	0.23	0.27	0.62	0.26

Source: Reproduced from Chen, M. R. et al. *Journal of Zhejiang University, Science A* 8 (12): 1905–1911, 2007. With permission.

perform better than NSGA-II on the problems Spindle concerning the C metric. It is interesting to note that MOEO outperforms SPEA2 on all the problems in terms of the C metric. It is also clear from Table 6.10 that MOEO significantly outperforms PAES on problems Two Bar and Spindle, with respect to the C metric.

For illustration, we also show 1 of 50 runs of all the algorithms on each problem in Figures 6.18 through 6.20, respectively. The figures show the trade-off fronts obtained by MOEO, NSGA-II, SPEA2, and PAES. The insets in all the figures show the parts which may be unclear in the main plots. From Figures 6.18 through 6.20, we can see that MOEO is able to converge nearly as well as or better than the other three algorithms on all the problems. Moreover, MOEO can find a well-distributed set of nondominated solutions on all the problems.

Note that the problem Spindle has two continuous and two discrete decision variables. As can be observed from Tables 6.9 and 6.10 and Figure 6.20, MOEO seems to be the best performer in both aspects of convergence and diversity of solutions on this problem. Two extreme nondominated solutions found by MOEO are $(4.767 \times 10^5, 0.0346)$ and $(1.607 \times 10^6, 0.01465)$, respectively, which are better than those found by other techniques listed in the literature (Baykasoglu, 2006). Consequently, MOEO may be a good choice to solve those problems with mixed continuous or discrete decision variables.

6.6.4 Conclusions

In this section, the MOEO algorithm is extended to solve the mechanical components design MOPs. The proposed approach is validated by three mechanical components design problems. The simulation results indicate that MOEO is highly competitive with three state-of-the-art MOEAs, that is, NSGA-II, SPEA2, and PAES in terms of convergence and diversity of solutions. Thus, MOEO may be well-suited for handling those mechanical components design problems with multiple objectives.

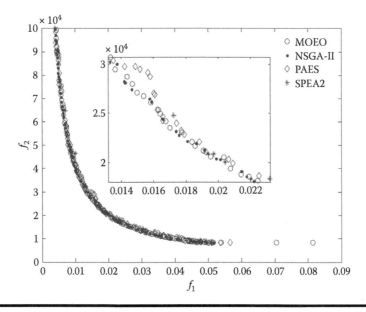

Figure 6.18 Approximate Pareto fronts for Two Bar. (Reproduced from Chen, M. R. et al. *Journal of Zhejiang University, Science A* **8 (12): 1905–1911, 2007. With permission.)**

6.7 Portfolio Optimization with MOEO

Portfolio optimization plays an important role in determining portfolio strategies for investors. What investors wish to achieve from portfolio optimization is to maximize portfolio return and minimize portfolio risk. However, the risk is often high when the return is maximized, and the return is often low when the risk is minimized. Thus, investors have to balance the risk-return trade-off for their investments. Hence, the portfolio optimization problem is intrinsically a MOP which has two competing objectives.

6.7.1 Portfolio Optimization Model

Originally the portfolio optimization problem attempts to maximize the return and minimize the risk simultaneously. However, the risk and the return conflict with each other. Therefore, the portfolio optimization problem is intrinsically a MOP. This work utilizes the following portfolio optimization model (Armañanzas and Lozano, 2005) as the mathematical model, which is based on the well-known Markowitz' model:

$$\text{minimize } \sigma_p^2 = \sum_{i=1}^{N}\sum_{j=1}^{N} w_i w_j \sigma_{ij} \tag{6.16}$$

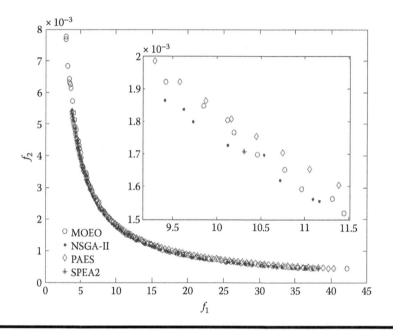

Figure 6.19 Approximate Pareto fronts for Welded Beam. (Reproduced from Chen, M. R. et al. *Journal of Zhejiang University, Science A* **8 (12): 1905–1911, 2007. With permission.)**

$$\text{maximize } r_p = \sum_{i=1}^{N} w_i \mu_i \tag{6.17}$$

$$\text{subject to } \sum_{i=1}^{N} w_i = 1 \tag{6.18}$$

$$\sum_{i=1}^{N} x_i = K \tag{6.19}$$

$$\varepsilon_i x_i \leq w_i \leq \delta_i x_i, \quad i = 1, \ldots, N \tag{6.20}$$

$$x_i \in \{0, 1\}, \quad i = 1, \ldots, N \tag{6.21}$$

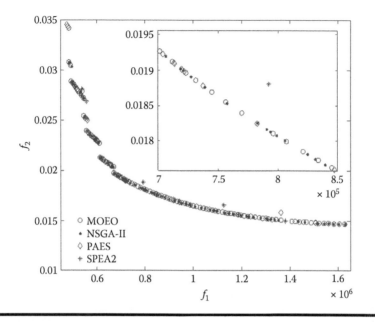

Figure 6.20 Approximate Pareto fronts for Spindle. (Reproduced from Chen, M. R. et al. *Journal of Zhejiang University, Science A* 8 (12): 1905–1911, 2007. With permission.)

where N is the number of available assets, σ_p^2 the risk of the portfolio assets, σ_{ij} the covariance between the ith and the jth asset, r_p the expected return of the portfolio assets, μ_i the expected return of the ith asset, K the number of assets to invest in ($K \le N$), ε_i the minimum investment ratio allowed in the ith asset, and δ_i the maximum investment ratio allowed in the ith asset.

There are two decision variables in this model: one is $X = (x_1, x_2, \ldots, x_i, \ldots, x_N) \in \{0, 1\}^N$, where $x_i = 1$, if the ith asset is chosen, otherwise $x_i = 0$; and the other is $W = (w_1, w_2, \ldots, w_j, \ldots, w_N)$, $w_j \in [0, 1]$, $j = 1, \ldots, N$, where w_j is the money ratio invested in the jth asset.

Equation 6.16 minimizes the total risk associated with the portfolio of assets. Equation 6.17 maximizes the return associated with the portfolio of assets. Equation 6.19 fixes the number of assets to invest in at K. Equation 6.20 imposes the maximum and minimum investment ratio allowed for each asset.

6.7.2 MOEO for Portfolio Optimization Problems

In order to further extend MOEO to solve portfolio optimization problems, Chen et al. (2009) presented an extended MOEO algorithm as shown in Figure 6.21. There are two decision variables, that is, X and W, in the above portfolio

1. Randomly select K assets from N assets, and then generate an initial feasible solution $W = (w_1, w_2, ..., w_j, ..., w_N)$, $w_j \in [0, 1]$, $j = 1, ..., N$. Repair W using repair strategy Repair (W). Add W to the external archive. Set *iteration* = 0.
2. Generate N offspring of the current solution W by performing mutation on each component one by one, while keeping other components unchanged.
3. Perform dominance ranking on the N offspring and then obtain their rank numbers, i.e., $r_j \in [0, N-1]$, $j \in \{1, ..., N\}$.
4. Assign the fitness $\lambda_j = r_j$ for each component w_j, $j \in \{1, ..., N\}$.
5. If there is only one component with fitness value of zero, the component will be considered the worst one. Otherwise, if there are two or more components with fitness value of zero, the diversity preservation mechanism is invoked to decide the worst component. Assuming that the kth component is the worst one, and W_k is the corresponding offspring generated by performing mutation only on the kth component, $k \in \{1, ..., N\}$.
6. Accept $W = W_k$ unconditionally.
7. Apply the archiving logic to updating the external archive. Otherwise, go to Step 8.
8. If the iterations reach the predefined maximum number of the generations, go to Step 9; otherwise, set *iteration* = *iteration* + 1, and go to Step 2.
9. Output the external archive as the Pareto-optimal set.

Figure 6.21 Pseudo-code of extended MOEO for portfolio optimization problems. (Chen, M. R. et al. Multi-objective extremal optimization for portfolio optimization problem. *Proceedings of 2009 IEEE International Conference on Intelligent Computing and Intelligent Systems (ICIS 2009)***, pp. 552–556. © 2009 IEEE.)**

optimization model. However, it is noticed that $x_i = 1$ when $w_i \neq 0$ and $x_i = 0$ when $w_i = 0$. Thus, if W is determined, then X will be determined exclusively. As a consequence, for convenience of encoding, there is only one decision variable W in the extended MOEO algorithm.

Note that fitness assignment, diversity preservation, and external archive in the extended MOEO for portfolio optimization problems are the same as those in the numerical MOEO. Different from the numerical MOEO, the extended MOEO for portfolio optimization problems adopts the following mutation operation and repair strategy.

6.7.2.1 Mutation Operation

Assuming that the kth component is the current component, perform mutation on the current component, that is, set $w_k = 0$ when its sign function $\text{sgn}(w_k) = 1$; otherwise, set w_k to be a Gaussian random number. At the same time, randomly choose another component with the sign function value different from that of the current component, and perform mutation on it. Thus, the total number of the chosen assets will be still kept at K after mutation. Finally, in order to ensure $\sum_{i=1}^{N} w_i = 1$ and $\varepsilon_i x_i \leq w_i \leq \delta_i x_i$, $i = 1, ..., N$, carry out repair strategy Repair (W) on the decision variable W. This mutation operation can ensure the new offspring to be a feasible solution.

Procedure Repair (W)
begin

$$L = 1 - \sum_{i=1}^{N} w_i$$

if $L \neq 0$ then
   for each $\alpha_i \in \mathfrak{I}$
     if $L > 0$ then

$$d_i = \delta_i - w_i$$

     else

$$d_i = w_i - \delta_i$$

     end if
   end for
   for each $\alpha_i \in \mathfrak{I}$

$$m_i = \frac{d_i}{\sum_{j=1}^{K} d_j}$$

$$w_i = w_i + L \cdot m_i$$

   end for
end if
   return W
end

Figure 6.22 **Pseudo-code of repair strategy. (Chen, M. R. et al. Multi-objective extremal optimization for portfolio optimization problem.** *Proceedings of 2009 IEEE International Conference on Intelligent Computing and Intelligent Systems (ICIS 2009)***, pp. 552–556. © 2009 IEEE.)**

6.7.2.2 Repair Strategy

The extended MOEO uses the repair strategy to repair the new generated W to ensure that $\sum_{i=1}^{N} w_i = 1$ and $\varepsilon_i x_i \leq w_i \leq \delta_i x_i$, $i = 1, \ldots, N$. The pseudo-code of repair strategy is shown in Figure 6.22.

In Figure 6.22, N is the number of available assets, K the number of assets to invest in ($K \leq N$), $\mathfrak{I}$ an asset subset of the problem, α_i the asset of the ith index, $\alpha_i \in \mathfrak{I}$; L the solution's leftover; d_i the deviation of the α_i asset investment ratio with respect to its high δ_i or low ε_i restriction; and m_i the weight modifier for α_i.

6.7.3 Experimental Settings

In the literature (Chen et al., 2009), the extended MOEO for portfolio optimization problems was tested by five popular stock indexes which come from OR-Library (Beasley, 1990). The five indexes are named port1 to port5, respectively. Each index corresponds to a different stock market of the world. The first index is the Hong Kong *Hang Seng*, the second one is the German *DAX 100*, the third one is the

British *FTSE 100*, the fourth one is the U.S. *S&P 100*, and the fifth one is the Japanese *Nikkei 225*. The values included for each index were collected from March 1992 to September 1997. The data package contains the complete identification list of the assets included. The data files are composed of 31, 85, 89, 98, and 225 assets, respectively. For each asset i, the average return μ_i and the individual risk σ_i is included. For each pair of assets i and j, the correlation ρ_{ij} between them is also included. The risk of investing in an asset having invested in another one simultaneously is modeled by the covariance between both. Then, the risk (or covariance) between two assets, i and j, is given by the expression $\sigma_{ij} = \rho_{ij} \cdot \sigma_i \cdot \sigma_j$.

The simulation results of MOEO (Chen et al., 2009) were compared with those of NSGA-II, SPEA2, and PAES when $K = 5$. In addition, to conduct a fair comparison, all algorithms were performed under the same FFE. All the algorithms were run for a maximum of 31,000, 85,000, 89,000, 98,000 and 225,000 FFE for problems port1 to port5, respectively. For MOEO, it was encoded in the floating point representation and an archive of size 100 was used. For NSGA-II, the population size was 100. For SPEA2, a population of size 80 and an external population of size 20 were adopted, so that the overall population size became 100. For PAES, a depth value d equal to 4 and the archive size of 100 were used. NSGA-II and SPEA2 adopt uniform crossover operator and Gaussian mutation, while PAES uses Gaussian mutation. The crossover probability of $p_c = 0.9$ and a mutation probability of $p_m = 0.1$ were used. Additionally, all the algorithms were executed 50 times independently on each problem. In order to consider a search space as wide as possible, the values of the maximum and minimum investments were set at extreme values. For all the experiments, the values of ε_i and δ_i are 0.001 and 1.0, respectively. Additionally, two performance metrics, that is, *FS* and coverage of two sets (*C* metric) were used to assess the performance of MOEO.

6.7.4 Experimental Results and Discussion

Table 6.11 shows the mean and standard deviation of the *FS* metric obtained using four algorithms, that is, MOEO, NSGA-II, SPEA2, and PAES. The bold numeral in Table 6.11 means it is the best optimum among all the optima. As can be seen from Table 6.11, MOEO is capable of finding a wider spread of solutions than any other algorithms on all the problems. This indicates that MOEO has the ability to find a wider distributed set of nondominated solutions than the other three algorithms. In all cases with MOEO, the standard deviation of *FS* metric in 50 runs is also small.

The direct comparison of the different algorithms based on the metric *C* is shown in Figure 6.23. It is clear that MOEO significantly outperforms SPEA2 and PAES on all the problems in terms of the metric *C*. It can be also observed from Figure 6.23 that MOEO performs a little worse than NSGA-II on all the problems concerning the metric *C*. This illustrates that MOEO is capable of performing nearly as well as or better than the other three algorithms with respect to the convergence of solutions.

Table 6.11 Comparisons of *FS* Metric

Algorithms	port1		port2		port3		port4		port5	
	Mean	St. Dev.	Mean	St. Dev.	Mean	St. Dev.	Mean	St. Dev.	Mean	St. Dev.
MOEO	**8.59E–3**	4.17E–4	**7.78E–3**	6.46E–4	**5.30E–3**	3.71E–4	**7.30E–3**	2.57E–4	**3.98E–3**	2.88E–4
NSGA-II	8.06E–3	2.80E–4	6.99E–3	1.07E–4	4.89E–3	1.20E–4	6.90E–3	2.70E–4	3.67E–3	1.92E–4
SPEA2	8.46E–3	3.46E–4	7.20E–3	2.06E–4	5.14E–3	1.13E–4	7.20E–3	1.85E–4	3.92E–3	1.85E–4
PAES	2.13E–3	9.55E–4	2.38E–3	2.45E–3	2.20E–3	1.52E–3	3.99E–3	2.07E–3	1.73E–3	7.30E–4

Source: Chen, M. R. et al. 2009. Multi-objective extremal optimization for portfolio optimization problem. *Proceedings of 2009 IEEE International Conference on Intelligent Computing and Intelligent Systems (ICIS 2009)*, pp. 552–556. © 2009 IEEE.

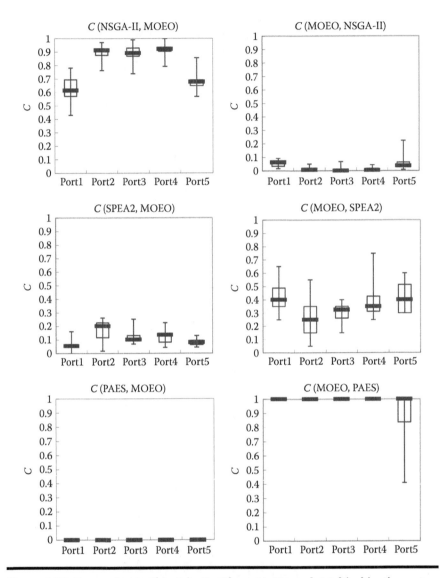

Figure 6.23 Comparisons of metric *C*. (Chen, M. R. et al. Multi-objective extremal optimization for portfolio optimization problem. *Proceedings of 2009 IEEE International Conference on Intelligent Computing and Intelligent Systems (ICIS 2009)*, **pp. 552–556. © 2009 IEEE.)**

For illustration, we also show 1 of 50 runs of all the algorithms on each problem when *K* = 5 in Figure 6.24a–e, respectively. The real curve represents the unconstrained Pareto front. Figure 6.24 shows the trade-off fronts obtained by MOEO, NSGA-II, SPEA2, and PAES. From Figure 6.24, we can see that MOEO is able to converge nearly as well as or better than the other three algorithms on all the

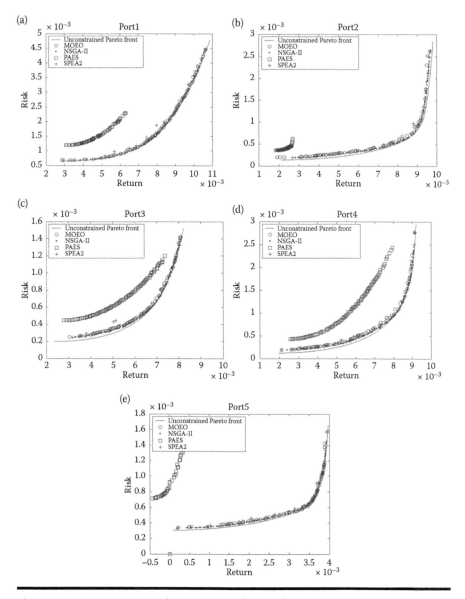

Figure 6.24 (a)–(e) Approximate Pareto fronts for port1–port5 ($K = 5$). (Chen, M. R. et al. Multi-objective extremal optimization for portfolio optimization problem. *Proceedings of 2009 IEEE International Conference on Intelligent Computing and Intelligent Systems (ICIS 2009)*, pp. 552–556. © 2009 IEEE.)

problems. Moreover, MOEO can find a well-distributed set of nondominated solutions on all the problems.

6.7.5 Conclusions

The above simulation results indicate that MOEO is highly competitive with three state-of-the-art MOEAs, that is, NSGA-II, SPEA2, and PAES in terms of convergence and diversity of solutions. The results of this study are encouraging. As a consequence, MOEO may be a good alternative to deal with the portfolio optimization problem.

6.8 Summary

EO is a general-purpose local-search heuristic algorithm, which has been successfully applied to some NP-hard COPs such as graph bipartitioning, TSP, graph coloring, spin glasses, MAXSAT, and so on. In this chapter, in order to extend EO to solve MOPs in an efficient way, a novel elitist Pareto-based multiobjective algorithm, called MOEO, is investigated in detail. The MOEO algorithm and its variation versions were used to solve several benchmark MOPs, such as numerical unconstrained/constrained MOPs, multiobjective mechanical components design problems, multiobjective portfolio optimization problems, and MOKP. Experimental results indicate that MOEO and its variation versions are highly competitive to the state-of-the-art MOEAs, such as NSGA-II, SPEA2, PAES, NSGA, SPEA, and NPGA. Thus, MOEO may be a good alternative to deal with MOPs. We will explore the efficiency of MOEO on those problems with large number of decision variables in the future. It is desirable to further apply MOEO to solving those complex engineering MOPs in the real world.

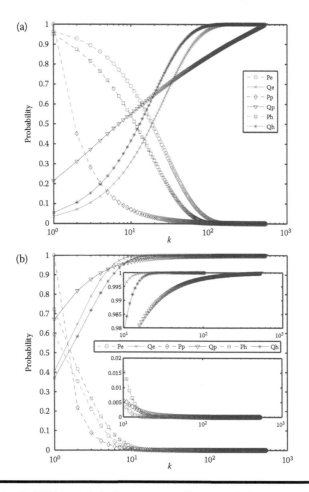

Figure 4.1 Probabilities and their corresponding cumulative ones of power-law, exponential, and hybrid distributions for $n = 532$, of which (a) $\mu = 0.037$, $\tau = 1.15$, and $h = 0.05$; (b) $\mu = 0.52$, $\tau = 2.20$, and $h = 0.325$. (Reprinted from *Physica A*, 389 (9), Zeng, G. Q. et al. Study on probability distributions for evolution in modified extremal optimization, 1922–1930, Copyright 2010c, with permission from Elsevier.)

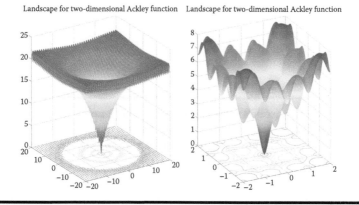

Figure 5.6 Landscape for two-dimensional Ackley function; left: surface plot in an area from −20 to 20, right: focus around the area of the global optimum at [0, 0] in an area from −2 to 2.

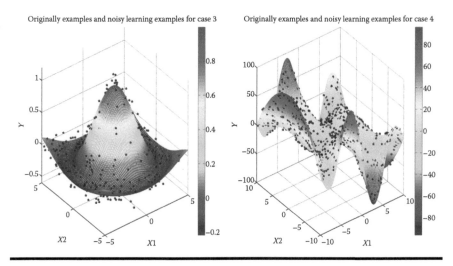

Originally examples and noisy learning examples for case 3   Originally examples and noisy learning examples for case 4

Figure 7.9 Original function examples and noisy learning examples (★) for case 3 (noise level $N(0, 0.01)$) and case 4 (noise level $N(0, 5)$). (Reproduced from Chen, P. and Lu, Y. Z., *Journal of Zhejiang University, Science C* 12: 297–306, 2011b. With permission.)

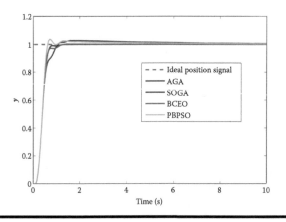

Figure 7.19 Comparison of step response for plant 1 under different algorithms-based PID controllers. (Reprinted from *Neurocomputing*, 138, Zeng, G. Q. et al., Binary-coded extremal optimization for the design of PID controllers, 180–188, Copyright 2014, with permission from Elsevier.)

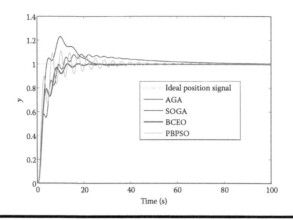

Figure 7.20 Comparison of step response for plant 2 under different algorithms-based PID controller. (Reprinted from *Neurocomputing*, 138, Zeng, G. Q. et al., Binary-coded extremal optimization for the design of PID controllers, 180–188, Copyright 2014, with permission from Elsevier.)

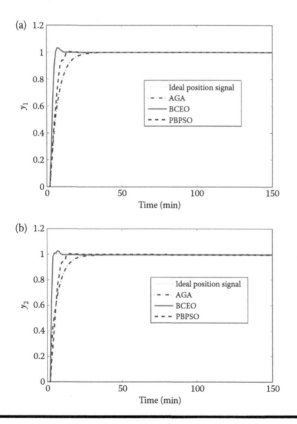

Figure 7.21 Comparison of output y_1 (a) and y_2 (b) under different algorithms-based PID controllers for multivariable plant with decoupler. (Reprinted from *Neurocomputing*, 138, Zeng, G. Q. et al., Binary-coded extremal optimization for the design of PID controllers, 180–188, Copyright 2014, with permission from Elsevier.)

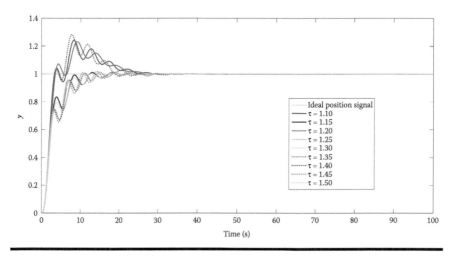

Figure 7.22 Adjustable parameter τ versus the step response for single-variable plant 2. (Reprinted from *Neurocomputing*, 138, Zeng, G. Q. et al., Binary-coded extremal optimization for the design of PID controllers, 180–188, Copyright 2014, with permission from Elsevier.)

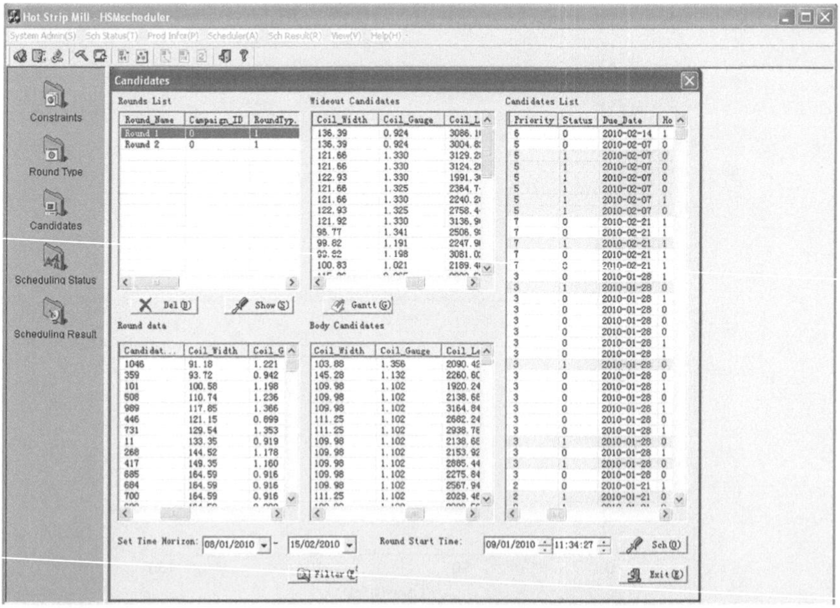

Figure 8.12 Main graphical user interface of the developed HSM-scheduling system. (Reprinted from *Computers and Operations Research*, 39 (2), Chen, Y. W. et al., Development of hybrid evolutionary algorithms for production scheduling of hot strip mill, 339–349, Copyright 2012, with permission from Elsevier.)

APPLICATIONS

III

Chapter 7

EO for Systems Modeling and Control

7.1 Problem Statement

There are many optimization problems in industrial process control, for example, system identification, controller parameter tuning, and online/real-time optimization. The rapid growth and huge progress in manufacturing business environment and technology leads to new requirements for control technology and system science. The primary mission of process control has extended from stability, safety, and quality to constrained optimization control for desired KPI (key performance index). As one of the popular advanced process control (APC) solutions, the model predictive control (MPC) has been widely applied in process industries. However, with serious uncertainty and competition in the global marketplace, a manufacturing enterprise faces critical challenges in making its business and production more flexible and dynamic. In other words, a manufacturing enterprise should be able to maximize its profits and provide customers with high-quality products/services under varying marketplace conditions. The issues mentioned above involve different types of optimization problems. Eventually, the traditional modeling, optimization, and control approaches could hardly be applied in solving those complex systems with highly nonlinear and highly uncertain natures. All those issues will encourage researchers and practitioners to devote themselves to the development of novel methods and new technology that is suitable for applications in process automations.

System modeling becomes more and more important in current system control and optimization. The state space of classical control theory in the time domain method, root locus method, frequency domain method, modern control theory analysis method, and the optimal control method is carried on the system analysis and design of the known mathematical system model. Therefore, the establishment of an accurate dynamic system model is very crucial. On the other hand, with the trends toward integration and large scale in industry, process control gradually developed from simple proportional-integral-derivative (PID) loops to large-scale multivariable complex nonlinear systems, and optimization problems in control systems have become more complex, resulting in MOO, increased process variables, correlation, and lots of constraints. Meanwhile market competition has put forward higher requirements on control strategies, which requires the control system to be more flexible to lower the consumption and deliver the qualified products on time.

Based on the ideas of memetic algorithms (MA), the efficiency of CI can be improved significantly by incorporating a local search procedure into the optimization. Considering the problems that occur in the application of evolution computation methods in control theory and application, this chapter further applies the two hybrid EO-based memetic algorithms, called "EO-LM" and "EO-SQP," to system identification and industrial optimal control.

7.2 Endpoint Quality Prediction of Batch Production with MA-EO

The BOF is one of the key processes in iron and steel production. It is a typical complex batch chemical reactor with sophisticated thermal and chemical reaction processes, which converts liquid pig iron into steel with desired grade (temperature and chemical compositions, such as carbon, etc.) under oxidation conditions. The BOF steelmaking endpoint quality control is always managed with the aid of either operator experience or a first-principle charge model, therefore, often resulting in a relatively poor hit ratio of the bath endpoint. The accurate prediction of the endpoint quality is vital not only for BOF operation but also for "steelmaking-caster-hot rolling" production balance. Due to the imprecision of the BOF charge model, however, and lack of real-time measurements and process uncertainties, it is difficult to predict the BOF endpoint product precisely (Cox et al., 2002; Wang et al., 2006). In addition to the first-principle models, the supervised learning with a neural network (NN) has played an important role in BOF endpoint quality prediction because of its capability in nonlinear mapping and simplicity in implementation (Cox et al., 2002; Tao et al., 2002; Fileti et al., 2006; Wang et al., 2006; Feng et al., 2008). Due to the inherent defects, the popular BP NN training algorithms based on GS often suffer from local minima, being sensitive to initial weights and poor generalization (Haykin, 1994; Yao and Islam, 2008).

In this section, the proposed hybrid EO-LM algorithm is applied to a practical engineering problem for a production-scale BOF in steelmaking (Chen and Lu, 2010a). As mentioned above, BOF is a complex multiphase batch chemical reactor with more than a dozen of input variables to produce the desired steel grade. In this section, the former proposed EO-LM algorithm is applied in multilayer perceptron (MLP) network training for BOF endpoint quality prediction with the abilities of avoiding local minimum and performing detailed local search. Based on the mechanism of the BOF process, a feed-forward MLP network {13, 10, 2} is adopted for real applications, which consists of 1 hidden layer with 10 hidden nodes. As shown in Figure 7.1, the inputs are 13 variables, including uncontrollable hot metal information (hot metal temperature, weight, chemical compositions, etc.) and the operational receipt variables (oxygen, scrap, a variety of recipes, etc.). Two outputs are defined as the endpoint temperature and the carbon content, respectively. The

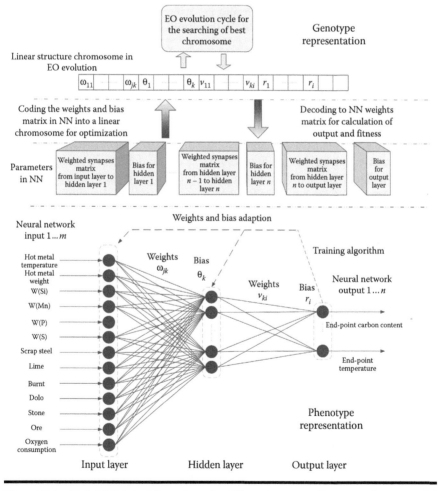

Figure 7.1 Hybrid EO-LM learning algorithm structure diagram of NN model for BOF endpoint prediction. (Reproduced from Chen, P. and Lu, Y. Z., *Journal of Zhejiang University, Science A* 11: 841–848, 2010a. With permission.)

Table 7.1   Comparison between the EO-LM and LM on BOF Endpoint Prediction Model

Output	Perfor-mance	EO-LM	Conventional LM	Improvement (%)
Endpoint temperature (°C)	RMSE	17.5232	19.1821	8.65
	ME	−1.1372	−1.7186	33.83
Endpoint carbon content (%)	RMSE	0.0129	0.0150	14
	ME	−7.7257e − 004	−0.0030	74

Source: Reproduced from Chen, P. and Lu, Y. Z., *Journal of Zhejiang University, Science A* 11: 841–848, 2010a. With permission.

EO-LM learning is executed in two parallel phases: the genotype for EO-LM and the phenotype for MLP network. The synaptic weights and biases are encoded as a real-valued chromosome, to be evolved during EO-LM iterations.

To evaluate the effectiveness of the proposed hybrid EO-LM for the predictions of endpoint temperature and carbon content, the simulation experiment is performed using real industry data. Over 1600 pairs of data are gathered from the steel plant database; among them, 800 pairs are selected randomly as training data, 480 pairs as validation data, and the balance 320 pairs as test data. In order to evaluate the performance of the proposed EO-LM-based NN model, the conventional LM algorithm is applied with the same test data set as used in the EO-LM model.

The simulation performance of the EO-LM model is evaluated in terms of root mean square error (RMSE), mean error (ME), and correlation coefficient. Table 7.1 gives the RMSE and ME values of the two different models on test data set. Compared with the conventional LM algorithm, the proposed algorithm reduces the prediction RMSE by 8.65% and 14% for endpoint temperature and carbon content, respectively.

The scatter diagram in Figure 7.2 shows the extent of the match between the measured and predicted values by EO-LM and LM learning algorithms. It can be seen that the EO-LM model shows a better agreement with the target than those by conventional LM learning algorithm.

The comparison of prediction error distributions for endpoint temperature and carbon content between the hybrid EO-LM algorithm and conventional LM algorithm are shown in Figure 7.3. It can also be seen that the range of prediction residuals are reduced by the hybrid EO-LM algorithm compared to the conventional LM algorithm.

The BOF steelmaking is a highly complex process and difficult to model and control. In this section, a novel hybrid NN training method with the integration of EO and LM is presented for BOF endpoint quality prediction. The main advantage of the proposed EO-LM algorithm is to utilize the superior features of EO and

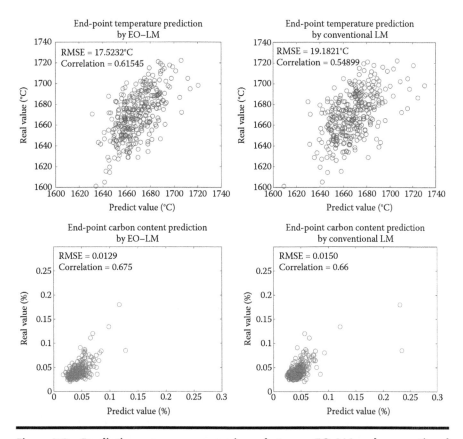

Figure 7.2 Prediction accuracy comparisons between EO-LM and conventional LM. (Reproduced from Chen, P. and Lu, Y. Z., *Journal of Zhejiang University, Science A* 11: 841–848, 2010a. With permission.)

LM in global and local search, respectively. As a result, the application of the proposed EO-LM algorithm in NN learning may create a BOF endpoint prediction model with better performance, the experimental results indicate that the proposed EO-LM can easily avoid the local minima, overfitting and underfitting problems suffered by traditional GS-based training algorithms, and provide better prediction results.

7.3 EO for Kernel Function and Parameter Optimization in Support Vector Regression

Based on statistical learning theory and the structural risk minimization principle, one of the most critical statistical learning solutions, called the support vector

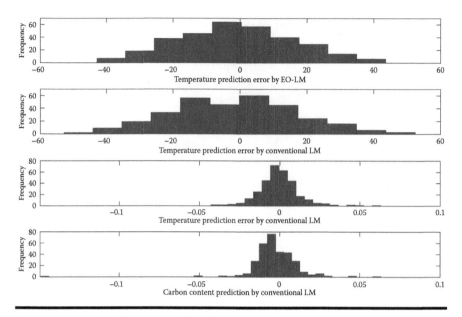

Figure 7.3 **Prediction error distribution comparisons between EO-LM and conventional LM. (Reproduced from Chen, P. and Lu, Y. Z., *Journal of Zhejiang University, Science A* 11: 841–848, 2010a. With permission.)**

machine (SVM), was first proposed by Vapnik to solve pattern recognition problems (Vapnik, 1995; Burges, 1998). In the last decade, SVM-based algorithms have been developed rapidly and employed in many real-world applications, such as handwriting recognition, identification, bioinformatics, classification, regression, etc.

The performance of support vector regression (SVR) model is sensitive to the kernel type and its parameters. The determination of appropriate kernel type and associated parameters for SVR is a challenging research topic in the field of support vector learning. In this section, a novel method is presented for simultaneous optimization of SVR kernel function and its parameters, which is formulated as a mixed integer optimization problem and solved by the recently proposed heuristic "extremal optimization (EO)." We will present the problem formulation for the optimization of the SVR kernel and parameters, EO-SVR algorithm, and experimental tests with five benchmark regression problems (Chen and Lu, 2011b). The comparison results with other traditional approaches show that the proposed EO-SVR method can provide better generalization performance through successfully identifying the optimal SVR kernel function and its parameters.

7.3.1 Introduction

High generalization capability and global optimal solution constitute the major advantages of SVM over other machine learning approaches. However, the performance of SVM strongly depends on the embedded kernel type (Ali and Smith, 2006) and the associated parameter values (Min and Lee, 2005; Jeng, 2006; Hou and Li, 2009). Therefore, the selection of the SVM kernel function and its parameters should be conducted carefully on a case-to-case basis as different functions and parameters may lead to widely varying performance. Up to now, several kernels have been proposed by researchers, but there are no effective guidelines or systematic theories concerning how to choose an appropriate kernel for a given problem. The empirical search for the SVM kernel and parameters through a trial-and-error approach has been proven to be often time consuming, imprecise, and unreliable (Zhang et al., 2004; Lorena and De Carvalho, 2008).

New parameter optimization techniques for SVM have been proposed and investigated by researchers in recent years. Among them, the most commonly used methods are grid search (Hsu et al., 2004) and CI (Engelbrecht, 2007). The former is time consuming (Engin, 2009) and can only adjust a few parameters (Friedrichs and Igel, 2005). On the contrary, CI methods have shown high suitability for constrained nonlinear optimization problems, and are able to avoid local minima inherently.

So far, the optimization of SVR parameters based on CI methods has been realized in many studies and applications (Zhang et al., 2004; Friedrichs and Igel, 2005; Mao et al., 2005; Pai and Hong, 2005; Engin, 2009; Tang et al., 2009; Wu, 2010). However, there has been relatively limited research published on how to determine an appropriate SVR kernel function automatically (Howley and Madden, 2005; Thadani et al., 2006) and only a few of the existed work was devoted to simultaneous optimization of the SVR kernel and its associated parameters (Wu et al., 2009). Therefore, we will introduce a novel SVR tuning algorithm based on EO to improve the predictive accuracy and generalization capability. Inspired by far-from-equilibrium dynamics, EO provides a new philosophy to optimization using nonequilibrium statistical physics and the capability to elucidate the properties of phase transitions in complex optimization problems (Boettcher and Percus, 1999). In this section, we will, for the first time, apply EO in the simultaneous optimization of the SVR kernel and its parameters. Considering the complexity of SVR tuning, a novel EO-SVR algorithm with carefully designed chromosome structure and fitness function is proposed to deal with this hard issue.

7.3.2 Problem Formulation

In this section, the basic idea of SVR is first introduced, and then the determination of the optimal SVR kernel function and parameters is formulated into a mixed integer optimization problem.

7.3.2.1 Support Vector Regression

SVR is a technique to build SVM for regression learning. Suppose we have a set of learning samples (x_1, y_1), ..., (x_l, y_l), where $x_i \subset R^n$ represents the input vector and has a corresponding output $y_i \subset R$ for $i = 1, ..., l$, where l denotes the size of the training data. The SVR regression function can be generally described by the form

$$f(x) = (w * \Phi(x)) + b \tag{7.1}$$

where $w \subset R^n$, $b \subset R$, and Φ denotes a nonlinear transformation from low-dimensional space of R^n to high-dimensional feature space. The training of SVR is to find w and b values by minimizing the regression risk:

$$R_{reg}(f) = \gamma \sum_{i=1}^{l} \Gamma(f(x_i) - y_i) + \frac{1}{2}\|w\|^2 \tag{7.2}$$

where $\Gamma(\cdot)$ is a cost function and γ is a constant that determines penalties to regression errors and the vector w can be written in terms of data points as

$$w = \sum_{i=1}^{l} (\alpha_i - \alpha_i^*)\Phi(x_i) \tag{7.3}$$

Only the most critical samples associated with nonzero Lagrange multipliers account in the solution, called support vectors. More detailed descriptions for the training process of SVR can be found in Steve's work (Steve, 1998).

Substituting Equation 7.3 into Equation 7.1, the general form can be rewritten as

$$f(x) = \sum_{i=1}^{l} (\alpha_i - \alpha_i^*)(\Phi(x_i) * \Phi(x)) + b = \sum_{i=1}^{l} (\alpha_i - \alpha_i^*)k(x_i, x) + b \tag{7.4}$$

The dot product in Equation 7.4 can be replaced by the kernel function $k(x_i, x)$, which provides an elegant way of dealing with nonlinear mapping in the feature space, thereby avoiding all difficulties inherently in high dimensions. There are several commonly used kernel types in SVR: linear, polynomial, radial basis function (RBF), and MLP, as listed in Equations through 7.8.

1. Linear kernel function:

$$k(x_i, x) = (x_i * x) \tag{7.5}$$

2. Polynomial kernel function:

$$k(x_i, x) = (x_i * x + t)^d \tag{7.6}$$

3. Radial basis function (RBF):

$$k(x_i, x) = \exp\left(\frac{-\|x_i - x\|^2}{2\sigma^2}\right) \tag{7.7}$$

4. Multilayer perceptron (MLP) function:

$$k(x_i, x) = \tanh(s * x_i^T * x + t^2) \tag{7.8}$$

Here, γ, t, d, σ, and s are adjustable parameters for above-mentioned kernel functions. To design an effective SVR model, the kernel function and parameters must be chosen carefully.

7.3.2.2 Optimization of SVR Kernel Function and Parameters

In this study, the selection of optimal kernel function and parameters is formulated into an optimization problem and then solved by EO. The kernels and parameters described in Equations 7.5 through 7.8 are first encoded into a chromosome and a predefined cost function J is used to assess the performance of candidate solutions during the optimization. In addition, the performance is assessed in a standard way: by learning different SVRs with a training data set and evaluating them on an independent validation data set, as described in the following equation:

$$J = \underset{\prod \text{Validation set}}{f} (\textit{Kernel type, Kernel parameters}) \tag{7.9}$$

where J is a function of *kernel type* (discrete variable) and *kernel parameters* (continuous variables). In this study, the mean square error (MSE) in Equation 7.10 is selected as the cost function:

$$
\begin{aligned}
J &= \underset{\prod \text{Validation set}}{f} (\textit{Kernel type, Kernel parameters}) \\
&= \underset{\prod \text{Validation set}}{MSE} (\textit{Kernel type, Kernel parameters}) \\
&= \frac{\sum_{i=1}^{N} (y_i - \hat{y}_i)^2}{N}
\end{aligned}
\tag{7.10}
$$

where N denotes the number of the validation data, y_i represents the actual output, and $\hat{y}_i$ is the SVR predicted value.

Based on the problem formulation described above, our goal is to employ the optimized procedure to explore a finite subset of possible values and determine the kernel type and associated parameters which can minimize the cost function J:

$$\text{Minimize} \quad \underset{\prod Validation\ set}{f} \quad (Kernel\ type,\ Kernel\ parameters)$$

(7.11)

subject to   Ranges of parameter values and kernel function types

7.3.3 Hybrid EO-Based Optimization for SVR Kernel Function and Parameters

In this section, the development of the proposed EO-SVR method is introduced. The detailed issues of applying EO to SVR kernel and parameter optimization are discussed, which include chromosome structure, fitness function, and EO-SVR workflow.

As described in Section 7.3.2, the simultaneous optimization of SVR kernel function and parameters falls into a mixed integer optimization problem, in which the discrete variable (kernel type) and the continuous variables (associated parameters) are parts of the same optimization problem. Traditional real coded EO algorithms are designed for the optimization of continuous variables and are unsuitable for this kind of problem. In this study, a carefully designed fitness function is proposed to deal with the mixed integer optimization problem, in which the real-valued gene (component) representing the SVR kernel type is rounded to the nearest integer for fitness calculation; thus, all the genes can be treated as real values in EO mutation/representation. The benefit of this design is that EO-SVR can operate independently of the variable types. There is no need for additional design of a mixed chromosome representation and mutation operations. The EO-based SVR kernel and parameters optimization is developed as follows:

7.3.3.1 Chromosome Structure

In the proposed EO-SVR optimization procedure, the kernel types and associated parameters can be represented by a real coded chromosome without considering the variable types (discrete or continuous). The SVR kernel and associated parameters are directly coded into the chromosome as $S = KT_i, P(1)_i, P(2)_i, P(3)_i$, where $KT_i, P(1)_i, P(2)_i,$ and $P(3)_i$ denote the kernel type and associated kernel parameters, respectively. The chromosome structure of the proposed EO-SVR is shown in Figure 7.4.

The SVR kernel functions and associated parameters to be optimized in this study are listed in Table 7.2.

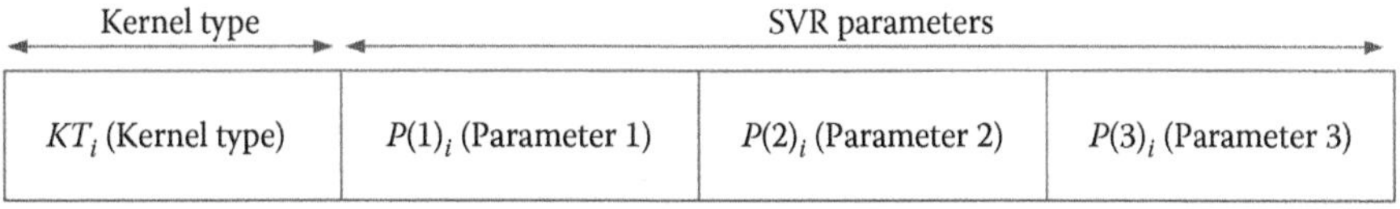

Kernel type	SVR parameters		
KT_i (Kernel type)	$P(1)_i$ (Parameter 1)	$P(2)_i$ (Parameter 2)	$P(3)_i$ (Parameter 3)

Figure 7.4 Chromosome structure of the proposed EO-SVR. (Reproduced from Chen, P. and Lu, Y. Z., *Journal of Zhejiang University, Science C* 12: 297–306, 2011b. With permission.)

One of the main issues in simultaneous optimization of SVR kernels and parameters is to deal with the different search space of parameter values for various kernel types. Therefore, we map all SVR parameters ($P(1)_i$, $P(2)_i$, and $P(3)_i$) to continuous values between 0 and 1 during EO evolution. When the fitness needs to be calculated, the parameters $P(j)_i$, $j = 1$, 2, 3 are converted to actual values $P(j)_{real}$ based on the upper bound $P_{UB}(j)_i$ and the lower bound $P_{LB}(j)_i$ associated with kernel types (KT_i), as described in Section 7.3.2.

7.3.3.2 Fitness Function

The EO-SVR in this study minimizes the cost function J in Equation 7.10 to establish an efficient SVR model. As shown in Figure 7.4, the chromosome in optimized procedure is represented by a real-valued string of genes:

$$S = \{KT_i, P(1)_i, P(2)_i, P(3)_i\} \tag{7.12}$$

Table 7.2 Kernel Types and Associated Parameters

KT Value	Kernel Type	P(1) (Parameter 1)	P(2) (Parameter 2)	P(3) (Parameter 2)
0	Linear kernel function	γ	–	–
1	Polynomial kernel function	γ	d	t
2	Radial basis function (RBF) kernel	γ	σ	–
3	Multilayer perceptron (MLP) kernel	γ	s	t

Source: Reproduced from Chen, P. and Lu, Y. Z., *Journal of Zhejiang University, Science C* 12: 297–306, 2011b. With permission.

Note: – Denotes no parameter needed.

As mentioned previously, the real-valued variable (KT_j) is converted into an integer representing the kernel type before fitness evaluation, the kernel parameters $(P(1)_j, P(2)_j,$ and $P(3)_j)$ are transformed to actual values, as described below:

$$\bar{S} = \{ROUND(KT_i), \quad CONVERT(P(1)_i), \quad CONVERT(P(2)_i),$$

$$CONVERT(P(3)_i)\}$$

$$= \{KT_{integer}, \quad P(1)_{real}, \quad P(2)_{real}, \quad P(3)_{real}\} \tag{7.13}$$

where $ROUND$ is a function that rounds to the nearest integer, and $CONVERT$ is a function that maps the $P(1)_i, P(2)_i,$ and $P(3)_i$ from [0, 1] to actual variable regions:

$$P(j)_{real} = CONVERT(P(j)_i) = P_{LB}(j)_i + (P_{UB}(j)_i - P_{LB}(j)_i) * P(j)_i, \quad j = 1, 2, 3 \tag{7.14}$$

where $P_{LB}(j)_i$ and $P_{UB}(j)_i$ are the variable bounds satisfying $P_{LB}(j)_i \leq P(j)_{real} \leq P_{UB}(j)_i$.

As described in Equation 7.10, to solve the SVR optimization problems, the global fitness can be defined as

$$Fitness_{global}(\bar{S}) = \underset{\Pi Validation\ set}{MSE}(\bar{S}) = \frac{\sum_{i=1}^{N}(y_i - \hat{y}_i)^2}{N} \tag{7.15}$$

Unlike GA, which works with a population of candidate solutions, EO evolves a single solution and makes local modification to the worst component. Each component needs to be assigned with a quality measure (i.e., fitness) called "local fitness." In this study, the local fitness λ_k is defined as an improvement in global fitness $Fitness_{global}$ made by the mutation imposed on the kth component of the best-so-far chromosome $\bar{S}$:

$$\lambda_k = Fitness_{local}(\bar{S'_k}) = \underset{S \to S'_k}{\Delta Fitness}_{global}(\cdot) = Fitness_{global}(\bar{S}) - Fitness_{global}(\bar{S'_k}) \tag{7.16}$$

7.3.3.3 EO-SVR Workflow

Figure 7.5 shows the workflow of our proposed EO-SVR method. The kernel function and associated parameters are optimized by EO-SVR with the initial chromosome randomly generated. Through always performing mutation on the worst component and its neighbors successively, the component in EO can evolve itself toward the global optimal solution generation by generation. The EO-SVR terminates when a predefined number of iterations is reached. Then, the optimal SVR kernel and parameter values can be finally determined.

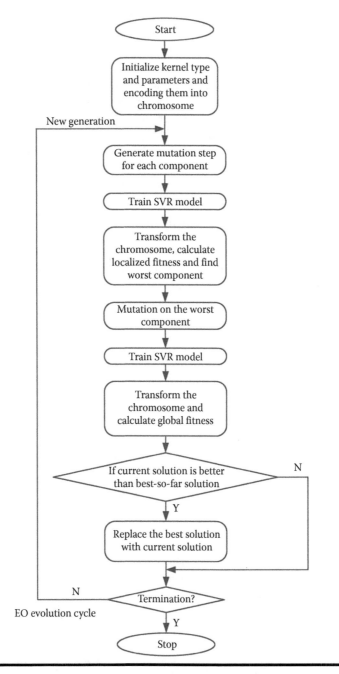

Figure 7.5 Optimization process of EO-SVR. (Reproduced from Chen, P. and Lu, Y. Z., *Journal of Zhejiang University, Science C* 12: 297–306, 2011b. With permission.)

7.3.4 Experimental Results

In this section, we consider five benchmark regression problems, which will cover the approximation of two single-variable functions and three multivariable functions. As mentioned above, we choose four commonly used SVR kernel types (linear, poly, MLP, RBF) listed in Table 7.2 as the candidate kernels. The programs are implemented in MATLAB and the experiments are carried out on a P4 E5200 (2.5) GHz PC with 2 GB RAM under WINXP platform.

For comparison, we showed the experimental results obtained by EO-SVR, RBF kernel with grid search and NN model, respectively. The index used for performance comparison is the MSE on test data set.

7.3.4.1 Approximation of Single-Variable Function

To verify the effectiveness of the proposed EO-SVR algorithm, we first consider two illustrative problems involving one input and one output. Figure 7.6 shows the original function examples and noisy learning examples (including the training and validation examples) for those two single-variable functions.

Case 1:
In this experiment, we approximate the following simple function (Chuang and Lee, 2009):

$$y = \frac{\sin x}{x}, \quad x \in [-10, 10] \tag{7.17}$$

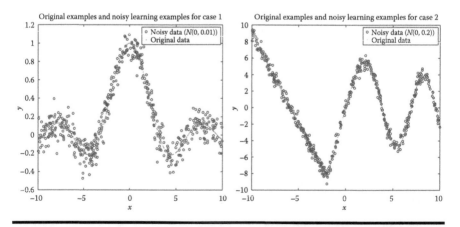

Figure 7.6 Original function examples ("·") and noisy learning examples ("o") for cases 1 and 2. (Reproduced from Chen, P. and Lu, Y. Z., *Journal of Zhejiang University, Science C* 12: 297–306, 2011b. With permission.)

We have 500 randomly generated examples, 200 of which are adopted as training examples, 200 as validation examples, and the balance 100 as testing examples. The learning samples are disturbed by additive noise ξ with zero mean and standard derivation δ, as described in the following equation:

$$\xi = N(0, \delta), \quad \delta \in \{0, 0.002, 0.005, 0.008, 0.01\} \tag{7.18}$$

Figure 7.7 illustrates the whole evolution process of MSE, the best-so-far result and the searching dynamics of kernel type during EO-SVR optimization with the noise level $\delta = 0.002$.

Figure 7.8 shows the predictions and the actual values of output y for test samples. As shown in Figure 7.8, the predictions by EO-SVR show good agreements with the actual outputs under different noise levels.

The performance comparisons between EO-SVR, RBF kernel with grid search, and NN model for case 1 are listed in Table 7.3. The best evolved model with optimal kernel function and associated parameters is highlighted in bold italic fonts. The comparisons imply that the predictive accuracy of the traditional SVR with grid search can be improved by simultaneous optimization of kernel type and parameters.

Case 2:
In this case, we consider a more complex example (Zhang et al., 2004):

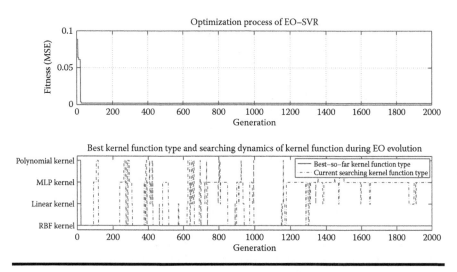

Figure 7.7 Evolution process, best kernel type, and searching dynamics during EO-SVR optimization (case 1, $\delta = 0.002$). (Reproduced from Chen, P. and Lu, Y. Z., *Journal of Zhejiang University, Science C* 12: 297–306, 2011b. With permission.)

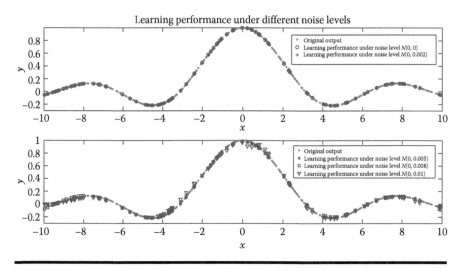

Figure 7.8 **EO-SVR-based predictions and actual values for test samples under different noise levels. (Reproduced from Chen, P. and Lu, Y. Z.,** *Journal of Zhejiang University, Science C* **12: 297–306, 2011b. With permission.)**

$$y = \begin{cases} -2.186x - 12.864 & -10 \le x < -2 \\ 4.246x & -2 \le x < 0 \\ 10e^{-0.05x-0.5} * \sin((0.03x + 0.7)x) & 0 \le x \le 10 \end{cases} \quad (7.19)$$

In this case, we randomly generate 500 examples, 200 of which are adopted as training examples, 200 as validation examples, and the balance 100 as testing examples. The learning samples are disturbed by an additive noise ξ with zero mean and standard derivation δ:

$$\xi = N(0, \delta), \quad \delta \in \{0, 0.02, 0.05, 0.1, 0.2\} \quad (7.20)$$

The improvement in generalization performance via EO-SVR is shown in Table 7.4. The best evolved model with optimal kernel function and associated parameters is highlighted in bold italic fonts. Compared with traditional methods, the prediction error can be further reduced by the proposed EO-SVR. Moreover, in many SVR applications, the RBF kernel function is supposed to have a better performance over other kernel functions; however, in this case, the MLP kernel with optimized parameters yields a lower prediction error and better generalization capability.

Table 7.3 Performance Comparisons between EO-SVR and Other Conventional Approaches (Case 1)

Noise Level	Method	Best Kernel	Parameter 1	Parameter 2	Parameter 3	MSE
$\delta = 0$	*EO-SVR*	*RBF*	*964.5230*	*0.2695*	–	*2.8266e – 09*
	RBF kernel with grid search	–	671.7264	0.2888	–	3.9785e – 09
	NN	–	–	–	–	2.3429e – 08
$\delta = 0.002$	*EO-SVR*	*RBF*	*998.8666*	*0.7968*	–	*5.8237e – 05*
	RBF kernel with grid search	–	262.8475	0.6111	–	6.1862e – 05
	NN	–	–	–	–	1.5466e – 04
$\delta = 0.005$	*EO-SVR*	*RBF*	*3.0704*	*0.1745*	–	*1.3436e – 04*
	RBF kernel with grid search	–	2.1483	0.1385	–	1.8660e – 04
	NN	–	–	–	–	4.4858e – 04
$\delta = 0.008$	*EO-SVR*	*RBF*	*8.2553*	*0.4723*	–	*2.1916e – 04*
	RBF kernel with grid search	–	4.9925	0.3983	–	2.3985e – 04
	NN	–	–	–	–	0.0010
$\delta = 0.01$	*EO-SVR*	*MLP*	*530.0677*	*3.9864*	*0.6656*	*8.6413e – 04*
	RBF kernel with grid search	–	4.8030	0.0960	–	8.9598e – 04
	NN	–	–	–	–	0.0017

Source: Reproduced from Chen, P. and Lu, Y. Z., *Journal of Zhejiang University, Science C* 12: 297–306, 2011b. With permission.

Note: – Denotes no parameter needed.

Table 7.4 Performance Comparisons between EO-SVR and Other Conventional Approaches (Case 2)

Noise Level	Method	Best Kernel	Parameter 1	Parameter 2	Parameter 3	MSE
$\delta = 0$	*EO-SVR*	*RBF*	*997.6446*	*0.0065*	–	*2.3383e − 04*
	RBF kernel with grid search	–	616.9396	0.0236	–	5.5786e − 04
	NN	–	–	–	–	0.0145
$\delta = 0.02$	*EO-SVR*	*MLP*	*1758.8*	*4.4702*	*2.0033*	*0.0034*
	RBF kernel with grid search	–	616.9396	0.0236	–	0.0042
	NN	–	–	–	–	0.0069
$\delta = 0.05$	*EO-SVR*	*MLP*	*1140.1*	*3.7743*	*1.6610*	*0.0045*
	RBF kernel with grid search	–	49.9279	0.0401	–	0.0061
	NN	–	–	–	–	0.0256
$\delta = 0.1$	*EO-SVR*	*MLP*	*1431.3*	*4.2647*	*1.8316*	*0.0106*
	RBF kernel with grid search	–	831.0181	0.0965	–	0.0124
	NN	–	–	–	–	0.0131
$\delta = 0.2$	*EO-SVR*	*MLP*	*1096.3*	*2.4390*	*1.5260*	*0.0216*
	RBF kernel with grid search	–	16.7501	0.1735	–	0.0304
	NN	–	–	–	–	0.0238

Source: Reproduced from Chen, P. and Lu, Y. Z., *Journal of Zhejiang University, Science C* 12: 297–306, 2011b. With permission.

Note: – Denotes no parameter needed.

7.3.4.2 *Approximation of Multivariable Function*

In this part, the effectiveness of EO-SVR is further validated on three benchmark functions involving multi-inputs and one output. Figure 7.9 shows the original function examples and noisy learning examples for cases 3 and 4.

Case 3:
In this case, a two-variable function is considered (Chuang and Lee, 2009):

$$y = \frac{\sin\left(\sqrt{x_1^2 + x_2^2}\right)}{\sqrt{x_1^2 + x_2^2}}, \quad -5 \le x_1, \quad x_2 \le 5 \tag{7.21}$$

We have 700 randomly generated examples, 400 of which are adopted as training examples, 200 as validation examples, and the balance 100 as testing examples. The learning samples are disturbed by an additive noise ξ with zero mean and standard derivation δ:

$$\xi = N(0, \delta), \quad \delta \in \{0, 0.002, 0.005, 0.008, 0.01\} \tag{7.22}$$

For the sake of comparison, the experimental results by EO-SVR, RBF kernel with grid search, and NN model for case 3 are summarized in Table 7.5. The best

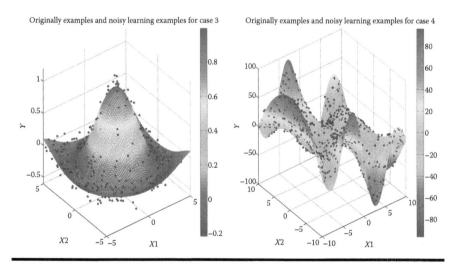

Figure 7.9 (See color insert.) Original function examples and noisy learning examples (★) for case 3 (noise level *N*(0, 0.01)) and case 4 (noise level *N*(0, 5)). (Reproduced from Chen, P. and Lu, Y. Z., *Journal of Zhejiang University, Science C* 12: 297–306, 2011b. With permission.)

Table 7.5 Performance Comparisons between EO-SVR and Other Conventional Approaches (Case 3)

Noise Level	Method	Best Kernel	Parameter 1	Parameter 2	Parameter 3	MSE
δ = 0	*EO-SVR*	*RBF*	*999.9703*	*0.9227*	–	*6.8732e − 08*
	RBF kernel with grid search	–	547.4999	0.8840	–	1.2564e − 07
	NN	–	–	–	–	1.4118e − 04
δ = 0.002	*EO-SVR*	*RBF*	*3.1471*	*1.0067*	–	*1.5906e − 04*
	RBF kernel with grid search	–	2.1308	0.7159	–	1.8197e − 04
	NN	–	–	–	–	4.3281e − 04
δ = 0.005	*EO-SVR*	*RBF*	*70.5141*	*2.5893*	–	*1.8091e − 04*
	RBF kernel with grid search	–	78.9656	2.7952	–	1.8725e − 04
	NN	–	–	–	–	2.6633e − 04
δ = 0.008	*EO-SVR*	*RBF*	*8.4046*	*1.4856*	–	*4.9897e − 04*
	RBF kernel with grid search	–	5.0120	1.2219	–	5.3518e − 04
	NN	–	–	–	–	0.0018
δ = 0.01	*EO-SVR*	*RBF*	*2.3983*	*1.2687*	–	*6.5814e − 04*
	RBF kernel with grid search	–	0.8969	0.7793	–	6.8268e − 04
	NN	–	–	–	–	0.0017

Source: Reproduced from Chen, P. and Lu, Y. Z., *Journal of Zhejiang University, Science C* 12: 297–306, 2011b. With permission.

Note: – Denotes no parameter needed.

evolved model with optimal kernel function and associated parameters is labeled in bold italic fonts.

Case 4:

This case is to consider a two-variable function as (Zhang et al., 2004):

$$y = (x_1^2 - x_2^2)\sin(0.5x_1), \quad -10 \le x_1 \le 10, \quad -10 \le x_2 \le 10 \qquad (7.23)$$

We have 700 randomly generated examples, 400 of which are adopted as training examples, 200 as validation examples, and the balance 100 as testing examples. The learning samples are disturbed by an additive noise ξ with zero mean and standard derivation δ:

$$\xi = N(0, \delta), \quad \delta \in \{0, 0.5, 1, 2, 5\} \qquad (7.24)$$

Table 7.6 shows the simulation results comparison between the proposed EO-SVR, RBF kernel with grid search, and NN model for case 4. The best evolved model with optimal kernel function and associated parameters is labeled in bold italic fonts. It can be seen that the prediction model, which is evolved by the proposed EO-SVM method, can always have satisfactory performances under different noise levels.

Case 5:

In this example, we approximate a widely used three-input nonlinear function (Qiao and Wang, 2008) to verify the effectiveness of the proposed EO-SVR:

$$y = (1 + x_1^{0.5} + x_2^{-1} + x_3^{-1.5})^2, \quad 1 \le x_1 \le 6, \quad 1 \le x_2 \le 6, \quad 1 \le x_3 \le 6 \qquad (7.25)$$

The training samples consist of 600 randomly generated data. Another 200 examples are independently generated as validation data set, and 100 examples as testing data set. The learning samples are disturbed by additive noise ξ with zero mean and standard derivation δ, as described in the following equation:

$$\xi = N(0, \delta), \quad \delta \in \{0, 0.02, 0.05, 0.08, 0.1\} \qquad (7.26)$$

Table 7.7 shows the forecasting accuracy on the testing data set with the proposed EO-SVR, RBF kernel with grid search, and NN model. The best evolved model with optimal kernel function and associated parameters is labeled in bold italic fonts. Obviously, the developed EO-SVR model yields more appropriate kernel and parameters, thus giving higher predictive accuracy and generalization capability.

Table 7.6 Performance Comparisons between EO-SVR and Other Conventional Approaches (Case 4)

Noise Level	Method	Best Kernel	Parameter 1	Parameter 2	Parameter 3	MSE
$\delta = 0$	*EO-SVR*	*RBF*	*999.9983*	*0.2437*	–	*0.0649*
	RBF kernel with grid search	–	557.4963	0.2478	–	0.0946
	NN	–	–	–	–	0.5883
$\delta = 0.5$	*EO-SVR*	*RBF*	*998.0115*	*0.5039*	–	*0.1722*
	RBF kernel with grid search	–	588.8470	0.4691	–	0.1875
	NN	–	–	–	–	0.5972
$\delta = 1$	*EO-SVR*	*RBF*	*998.0890*	*0.8163*	–	*0.3354*
	RBF kernel with grid search	–	434.7411	0.7442	–	0.3920
	NN	–	–	–	–	0.5844
$\delta = 2$	*EO-SVR*	*RBF*	*999.3974*	*0.8031*	–	*0.5222*
	RBF kernel with grid search	–	434.7411	0.7442	–	0.5471
	NN	–	–	–	–	1.1521
$\delta = 5$	*EO-SVR*	*RBF*	*879.9280*	*0.7545*	–	*0.9555*
	RBF kernel with grid search	–	215.5972	0.5907	–	1.0473
	NN	–	–	–	–	2.8953

Source: Reproduced from Chen, P. and Lu, Y. Z., *Journal of Zhejiang University, Science C* 12: 297–306, 2011b. With permission.

Note: – Denotes no parameter needed.

Table 7.7 Performance Comparisons between EO-SVR and Other Conventional Approaches (Case 5)

Noise Level	Method	Best Kernel	Parameter 1	Parameter 2	Parameter 3	MSE
$\delta = 0$	*EO-SVR*	*Polynomial*	*0.5282*	*8.7922*	*4.8795*	*1.0962e − 07*
	RBF kernel with grid search	–	559.3278	23.4082	–	2.5487e − 05
	NN	–	–	–	–	1.7727e − 07
$\delta = 0.02$	*EO-SVR*	*MLP*	*438.2326*	*0.1241*	*3.3632*	*4.4824e − 04*
	RBF kernel with grid search	–	650.2169	19.1805	–	0.0012
	NN	–	–	–	–	0.0018
$\delta = 0.05$	*EO-SVR*	*MLP*	*21.3813*	*0.0235*	*0.3436*	*8.1356e − 04*
	RBF kernel with grid search	–	24.6381	36.8418	–	0.0021
	NN	–	–	–	–	0.0064
$\delta = 0.08$	*EO-SVR*	*MLP*	*607.0087*	*0.0077*	*0.1461*	*0.0012*
	RBF kernel with grid search	–	13.4786	33.1361	–	0.0029
	NN	–	–	–	–	0.0044
$\delta = 0.1$	*EO-SVR*	*MLP*	*31.4823*	*0.0396*	*2.2208*	*0.0015*
	RBF kernel with grid search	–	32.6444	38.0491	–	0.0028
	NN	–	–	–	–	0.0121

Source: Reproduced from Chen, P. and Lu, Y. Z., *Journal of Zhejiang University, Science C* 12: 297–306, 2011b. With permission.

Note: – Denotes no parameter needed.

In this section, we demonstrate the effectiveness of the proposed EO-SVR through experiments on five typical benchmark regression problems at different noise levels. As shown in the experimental results, the optimal kernel and parameters vary with the problems to be solved and the noise level, the proposed EO-SVR can successfully determine the optimal SVR kernel type and associated parameter values consistently for all the five experimental cases, leads to a better generalization performance, and a lower forecasting error. On the other side, it is also shown in our experimental results that the most frequently used SVR kernel function, the RBF kernel, may not be the best choice for nonlinear regression in some cases: the MLP kernel can offer a better performance if the associated parameters are well tuned.

7.4 Nonlinear Model Predictive Control with MA-EO

MPC denotes a broad range of control strategies that use a process model to predict and optimize the effect of future control actions on controlled variables over a finite horizon. The MPC has become one of the most popular APC solutions for its ability to control multivariable systems in the presence of constraints. The application areas have extended from chemical, petroleum refinery, pulp and paper, and metal to automotive, aerospace, power industries, etc. (Qin and Badgwell, 2003).

Since the MPC is a model-based controller, a reasonably accurate process model over the whole prediction horizon is essentially required for the controller to make effective control decisions. In industrial applications, the MPC normally relies on linear dynamic models, even though most processes are nonlinear (Al Seyab and Cao, 2006). Therefore, it is generally referenced as "linear model predictive control (LMPC)." The assumption of process linearity greatly simplifies model development and controller design. However, LMPC is inadequate for processes which are highly nonlinear or have large operating regimes. This has led to the development of nonlinear model predictive control (NMPC), in which a more accurate nonlinear model is used for prediction and optimization (Henson, 1998).

While NMPC offers the potential for improved performance, it brings both theoretical and practical problems which are considerably more complicated than LMPC. First, the development of an accurate nonlinear model may be difficult and there is not a general model form that is clearly suitable to represent the nonlinear process, either from input/output data correlation or first principles derived from well-known mass and energy conservation laws. Second, besides the difficulties associated with nonlinear modeling, the future manipulation variables have to be obtained in a prescribed control interval; thus, the computational cost to find optimal (or near-optimal) control actions online is another critical issue. According to Camacho and Bordons (1995), the practical usefulness of NMPC is hampered by the unavailability of suitable optimization tools. From a theoretical point of view, the usage of nonlinear models changes the online optimization from convex quadratic programming (QP) problems to nonconvex nonlinear programming (NLP) problems, which gives rise to computational difficulties related to the expense and reliability of solving the NLP problems online. The classical analytical and numerical optimization techniques are very sensitive to the initial conditions and usually lead to unacceptable solutions due to the convergence to local optima (Onnen et al., 1997; Venkateswarlu and Reddy, 2008), or are not even able to guarantee a feasible solution because of the complexity of optimization problems.

Consequently, new optimization techniques have been proposed and investigated by researchers for NMPC solution development in recent years. Among them, methods of CI (e.g., EAs, EO, and ACO) (Engelbrecht, 2007) have shown notable suitability for constrained nonlinear optimization problems, and are able to avoid local minima inherently. Therefore, combinations of NMPC and various computation intelligence methods as online optimizers have been successfully applied to several typical nonlinear processes (Onnen et al., 1997; Martinez et al., 1998; Potocnik and Grabec, 2002; Song et al., 2007; Venkateswarlu and Reddy, 2008). However, CI methods have the weakness of slow convergence and are unable to provide precise solutions because of the failure to exploit local information (Tahk et al., 2007). Moreover, for the constrained optimization problems involving a number of constraints which the optimal solution must satisfy, CI methods often lack an explicit mechanism to bias the search in constrained search space (Runarsson and Yao, 2000; Zhang et al., 2008).

In NMPC, the system performance is greatly dependent upon predictive model accuracy and online optimization algorithm efficiency. This section introduces a novel NMPC with the integration of SVM and a hybrid optimization algorithm called "EO-SQP," which combines recently proposed heuristic EO and deterministic sequential quadratic programming (SQP) under the framework of MA. Inheriting the advantages of the two approaches, the proposed EO-SQP algorithm is capable of solving nonlinear programming (NLP) problems effectively. Furthermore, the hybrid EO-SQP algorithm is employed as the online solver of a multistep-ahead SVM model-based NMPC. Simulation results on a nonlinear multi-input multi-output (MIMO) continuous stirred tank reactor (CSTR) reactor show considerable performance improvement obtained by the proposed NMPC. This work focused on the development of a novel NMPC framework, which combines a multistep-ahead SVM predictive model with an EO-SQP-based online optimizer.

7.4.1 Problem Formulation for NMPC Based on SVM Model

Based on statistical learning theory and the structural risk minimization principle, SVM was first proposed by Vapnik to solve pattern recognition problems in 1995 (Vapnik, 1995). In the last decade, SVM-based algorithms have been rapidly developed and applied to many areas, such as classification and regression. Global optimal solution and higher generalization capability constitute the major advantages of SVM over other regression techniques. Consequently, in this study, SVM is employed to build an accurate input/output predictive model.

Discrete-time models are most appropriate for engineering applications because plant data (learning samples) are available at discrete-time instants; thus, NMPC is most naturally formulated in discrete time. The model form considered in this study is a nonlinear autoregressive moving average model with exogenous inputs (NARMAX):

$$\hat{y}(k) = g[y(k-1), \ldots, y(k-n_y), u(k-1), \ldots, u(k-n_u)] \tag{7.27}$$

Here, k refers to the sampling time point; y and u are the output and input variables; and n_y and n_u refer to input and output lags, respectively.

The primary purpose of MPC is to deal with complex dynamics over an extended horizon. Thus, based on the currently measured system outputs and future inputs, the model in MPC must predict the future outputs over a predefined prediction horizon. However, the NARMAX model described in Equation 7.27 can only provide a one-step-ahead prediction. If a wider prediction horizon is required, this problem can be handled by a procedure known as multistep-ahead prediction: the input vector's components, previously composed of actual sample points of the time series, are gradually replaced by predicted values. By feeding back the model outputs, the one-step-ahead predictive model can be recurrently cascaded to itself to generate future predictions for process output, as described below:

$$\hat{y}_{k+i} = \hat{y}(k+i\,|\,k)$$

$$= \begin{cases} e(k+i\,|\,k) + g[\hat{y}(k+i-1), \ldots, \hat{y}(k+i-n_y), u(k+i-1), \ldots, u(k+i-n_u)] \\ \quad n_y \le i \\[6pt] e(k+i\,|\,k) + g[\hat{y}(k+i-1), \ldots, y(k+i-n_y), u(k+i-1), \ldots, u(k+i-n_u)] \\ \quad 1 < i < n_y \\[6pt] e(k+i\,|\,k) + g[y(k+i-1), \ldots, y(k+i-n_y), u(k+i-1), \ldots, u(k+i-n_u)] \\ \quad i = 1 \end{cases}$$

$$\tag{7.28}$$

where $e(k+i\,|\,k)$ represents the model uncertainty and unmeasured process disturbances at time step $(k+i)$.

In this study, the RBF is adopted as the SVM kernel function. The dynamic modeling task in NMPC can then be formulated into the problem of finding an SVM model that can be used to predict the future trajectory of the plant over a prediction horizon:

$$\hat{y}_{k+i} = \sum_{j=1}^{\#SV} \alpha_j \exp\left(-\frac{d_{j,k+i}}{2\sigma^2}\right) + b + e(k+i\,|\,k), \quad i = 0, 1, 2, \ldots, P-1 \tag{7.29}$$

The #*SV* in Equation 7.29 represents the support vectors. P is the prediction horizon in NMPC.

The $d_{j,k+i}$ in Equation 7.29 can be further described in the following equation:

$$d_{j,k+i} = \sum_{n=0}^{\min(i,n_y)} (x_{j,n_u+n+1} - \hat{y}_{k+i-n})^2 + \sum_{n=i+1}^{n_y} (x_{j,n_u+n+1} - y_{k+i-n})^2$$

$$+ \sum_{n=0}^{n_u} \begin{cases} (x_{j,n+1} - u_{k+i-n})^2, & i - n_u < n \\ (x_{j,n+1} - u_{k+n_u})^2, & i - n_u > n \end{cases} \tag{7.30}$$

The NMPC based on SVM can then be formulated into an online constrained optimization problem as follows:

$$\min_{\overline{U}(k)} J(\overline{U}(k), \overline{Y}(k))$$

subject to:

$$\overline{U}(k) = [u_k, u_{k+1|k}, \ldots, u_{k+i|k}]$$

$$\overline{Y}(k) = [y_k, y_{k+1|k}, \ldots, y_{k+i|k}]$$

$$\hat{y}_{k+i|k} = g(U_{k+i}, Y_{k+i-1})$$

$$i \in [0, P-1],$$

$$\text{if } i > M-1, u_{k+i|k} = u_{k+M-1|k} \tag{7.31}$$

and

$$u_{\min} \le u_k \le u_{\max}$$

$$\Delta u_{\min} \le \Delta u_k \le \Delta u_{\max}$$

$$y_{\min} \le y_k \le y_{\max}$$

where $g(\cdot)$ in Equation 7.31 represents the process model, as indicated in Equations 7.27 and 7.28. P and M are prediction horizon and control horizon with $M \le P$.

The cost function J in Equation 7.31 is usually defined in NMPC as an objective function which minimizes the differences between model predictions and the desired trajectory over prediction horizon and control energy over control horizon:

$$J = \sum_{i=1}^{P} \|y_{k+i} - r\|_{Q_i} + \sum_{j=1}^{M-1} \| \Delta u_{k+j}\|_{R_j} \qquad (7.32)$$

where r represents the desired set point. The relative importance of the objective function contributions is controlled by setting the weighting matrices Q_i and R_j.

7.4.2 Real-Time NMPC with SVM and EO-SQP

The NMPC formulation described in Section 7.4.1 requires a nonconvex NLP problem to be solved online at each time step. Based on the problem described by Equations 7.27 through 7.32, this section formulates a novel NMPC with the integration of multistep-ahead SVM model and real-time EO-SQP-based optimizer. The structure of the proposed NMPC workflow is demonstrated in Figure 7.10. The components of the framework include a nonlinear process, a predictive model, and an online optimizer. The optimizer operates based upon the predictive model, the cost function, and process constraints.

According to "No-Free-Lunch" theorem by Wolpert and Macready (1997), a search algorithm strictly performs in accordance with the amount and quality of the problem knowledge it incorporates. Consequently, an optimization algorithm only unfolds its full capabilities if it is designed properly for a particular application. Therefore, based on the hybrid EO-SQP algorithm in Chapter 5, some practical issues must be considered to deal with the specific properties of the optimization in NMPC. Those issues include the workflow of the proposed NMPC, encoding strategy, selection of the initial population, termination criteria, and mutation strategy.

7.4.2.1 Workflow of Proposed NMPC

As shown in Figure 7.10, the structure of the proposed NMPC is divided into two phases: *off-line modeling* and *online optimization*.

7.4.2.1.1 Off-line modeling

The quality of an empirical inputs/outputs model depends strongly on the input sequence (learning examples) used for model identification since it determines the process response and the model-relevant information contained in this response. A "good" input sequence means one whose amplitude is large enough to elicit significant information from process nonlinearities (Pearson, 2006). In this study, random step input sequences described in Equation 7.33 is used to generate the learning examples which contain informative process responses.

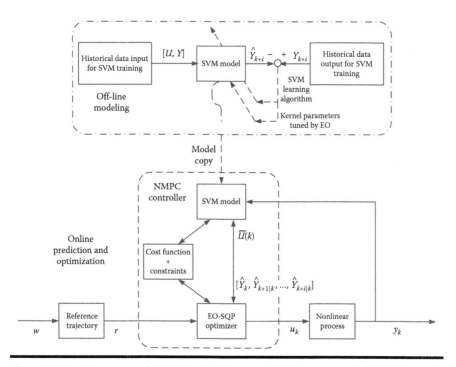

Figure 7.10 Structure of NMPC with integration of SVM model and EO-SQP optimization. (Adapted from Chen, P. and Lu, Y. Z. 2010b. Nonlinear model predictive control with the integration of support vector machine and extremal optimization. *Proceedings of the 8th World Congress on Intelligent Control and Automation,* **pp. 3167–3172.)**

$$u_i(k) = \begin{cases} z_i(k), & \text{with probability } p_s \\ u_i(k-1), & \text{with probability } 1 - p_s \end{cases}, k > 1 \qquad (7.33)$$

$$u_i(1) = z_i(1)$$

where $z_i(k)$ is a random step input sequence with any desired distribution (e.g., uniform, Gaussian) and $z_i(k)$ are fed into the plant to obtain the open-loop inputs/ outputs time sequence $y(k)$, $u(k)$.

Min and Lee (2005) stated that the optimal parameter search on SVM plays a crucial role in building a prediction model with high accuracy and stability. Therefore, to construct an efficient SVM model with RBF kernel, two extra parameters, sigma and gamma, have to be carefully tuned. This study proposes an EO-based SVM multistep predictive model, namely, EO-SVM, in which EO is adopted to search for the optimal values of sigma and gamma. According to the multistep-ahead prediction procedure described in Section 7.4.1, the generalization ability of the SVM model is evaluated in terms of the multistep-ahead MSE in parameter tuning process, defined as follows:

$$MSE_k = \sum_{i=1}^{P-1} e_{k+i}{}^2 = \sum_{i=1}^{P-1} (\hat{y}_{k+i} - y_{k+i})^2 \tag{7.34}$$

where $\hat{y}_{k+i}$ represents model prediction trajectory over the whole prediction horizon as defined in Equation 7.28. The tuning process of proposed EO-SVM is illustrated in Figure 7.11.

7.4.2.1.2 Online optimization

Generally, MPC is formulated as online optimization of a finite moving (receding) horizon open-loop optimal control problem subject to system dynamics and inputs/ states constraints, as presented in Figure 7.10 and Equation 7.31. The EO-SQP algorithm for NMPC online optimization works as follows:

1. At time step k, simultaneously obtain measurements of process inputs and outputs and initialize the EO-SQP optimizer.
2. Start with the previous calculated control input sequence, use EO-SQP to compute an updated optimal input sequence by minimizing a given cost function described in Equation 7.31 over prediction horizon based on the predictive model.
3. Repeat step 2 until the termination criteria of the optimization are satisfied.
4. Only the first input vector of the optimal control sequence is applied to control the process.
5. Repeat the entire procedure for each time step.

Model uncertainty and unmeasured process disturbances are handled by calculating an unchanged disturbance defined as the difference between the actual output value $y(k)$ and the predicted system output $\hat{y}(k)$ at the current time step, as described in Equation 7.28. It is assumed that the value of $e(k + i \mid k)$ will remain unchanged during the prediction horizon.

7.4.2.2 Encoding Strategy

In EO communities, there are several encoding strategies such as binary encoding strategy and real encoding strategy. The former uses a string of bits and imposes a precision to the variables. On the contrary, real coded EO works directly with real-valued variables, encoding solutions as a real vector, which can reduce calculation complexity, increase the solution precision and the operation speed without encoding/decoding error. Consequently, the real coded EO is adopted in this study.

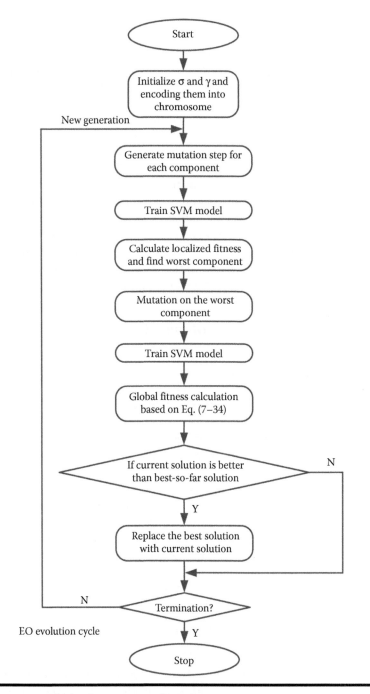

Figure 7.11 Optimization process of EO-SVM.

7.4.2.3 Selection of the Initial Population

The computational cost is a critical issue for NMPC online optimization. In order to provide fast convergence, suitable initialization procedure should be specified at the beginning of each NMPC time step. If no additional information is available, a random initialization is often used and the convergence may be slower. Fortunately, the receding horizon principle is adopted in NMPC, which implies that a time series of controller outputs have to be calculated at each time step and only the first control action in the sequence will be implemented. Consequently, the optimization can be accelerated by initializing the solution based on the interevolution steady-state principle (ISS). By doing so, the optimal solution of the previous evolution is recalled, shifted by one position, and reintroduced to the initialization of current evolution, with a randomly initialized value at the end of control horizon. The shifted solution might be a very good guess for the solution of the next optimization problem, especially if the system is close to the desired steady state. Heuristic creation of the initial solution based on the exploitation of previously accumulated knowledge results in a significant reduction in computational time.

7.4.2.4 Termination Criteria of the NLP Solver

In order to deal with real-time constraints in NMPC application, the termination criteria are proposed to stop the optimization loop when a predefined level of optimality is reached. This study introduces a hybrid convergence measurement to terminate the NLP solver in order to compromise between solution quality and computational time. Three different indexes are investigated and combined to serve as the criteria to terminate the optimization.

Convergence rate ($C1$)—the rate of convergence of the best-so-far solution fitness over the entire evolution process. If the fitness of best-so-far solution is unchanged for a given number of generations, this criterion is active.

Convergence rate of the first gene ($C2$)—this condition is quite similar to the previous one. The criterion is active once the first gene is unchanged for N generations.

Optimization time limit ($C3$)—a predefined parameter to restrict the time that NLP solver spends on the optimization of the optimum controller outputs series. The timeout is activated if the optimization timeout limit has been exceeded.

The EO-SQP-based NLP solver will stop if ($C1 \cap C2$) $\cup$ $C3$ is true.

7.4.2.5 Horizon-Based EO Mutation for NMPC Online Optimization

The conventional EO algorithm updates the solution by a "component-based mutation strategy," which begins with sequentially adding a randomly generated number $z \in [lb_j, ub_j]$ to every decision variable to get neighborhood chromosomes. However, the component-based mutation strategy is not suitable for the online

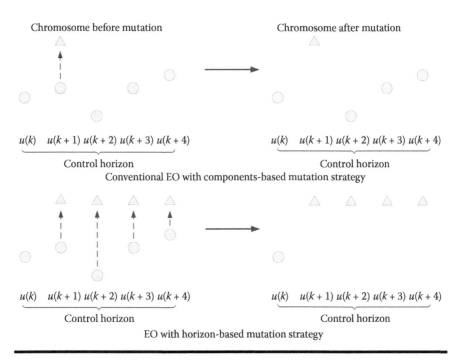

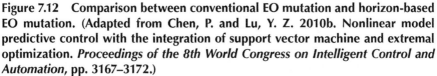

Figure 7.12 **Comparison between conventional EO mutation and horizon-based EO mutation. (Adapted from Chen, P. and Lu, Y. Z. 2010b. Nonlinear model predictive control with the integration of support vector machine and extremal optimization.** *Proceedings of the 8th World Congress on Intelligent Control and Automation,* **pp. 3167–3172.)**

optimization of NMPC, due to the fact that the system dynamics depend heavily on the input and the states series, as described in Equation 7.28. Thus, the mutation on a single component (controller output at a single time instant) does not help much to improve the system performance. For this reason, this study proposes a novel "horizon-based EO mutation" which is specially designed for NMPC. The horizon-based mutation updates all the components within a flexible window, whose width is decided by the position of the currently changed decision variable and control horizon P. For example, the decision variables between the current mutation position and the position at the end of the control horizon will update simultaneously, as described in Figure 7.12. This strategy will help to reduce the unnecessary variations of optimized control variables and consequently suppress the oscillation of the system outputs.

7.4.3 Simulation Studies

To demonstrate the effectiveness of the proposed NMPC, the nonlinear system considered here is an isothermal CSTR with first-order reaction $A + B \rightarrow P$, in the presence of excess concentration of A. The corresponding MIMO CSTR model is described as follows (Martinsen et al., 2004):

$$\begin{cases} \dfrac{dx_1}{dt} = u_1 + u_2 - k_1\sqrt{x_1} \\[2mm] \dfrac{dx_2}{dt} = (C_{B1} - x_2)\dfrac{u_1}{x_1} + (C_{B2} - x_2)\dfrac{u_2}{x_1} - \dfrac{k_2 x_2}{(1+x_2)^2} \end{cases} \tag{7.35}$$

$$y = \begin{pmatrix} 1 & 0 \\ 0 & 1 \end{pmatrix}[x_1\ x_2]^T$$

$$s.t. \quad \begin{cases} 0 \le u_1 \le 2 \\ 0 \le u_2 \le 2 \end{cases} \tag{7.36}$$

where $k_1 = 0.2$, $k_2 = 1$, $C_{B1} = 24.9$, and $C_{B2} = 0.1$. The plant has two inputs and two outputs and possesses strong nonlinearity.

Here, the NARMAX structure is used to model the nonlinear dynamical system in Equation 7.35 and 7.36, it can be mathematically represented as follows:

$$\hat{y}_i(k) = g(y_1(k-1), y_2(k-1), u_1(k-1), u_2(k-1)) \quad i = 1,2 \tag{7.37}$$

In this case, two separated SVM models are employed to build the nonlinear mapping $g(\cdot)$ between inputs and outputs for y_1 and y_2, respectively. Based on Equation 7.33, 500 observation pairs are generated and 400 of them are used as learning data, the other 100 are used as test data set. Since the generalization capability of the SVM model greatly depends on some essential parameters, the SVM parameters are tuned by EO to guarantee multistep-ahead-prediction accuracy of the predictive model, as described in Section 7.4.1 and Equation 7.34. The generalization capability of the optimized SVM predictive model is shown in Figure 7.13. It can be seen that the long-term recursive prediction by EO-SVM shows a good agreement with actual output.

The trained EO-SVM models, which show accurate predictions and good generalization properties over the whole prediction horizon, will be employed as the internal prediction model of the NMPC.

The effectiveness of the proposed NMPC framework is tested by simulating various control problem scenarios involving set-point tracking (cases 1–4 for Y1 and cases 1, 3 for Y2) and disturbance rejection (cases 2, 4 for Y2), and compared with the performance of a LMPC controller, which piecewise linearizes the plant model at each optimization loop and chooses QP algorithm as online optimizer. The simulation starts from $(y1(0), y2(0)) = (100, 6.637)$ with input constraints

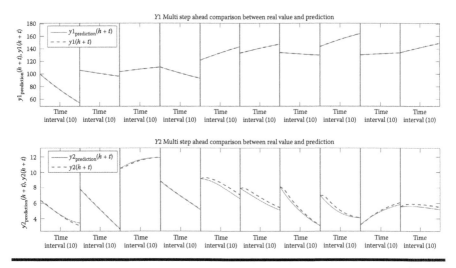

Figure 7.13 **Long-term recursive prediction (10 steps ahead) of the nonlinear process by EO-SVM model, starting from various initial states.**

$(0 \leq u1 \leq 2, \; 0 \leq u2 \leq 2)$, control horizon $M = 2$, prediction horizon $P = 8$. The programs are implemented in MATLAB and the experiments are carried out on a P4 E5200 (2.5) GHz PC with 2 GB RAM under WINXP platform. The times correspond to a complete simulation period of 3600 s with a sample time of 3 s. The results of these simulations are shown in Figures 7.14 and 7.15.

Table 7.8 shows the quantitative comparison of the proposed NMPC and LMPC under set-point tracking cases and disturbance rejection (regulation) cases as shown in Figures 7.14 and 7.15, based on three popular performance indexes used in control engineering: overshooting, settling time, and integral of time and absolute error (ITAE) specified in the following equation:

$$ITAE = \int_{0}^{n} t|(e(t))| \, dt \qquad (7.38)$$

As shown in Figures 7.14 and 7.15 and Table 7.8, EO-SQP-based NMPC performs better than LMPC in most cases, as it settles faster, has smaller overshooting, and lower ITAE values. These results are expected, considering that the long-term accurate predictions of the EO-SVM model and the global search capability of the online EO-SQP optimizer will result in more effective and efficient control actions. Moreover, Figure 7.15 shows that the manipulated variables of the LMPC controller tend to oscillate, while their NMPC counterparts stay close to the reference values, due to the fact that the NMPC controller can handle nonlinear process dynamics. Although the proposed EO-SQP-based NMPC requires more computational

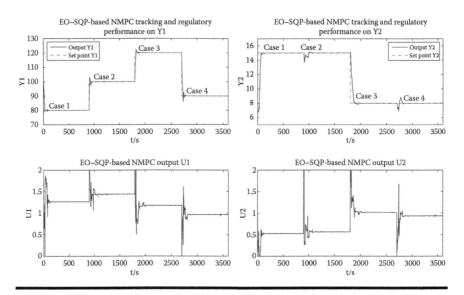

Figure 7.14 Performance of NMPC for tracking and disturbance rejection. (Chen, P. and Lu, Y. Z. Memetic algorithms based real-time optimization for nonlinear model predictive control. *International Conference on System Science and Engineering*, Macau, China, pp. 119–124. © 2011 IEEE.)

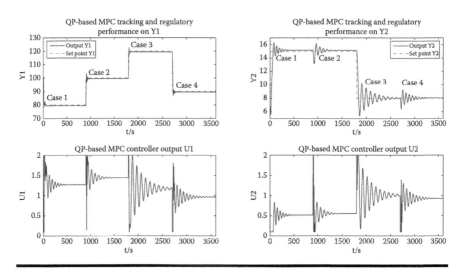

Figure 7.15 Performance of LMPC for tracking and disturbance rejection. (Chen, P. and Lu, Y. Z. Memetic algorithms based real-time optimization for nonlinear model predictive control. *International Conference on System Science and Engineering*, Macau, China, pp. 119–124. © 2011 IEEE.)

Table 7.8 Quantitative Comparison of NMPC and LMPC under Tracking and Regulation Cases

Control Type	Optimization Time (s)	Variable	Performance Index	Case 1	Case 2	Case 3	Case 4
NMPC	1.131	Y1		(Tracking)	(Tracking)	(Tracking)	(Tracking)
			Overshooting (%)	4.9%	14.6%	11.8%	13.1%
			Settling time (s)	24	42	42	42
			ITAE	4.499×10^3	5.031×10^3	5.792×10^3	7.021×10^3
		Y2		(Tracking)	(Regulation)	(Tracking)	(Regulation)
			Overshooting (%)	0.5	–	1.8	–
			Settling time (s)	60	18	63	87
			ITAE	8.548×10^3	4.476×10^3	7.289×10^3	4.532×10^3
LMPC	0.0032	Y1		(Tracking)	(Tracking)	(Tracking)	(Tracking)
			Overshooting (%)	6.4%	16%	12.9%	10.6%
			Settling time (s)	36	33	30	33
			ITAE	1.841×10^4	2.054×10^4	2.245×10^4	1.959×10^4
		Y2		(Tracking)	(Regulation)	(Tracking)	(Regulation)
			Overshooting (%)	15.2	–	37.2	–
			Settling time (s)	204	72	552	255
			ITAE	7.63×10^4	6.273×10^4	10.698×10^4	3.993×10^4

Source: Chen, P. and Lu, Y. Z. Memetic algorithms based real-time optimization for nonlinear model predictive control. *International Conference on System Science and Engineering*, Macau, China, pp. 119–124. © 2011 IEEE.

Note: – Denotes no statistical content.

time than the traditional LMPC, it is acceptable in this case for the reason that the average NMPC optimization time (1.131 s) is much smaller than the sample time (3 s). The dramatic improvement in tracking and regulatory performance makes it appealing and promising for processes with high nonlinearity.

7.5 Intelligent PID Control with Binary-Coded EO

It has been widely recognized that PID control is still one of the simplest but most efficient control strategies for many real-world control problems (Ang et al., 2005; Hsu et al., 2007; Ye, 2008; Chaiyatham and Ngamroo, 2012), although a variety of advances have been gained in control theories and practices. How to design and tune an effective and efficient single-variable and especially multivariable PID controller to obtain high-quality performances such as high stability and satisfied transient response is of great theoretical and practical significance. This issue has attracted considerable attention of some researchers using EAs (Iruthayarajan and Baskar, 2009; Meza et al., 2012), such as GA (Chang, 2007; Zhang et al., 2009), particle swarm optimization (PSO) (Kim et al., 2008; Menhas et al., 2012; Djoewahir et al., 2013; Li et al., 2014), differential evolution (DE) (Coelho and Pessôa, 2011; Davendra et al., 2010), and MOO algorithms (Zhao et al., 2011; Meza et al., 2012). However, the issue of designing and tuning PID controllers efficiently and adaptively is still open. To the best of our knowledge, there is only some reported research concerning the design of PID controllers based on EO. In Huo et al. (2007), an improved generalized EO algorithm is proposed for designing a two-degree-of-freedom PID regulator. As a consequence, this section focuses on addressing this issue by adopting another novel optimization algorithm called binary-coded extremal optimization (BCEO) (Zeng et al., 2014) in an attempt to obtain better performances.

7.5.1 PID Controllers and Performance Indices

A standard control system with a PID controller $D(s)$ and controlled plant $G(s)$ is shown as Figure 7.16. Let us consider first the simplest case, single-input and single-output control system. The transfer function $D(s)$ of a standard single-variable PID controller (Ang et al., 2005) is generally expressed in the following form:

$$D(s) = K_P\left(1 + \frac{1}{T_I s} + T_D s\right) = K_P + K_I \frac{1}{s} + K_D s \qquad (7.39)$$

where T_I and T_D are the integral time constant and derivative time constant, respectively, K_P, K_I, and K_D are proportional gain, integral gain, and derivative gain, respectively, $K_I = K_P/T_I$, and $K_D = K_P T_D$.

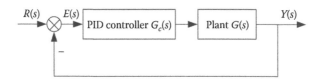

Figure 7.16 A control system with PID controller. (Reprinted from *Neurocomputing*, **138, Zeng, G. Q. et al., Binary-coded extremal optimization for the design of PID controllers, 180–188, Copyright 2014, with permission from Elsevier.)**

The output $U(s)$ of PID controller is described as follows:

$$U(s) = D(s)E(s) = K_P \left(1 + \frac{1}{T_I s} + T_D s\right) E(s) = K_P E(s) + K_I \frac{1}{s} E(s) + K_D s E(s)$$

$$(7.40)$$

where $E(s)$ is the transfer function of the system error $e(t)$. Furthermore, the continuous-time form of $U(s)$ is also written as the following equation:

$$u(t) = K_P e(t) + K_I \int_0^t e(t)\,dt + K_D \frac{de(t)}{dt} \qquad (7.41)$$

The discrete PID controller is described as follows:

$$u(k) = K_P e(k) + K_I T_s \sum_{j=0}^{k} e(j) + \frac{K_D}{T_s}[e(k) - e(k-1)] \qquad (7.42)$$

where T_s is the sampling time.

Then, consider more complex case, an $n \times n$ multivariable plant $G(s)$ in Figure 7.16, which is given as follows:

$$G(s) = \begin{bmatrix} g_{11}(s) & \cdots & g_{1n}(s) \\ \vdots & \ddots & \vdots \\ g_{n1}(s) & \cdots & g_{nn}(s) \end{bmatrix} \qquad (7.43)$$

The corresponding $n \times n$ multivariable PID controller is given as follows:

$$D(s) = \begin{bmatrix} d_{11}(s) & \cdots & d_{1n}(s) \\ \vdots & \ddots & \vdots \\ d_{n1}(s) & \cdots & d_{nn}(s) \end{bmatrix} \qquad (7.44)$$

where the form of $d_{ij}(s)$ is characterized as the following equation:

$$d_{ij}(s) = K_{Pij} + K_{Iij}\frac{1}{s} + K_{Dij}s, \quad \forall i, j \in \{1, 2, ..., n\} \tag{7.45}$$

In most of previous research work, the integral of absolute error (IAE) and the integral of time-weighted absolute error (ITAE) are generally used as the indexes measuring the performances of PID controllers (Ang et al., 2005). However, these indexes are not still sufficient to evaluate the control performances comprehensively (Zhang et al., 2009). Here, another much more reasonable performance index is presented by considering the following additional factors. The first one is the introduction of the square of the controllers' output, that is, $\int_0^\infty w_2 u^2(t)\,dt$ in order to avoid exporting a large control value. Second, the rising time $w_3 t_u$ is used to evaluate the rapidity of the step response of a control system. The third one $\int_0^\infty w_4|\Delta y(t)|\,dt$ is added to avoid a large overshoot value.

The objective function (also called fitness) evaluating the control performance of a single-variable PID controller is defined as follows (Zhang et al., 2009):

$$\min J = \min \begin{cases} \displaystyle\int_0^\infty (w_1|e(t)| + w_2 u^2(t))dt + w_3 t_u, & \text{if } \Delta y(t) \geq 0 \\[4mm] \displaystyle\int_0^\infty (w_1|e(t)| + w_2 u^2(t) + w_4|\Delta y(t)|)dt + w_3 t_u, & \text{if } \Delta y(t) < 0 \end{cases} \tag{7.46}$$

where $e(t)$ is the system error, $\Delta y(t) = y(t) - y(t - \Delta t)$, $u(t)$ is the control output at the time t, t_u is the rising time, w_1, w_2, w_3, w_4 are weight coefficients, and $w_4 \gg w_1$.

The objective function that evaluates the control performance of a multivariable PID controller is defined as follows:

$$\min J =$$

$$\min \begin{cases} \displaystyle\int_0^\infty \left(w_1 \sum_{i=1}^n |e_i(t)| + w_2 \sum_i^n u_i^2(t) \right) dt + w_3 \sum_{i=1}^n t_{ui}, & \text{if } \Delta y_i(t) \geq 0 \\[6mm] \displaystyle\int_0^\infty \left(w_1 \sum_{i=1}^n |e_i(t)| + w_2 \sum_i^n u_i^2(t) + w_4 \sum_{i=1}^n |\Delta y_i(t)| \right) \\[4mm] \qquad dt + w_3 \sum_{i=1}^n t_{ui}, & \text{if } \Delta y_i(t) < 0 \quad (7.47) \end{cases}$$

where $e_i(t)$ is the ith system error, $\Delta y_i(t) = y_i(t) - y_i(t - \Delta t)$, $u_i(t)$ is the ith control output at the time t, t_{ui} is the rising time of the ith system output y_i, w_1, w_2, w_3, w_4 are weight coefficients, and $w_4 \gg w_1$.

7.5.2 BCEO Algorithm

The fitness of any configuration S, where S is a binary string illustrated in Figure 7.17 with length $L = \sum_{j=1}^{M} l_j$ that encodes M PID design real parameters $\{K_{Pi}, K_{Ii}, K_{Di}, i = 1, 2, ..., n, M = 3n\}$ and the length l_j of jth binary substring encodes the jth parameter for an n-variable PID controller applied in a multivariable control system, is used to evaluate its effect on the control performances and defined as follows (Zeng et al., 2014):

$$C(S) = \sum_{i=1}^{n} \left(\sum_{\Delta y_i(k) \geq 0} (w_1 |e_i(k)| + w_2 u_i^2(k)) \right) + \sum_{\Delta y_i(k) < 0} (w_4 |\Delta y_i(k)|) + w_3 t_{ui}),$$

$$(k = 1, 2, ..., N) \tag{7.48}$$

$$e_i(k) = r_i(k) - y_i(k), \quad i = 1, 2, ..., n \tag{7.49}$$

$$u_i(k) = K_{Pi} e_i(k) + K_{Ii} T_s \sum_{j=0}^{k} e_i(j) + \frac{K_{Di}}{T_s} [e_i(k) - e_i(k-1)], \quad i = 1, 2, ..., n \tag{7.50}$$

$$\Delta y_i(k) = y_i(k) - y_i(k-1), \quad i = 1, 2, ..., n \tag{7.51}$$

$$t_{ui} = \min\{k\}^* T_s, \quad i = 1, 2, ..., n, \quad \text{if } 0.95 \leq y_i(k) \leq 1.05 \tag{7.52}$$

where T_s is predefined sampling period, N is the time number of sampling, $e_i(k)$ and $u_i(k)$ represent the system error and the control output for a specific ith PID controller at the kth sampling time, t_{ui} is the rising time of the system output y_i

l_1	l_2	l_3		l_{M-2}	l_{M-1}	l_M
1011...1101	1011...1101	1011...1101	...	1011...1101	1011...1101	1011...1101
K_{P1}	K_{I1}	K_{D1}	...	K_{Pn}	K_{In}	K_{Dn}

Figure 7.17 Binary encoding of multivariable PID real parameters. (Reprinted from *Neurocomputing*, 138, Zeng, G. Q. et al., Binary-coded extremal optimization for the design of PID controllers, 180–188, Copyright 2014, with permission from Elsevier.)

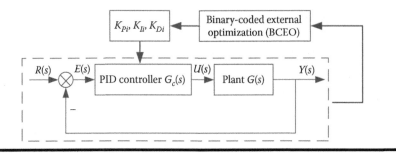

Figure 7.18 Block diagram of a control system with BCEO-based PID controller. (Reprinted from *Neurocomputing*, 138, Zeng, G. Q. et al., Binary-coded extremal optimization for the design of PID controllers, 180–188, Copyright 2014, with permission from Elsevier.)

corresponding to a specific ith PID controller, w_1, w_2, w_3, w_4 are weight coefficients, and $w_4 \gg w_1$. In addition, the relationship between $y_i(k)$ and $u_i(k)$ is obtained by the following steps: First, compute the equation $Y(z)/U(z) = Z[G(s)]$ by z-transform, and then obtain the difference equation describing the relationship between $y_i(k)$, $y_i(k-1)$, ... and $u_i(k)$, $u_i(k-1)$, ... by inverse z-transform. Clearly, the special case when $n = 1$ is just the fitness of single-variable PID controller.

The block diagram of a control system with BCEO-based PID controller is shown as Figure 7.18. The parameters K_{Pi}, K_{Ii}, and K_{Di} of a PID controller are optimized by the proposed BCEO algorithm after an iteration process. To be more precise, the basic idea behind the proposed BCEO-based PID design method is encoding the PID real-valued parameters, including K_{Pi}, K_{Ii}, and K_{Di}, into a binary string, evaluating the control performance by Equations 7.48 through 7.52, selecting bad binary elements according to a probability distribution, mutation of the selected elements by flipping the binary elements, and updating the current solution by accepting the new solution after mutation unconditionally. The detailed description of BCEO-based PID controller design algorithm is presented as follows (Zeng et al., 2014).

Input: A discrete-time model of controlled plant $G(s)$ with sampling period T_s, the number M of PID controllers, the length l_j of jth binary substring corresponding to jth parameter, the weight coefficients w_1, w_2, w_3, w_4 used for evaluating the fitness, the maximum number of iterations I_{max}, the control parameter of the probability distribution $P(k)$.

Output: The best solution S_{best} (the best PID parameters K_{PO}, K_{IO}, and K_{DO}) and the corresponding global fitness C_{best}.

1. Generate an initial configuration S randomly, where $S = [K_{P1}, K_{I1}, K_{D1}, ..., K_{Pi}, K_{Ii}, K_{Di}, ..., K_{Pn}, K_{In}, K_{Dn}]$ is a binary string with length $L = \sum_{j=1}^{M} l_j$ that encodes M PID design parameters and set $S_{best} = S$ and $C_{best} = C(S)$ based on Equations 7.48 through 7.52.

2. For configuration S,
   a. Generate the configuration S_i by flipping the bit i $(1 \leq i \leq L)$ and keeping the others unchanged, then compute the fitness $C(S_i)$ based on Equations 7.48 through 7.52.
   b. Evaluate the local fitness $\lambda_i = C(S_i) - C_{best}$ for each bit i and rank all the bits according to λ_i, that is, find a permutation Π_1 of the labels i such that $\lambda_{\Pi_1(1)} \geq \lambda_{\Pi_1(2)} \geq \cdots \geq \lambda_{\Pi_1(L)}$.
   c. Select a rank $\Pi_1(k)$ according to a probability distribution $P(k)$, $1 \leq k \leq L$ and denote the corresponding bit as x_j.
   d. Flip the value of x_j and set $S_{new} = S$ in which x_j value is flipped.
   e. If $C(S_{new}) \leq C(S_{best})$, then $S_{best} = S_{new}$.
   f. Accept S_{new} unconditionally.
3. Repeat step (2) until some predefined stopping criteria (e.g., the maximum number of iterations) are satisfied.
4. Obtain the best solution S_{best} and the corresponding global fitness C_{best}.

The power-law, exponential, or hybrid distribution can be chosen as the effective probability distributions $P(k)$ used in the above proposed BCEO method. Here, Zeng et al. (2014) chose the power law as the evolutionary probability distribution, where only parameter τ needs to be adjusted. The effects of other probability distributions on the performances will be analyzed in a similar way by readers. The parameter l_j determines the accuracy of binary encoding. The sampling period T_s plays an important role in effecting the stability of the control system. In general, the smaller the T_s, the more the stability of the control system. In real-world situations, the weight coefficients w_1, w_2, w_3, and w_4 used for evaluating the fitness are generally determined according to the experiential rules, which are similar to those in Zhang et al. (2009). More specifically, the parameters w_1 and w_2 are subject to the equation $w_1 + w_2 = 1$ and the value of w_1 is often much larger than w_2. Additionally, w_3 often ranges from 2 to 5 while w_4 is generally from 50 to 100 or larger values. In practice, the more precise values of these parameters for a specific instance are further determined by trial and error.

The proposed BCEO-based PID design algorithm has less adjustable parameters than other existing methods based on popular EAs such as adaptive GA (AGA) (Zhang et al., 2007), self-organizing genetic algorithm (SOGA) (Zhang et al., 2009), and probability-based binary PSO (PBPSO) (Menhas et al., 2012). The detailed description is shown in Table 7.9. More specially, except the maximum number of iterations and the length l_j of jth binary substring corresponding to jth PID parameter, only one parameter τ should be tuned in BCEO while three parameters, including population size, crossover probability, and mutation probability, need to be tuned in AGA at least, five adjustable parameters, including population size, selection pressure turning coefficient β, crossover rate P_c, mutation turning coefficient α, and mutation period T_c in SOGA, and more parameters, including population size, inertia weight w, acceleration factors c_1, c_2, need to be

Table 7.9 Adjustable Parameters Used in Different Optimization Algorithms-Based PID Controller Design Algorithms

Algorithm	Number of Parameters	Adjustable Parameters
AGA (Zhang et al., 2007)	5	Population size NP, maximum number of iterations I_{max}, binary-coded length l_j for each variable, crossover rate P_c, mutation rate P_m
SOGA (Zhang et al., 2009)	7	NP, I_{max}, l_j, the selection pressure turning coefficient β, crossover rate P_c, the mutation turning coefficient α, and the mutation period T_c
PBPSO (Menhas et al., 2012)	8	NP, I_{max}, l_j, inertia weight factor w_{max} and w_{min}, V_{max}, acceleration parameter c_1, c_2
BCEO (Zeng et al., 2014)	3	I_{max}, l_j, shaped parameter τ in power-law distribution

Source: Adapted from Zeng, G. Q. et al., *Neurocomputing* 138: 180–188, 2014.

tuned in PBPSO. Furthermore, BCEO has only selection and mutation operations from the perspective of EA. Therefore, BCEO-based PID design method is much simpler than the methods based on AGA, SOGA, and PBPSO. Additionally, the proposed BCEO is more effective and efficient than AGA, SOGA, and PBPSO, which has been demonstrated by the experimental results on some benchmark examples in Section 7.5.3.

7.5.3 Experimental Results

To demonstrate the effectiveness of the proposed BCEO algorithm, this section chooses some benchmark engineering instances as tested. It should be noted that all the following experiments have been implemented on by MATLAB software on a 3.10 GHz PC with I5-2400 processor and 4 GB RAM. In the following experiments, 20 runs of each algorithm are implemented for a benchmark example.

7.5.3.1 Single-Variable Controlled Plant

In this section, two single-variable plants (Zhang et al., 2009) are chosen to illustrate the superiority of the BCEO algorithm to the existing popular evolutionary optimization algorithms, such as AGA (Zhang et al., 2007), SOGA (Zhang et al., 2009), and PBPSO (Menhas et al., 2012). The transfer functions of the benchmark plants 1 and 2 denoted as $G_1(s)$ and $G_2(s)$ are as follows:

$$G_1(s) = \frac{1.6}{s^2 + 2.584s + 1.6} \tag{7.53}$$

$$G_2(s) = \frac{15}{50s^3 + 43s^2 + 3s + 1} \tag{7.54}$$

To compare performances of the proposed BCEO algorithm with the reported algorithms under the same criterion, the values of the weight coefficients used in the evaluation of fitness are predefined as follows: $w_1 = 0.999$, $w_2 = 0.001$, $w_3 = 2.0$, $w_4 = 50$, and sampling period T_s is set to 0.05 s. The parameters used in AGA (Zhang et al., 2007), SOGA (Zhang et al., 2009), PBPSO (Menhas et al., 2012), and BCEO (Zeng et al., 2014) for plants 1 and 2 are shown in Table 7.10.

The performances of these algorithms are evaluated by these indexes, including the best fitness (*BF*), overshoot σ%, rising time t_u (in seconds), steady-state error e_{ss}%, settling time 0.1%t_w (in seconds) with 0.1% error, and the average running time T (in seconds). The experimental results of AGA (Zhang et al., 2007), SOGA (Zhang et al., 2009), PBPSO (Menhas et al., 2012), and BCEO for single-variable controlled plants 1 and 2 are shown in Table 7.11.

The step responses for plant 1 under different algorithms-based PID controllers are compared as shown in Figure 7.19. Clearly, the indexes, including σ%, 0.1%t_w, and e_{ss}%, obtained by BCEO are all smaller than those by AGA, SOGA, and PBPSO, and t_u obtained by BCEO is equivalent to that by AGA and PBPSO and smaller than that by SOGA. This indicates that the transient and steady-state performances obtained by BCEO are better than those by AGA and SOGA. In addition, the *BF* and *T* of BCEO are also smaller than those of AGA, SOGA, and PBPSO. In other words, the proposed BCEO-based PID algorithm is more effective and efficient than AGA, SOGA, and PBPSO for plant 1.

Table 7.10 Adjustable Parameters Used in BCEO and Other Evolutionary Algorithms for Single-Variable Plants 1 and 2

Algorithm	Adjustable Parameters
AGA (Zhang et al., 2007)	$NP = 50$, $I_{max} = 100$, $I_j = 10$, $P_c = 0.6$, $P_m = 0.01$
SOGA (Zhang et al., 2009)	$NP = 50$, $I_{max} = 100$, $I_j = 10$, β = 0.7, $P_c = 0.6$, α = 4, $T_c = 50$
PBPSO (Menhas et al., 2012)	$NP = 50$, $I_{max} = 100$, $I_j = 16$, $w_{max} = 0.8$ and $w_{min} = 0.8$, $V_{max} = 50$, $c_1 = 2$, $c_2 = 2$
BCEO (Zeng et al., 2014)	$I_{max} = 100$, $I_j = 10$, τ = 1.3

Source: Reprinted from *Neurocomputing*, 138, Zeng, G. Q. et al., Binary-coded extremal optimization for the design of PID controllers, 180–188, Copyright 2014, with permission from Elsevier.

Table 7.11 Comparative Performances of BCEO with the Reported Popular Evolutionary Algorithms for Single-Variable Plant 1

Algorithm	K_P	K_I	K_D	BF	$\sigma\%$	t_u (s)	$e_{ss}\%$	$0.1\%t_w$ (s)	T (s)
AGA	15.2884	2.7566	3.4506	9.8644	0.21	0.65	0.05	5.80	52.2760
SOGA	19.3900	4.1190	5.1510	14.2753	2.81	0.90	0.51	> 10	51.3921
PBPSO	19.9990	4.2174	3.9562	14.1507	3.78	0.65	0.35	> 10	78.9850
BCEO ($\tau = 1.3$)	17.5171	2.6393	3.9296	9.7605	0.01	0.65	0.00	2.00	19.8290

Source: Reprinted from *Neurocomputing*, 138, Zeng, G. Q. et al., Binary-coded extremal optimization for the design of PID controllers, 180–188, Copyright 2014, with permission from Elsevier.

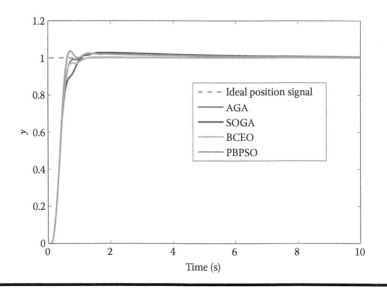

Figure 7.19 **(See color insert.) Comparison of step response for plant 1 under different algorithms-based PID controllers. (Reprinted from *Neurocomputing*, 138, Zeng, G. Q. et al., Binary-coded extremal optimization for the design of PID controllers, 180–188, Copyright 2014, with permission from Elsevier.)**

The step responses for plant 2 under different algorithms-based PID controllers are shown in Figure 7.20. From the experimental results of AGA (Zhang et al., 2007), SOGA (Zhang et al., 2009), PBPSO (Menhas et al., 2012), and BCEO (Zeng et al., 2014) for plant 2 shown in Table 7.12, it is obvious that the transient and steady-state performances obtained by BCEO for plant 2 are better than those by AGA, SOGA, and PBPSO. Moreover, the *BF* and *T* of BCEO are also smaller than those of AGA, SOGA, and PBPSO. In other words, the proposed BCEO-based PID algorithm is also more effective and efficient than AGA, SOGA, and PBPSO for plant 2.

7.5.3.2 Multivariable Controlled Plant

To illustrate the superiority of the proposed BCEO algorithm to AGA (Zhang et al., 2007) and PBPSO (Menhas et al., 2012) for the multivariable plants, the following binary distillation column plant with two input and two output (Wang et al., 1997) is chosen as a benchmark example.

$$G_m(s) = \begin{bmatrix} \dfrac{12.8e^{-s}}{1+16.7s} & \dfrac{-18.9e^{-3s}}{1+21s} \\ \dfrac{6.6e^{-7s}}{1+10.9s} & \dfrac{-19.4e^{-3s}}{1+14.4s} \end{bmatrix} \qquad (7.55)$$

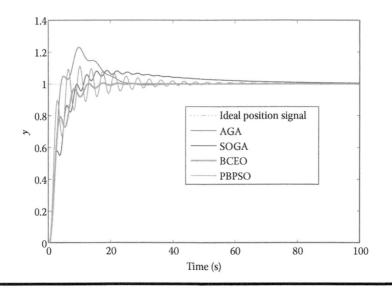

Figure 7.20 (See color insert.) Comparison of step response for plant 2 under different algorithms-based PID controller. (Reprinted from *Neurocomputing*, 138, Zeng, G. Q. et al., Binary-coded extremal optimization for the design of PID controllers, 180–188, Copyright 2014, with permission from Elsevier.)

The steady-state decoupling matrix of the above controlled plant model is given as follows:

$$D = G_m^{-1}(0) = \begin{bmatrix} 0.1570 & -0.1529 \\ 0.0534 & -0.1036 \end{bmatrix} \tag{7.56}$$

The experimental results on the comparative performances of BCEO (Zeng et al., 2014) with AGA (Zhang et al., 2007) and PBPSO (Menhas et al., 2012) for multivariable plant with decoupler are shown in Tables 7.13 and 7.14. In PBPSO, the original fitness is evaluated by IAE and the BF value is 159.5911 (Menhas et al., 2012). Here, for the convenience of comparison, the new BF value of PBPSO under the same values of PID parameters is obtained as 101.4572 by Equations 7.48 through 7.52. Figure 7.21 presents the step response of output y_1 and y_2 under different algorithm-based PID controllers. Although the overshoot $\sigma_1\%$ and $\sigma_2\%$ obtained by BCEO are larger than those by AGA and PBPSO, the performances indexes, including t_{u1}, $0.5\%t_{w1}$, t_{u2}, $0.5\%t_{w2}$, and $e_{ss2}\%$, are smaller than those by AGA and PBPSO, and $e_{ss1}\%$ obtained by BCEO is the same as that by AGA and smaller than PBPSO. The *BF* and *T* of BCEO are also smaller than those of AGA and PBPSO. By considering the above analyses comprehensively, it is clear that the proposed BCEO outperforms AGA and PBPSO for multivariable plant with decoupler in terms of simplicity, effectiveness, and computational efficiency.

Table 7.12 Comparative Performances of BCEO with Existing Evolutionary Algorithms for Single Plant 2

Algorithm	K_P	K_I	K_D	BF	$\sigma\%$	t_u (s)	$e_{ss}\%$	$0.1\% t_w$ (s)	T (s)
AGA	1.3294	0.1955	4.6921	108.0177	22.97	3.75	0.00	43.20	64.8280
SOGA	2.98	0.096	12.7	156.5308	8.45	8.45	0.45	>100	62.7620
PBPSO	4.0043	0.0355	10.0	165.4567	11.13	5.70	0.24	>100	125.1520
BCEO	1.7986	0.0196	6.3441	92.7998	0.50	7.75	0.00	30.0	41.7960

Source: Reprinted from *Neurocomputing,* 138, Zeng, G. Q. et al., Binary-coded extremal optimization for the design of PID controllers, 180–188, Copyright 2014, with permission from Elsevier.

Table 7.13 Multivariable PID Controller Parameters of AGA, PBPSO, and BCEO for Multivariable Plant with Decoupler

Algorithm	K_{P1}	K_{I1}	K_{D1}	K_{P2}	K_{I2}	K_{D2}	BF	T (s)
AGA	1.2199	0.0616	−0.7048	1.2199	0.3050	−0.6950	103.0780	181.631
PBPSO	1.9978	0.1121	−0.5443	1.9990	0.1485	−0.5619	101.5472	212.235
BCEO	2.0489	0.1095	−0.4692	2.0489	0.2561	−0.4878	78.6459	120.301

Source: Reprinted from *Neurocomputing*, 138, Zeng, G. Q. et al., Binary-coded extremal optimization for the design of PID controllers, 180–188, Copyright 2014, with permission from Elsevier.

Table 7.14 Comparative Performance of BCEO with AGA and PBPSO for Multivariable Plant with Decoupler

Algorithm	$\sigma_1\%$	t_{u1}	$e_{ss1}\%$	$0.5\%t_{w1}$	$\sigma_2\%$	t_{u2}	$e_{ss2}\%$	$0.5\%t_{w2}$
AGA	0.00	18.0	0.00	30	0.00	17.0	0.31	34.0
PBPSO	0.40	11.8	0.02	22	0.60	10.5	0.44	25.0
BCEO	2.98	6.0	0.00	11	2.49	4.0	0.27	10.0

Source: Reprinted from *Neurocomputing*, 138, Zeng, G. Q. et al., Binary-coded extremal optimization for the design of PID controllers, 180–188, Copyright 2014, with permission from Elsevier.

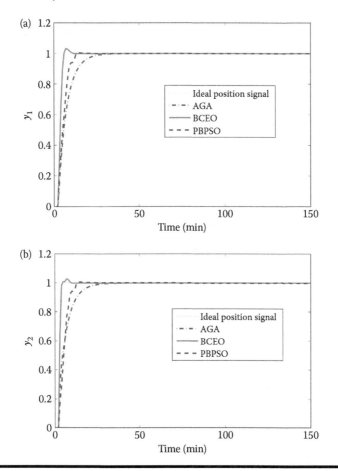

Figure 7.21 (See color insert.) Comparison of output y_1 (a) and y_2 (b) under different algorithms-based PID controllers for multivariable plant with decoupler. (Reprinted from *Neurocomputing*, 138, Zeng, G. Q. et al., Binary-coded extremal optimization for the design of PID controllers, 180–188, Copyright 2014, with permission from Elsevier.)

Table 7.15   Effects of Parameter τ on the Performances of BCEO for Single-Variable Plant 2

τ	K_P	K_I	K_D	BF	σ%	t_u (s)	e_{ss}%	$0.1\%t_w$ (s)	T (s)
1.10	2.1896	0.2933	6.4614	106.9520	24.12	3.15	0.00	28.60	43.1100
1.15	1.8768	0.0196	6.1877	94.0765	1.08	7.55	0.09	13.85	42.9370
1.20	1.7400	0.2346	5.3568	103.8372	23.19	3.35	0.07	27.25	43.1090
1.25	2.5024	0.0196	8.5826	97.8417	1.20	6.90	0.02	15.75	42.1880
1.30	1.7986	0.0196	6.3441	92.7998	0.50	7.75	0.00	18.40	41.7960
1.35	2.5024	0.4301	7.4682	109.8545	28.25	3.00	0.00	22.10	42.5630
1.40	2.5024	0.0196	8.5826	97.8417	1.20	6.90	0.02	15.75	42.9690
1.45	2.8152	0.0196	9.4233	101.9073	1.78	6.55	0.04	15.05	42.8900
1.50	2.5024	8.5142	8.2893	97.8486	1.23	6.85	0.03	15.85	43.6410

Source: Reprinted from *Neurocomputing*, 138, Zeng, G. Q. et al., Binary-coded extremal optimization for the design of PID controllers, 180–188, Copyright 2014, with permission from Elsevier.

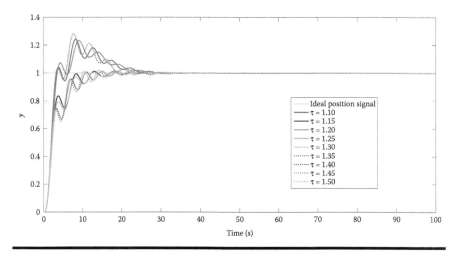

Figure 7.22 (See color insert.) Adjustable parameter τ versus the step response for single-variable plant 2. (Reprinted from *Neurocomputing*, 138, Zeng, G. Q. et al., Binary-coded extremal optimization for the design of PID controllers, 180–188, Copyright 2014, with permission from Elsevier.)

7.5.3.3 Parameters and Control Performances

As analyzed in Section 7.5.2, the adjustable parameters, such as I_{max}, l_j, and τ, are important to affect the performances of the proposed BCEO. However, the maximum number of iterations I_{max} and the length l_j of *j*th binary substring corresponding to *j*th PID parameter are often predefined as constants. This section focuses on the effects of the parameter τ on the performances of BCEO, which are illustrated for single-variable plant 2 shown in Table 7.15. In addition, Figure 7.22 presents the effects of τ on the step response for plant 2. In these experiments, the values of τ range from 1.10 to 1.50. Generally, the *BF* obtained by the BCEO algorithm is robust for the parameter τ. Despite the large fluctuation of overshoot σ% and the corresponding t_u and $0.1\%t_w$, the values of e_{ss}% maintains a much smaller variation. Not surprisingly, an appropriate value of σ% often leads to small value of t_u and t_w. The effects of the parameter τ on the control performance for other single-variable and multivariable plants is analyzed in a similar way.

7.6 Summary

As one of the most popular APC solutions, the MPC has been widely applied in process industries during recent years. Due to the rapid development of industry and technology, control performance requirements for large-scale, complicated, and uncertain systems keep rising. Basically, the issues mentioned above mainly involve solving various types of complicated nonlinear optimization problems. The LMPC, which usually relies on a linear dynamic model, is inadequate for the abovementioned problems and the limitation becomes more and more obvious. NMPC appears to be a perspective method. However, the introduction of the nonlinear predictive model brings some difficulties, such as the parameters/ structure selection of the nonlinear predictive model and the online receding horizon optimization. All these issues encourage researchers and practitioners to devote themselves to the development of novel optimization methods, which are suitable for applications in NMPC.

This chapter starts with the general review of two key research issues in modern control: the prediction model and online optimization. Based on the methods in Chapter 5, we first apply the proposed EO-LM methods in BOFs endpoint quality prediction. Then a new SVR tuning method, EO-SVR, is introduced for the automatic optimization of SVR kernel type and its parameters, to provide a better generalization performance and a lower forecasting error. Finally, a novel NMPC with the integration of EO-SVR prediction model and "EO-SQP"-based receding horizon optimization is proposed. Simulation results on a nonlinear MIMO CSTR reactor show considerable performance improvement obtained by the proposed NMPC. The main topics studied in this chapter are summarized as follows:

1. For a typical nonlinear industrial process-BOF-in steelmaking, the proposed NN model based on EO-LM learning algorithm is further applied to predict the endpoint product quality. The fundamental analysis reveals that the proposed EO-LM algorithm may provide superior performance in generalization, computation efficiency, and avoid local minima, compared to traditional NN learning methods. The experimental results with production-scale BOF data show that the proposed method can effectively improve the NN model for BOF endpoint quality prediction.

2. The performance of SVR model is sensitive to kernel type and its parameters. The determination of an appropriate kernel type and the associated parameters for SVR is a challenging research topic in the field of support vector learning. A novel method is presented for simultaneous optimization of SVR kernel function and its parameters, formulated as a mixed integer optimization problem and solved by the recently proposed heuristic "extremal optimization (EO)." This chapter presents the problem formulation for the optimization of SVR kernel and parameters, EO-SVR algorithm, and experimental tests with five benchmark regression problems. The comparison results with other traditional approaches show that the proposed EO-SVR method can provide better generalization performance by successfully identifying the optimal SVR kernel function and its parameters.

3. For the requirements of model identification and online receding horizon optimization in NMPC, the EO is extended and further applied to the SVR predictive model based on multistep-ahead error and the optimization of model parameters. Then, based on the predictive model, a "horizon-based mutation" EO-SQP is designed and serves as the online optimizer of a NMPC controller. The simulation results demonstrated improved tracking of multi-set-point profiles and good disturbance rejection in comparison with LMPC.

4. Furthermore, this chapter presents a novel design method for PID controllers based on the BCEO algorithm (Zeng et al., 2014). One of the most attractive advantages of BCEO is its relative simplicity comparing with the existing popular EAs, such as AGA (Zhang et al., 2007), SOGA (Zhang et al., 2009), and PBPSO (Menhas et al., 2012). More specially, only selection and mutation should be designed from the EAs' points of view and fewer adjustable parameters need to be tuned in BCEO. Furthermore, the experimental results on some benchmark instances have shown that the proposed BCEO-based PID design method is more effective and efficient than AGA, SOGA, and PBPSO. As a consequence, the proposed BCEO is a promising optimization method for the design of PID controllers. Nevertheless, the performances of BCEO may be improved further by a highly tailored method concerning the adaptive mechanism of τ or backbone-based learning algorithm. On the other hand, the BCEO can be extended to design PID or fractional-order PID controllers for more complex control systems.

Chapter 8

EO for Production Planning and Scheduling

8.1 Introduction

To make a manufacturing enterprise more competitive and profitable in the global marketplace, the profit- driven "make-to-order" or "make-to-stock" business model has been applied widely in manufacturing management. Among multidimensional business and production decisions, computer-aided production planning and scheduling to optimize desired business objectives subject to multiple sophisticated constraints has been one of the most important decisions. In general, many production-scheduling problems can be formulated as constrained combinatorial optimization models that are usually NP-hard problems, particularly for those large-scale real-world applications. This kind of scheduling problem is generally difficult to be solved with traditional optimization techniques. Consequently, many approximation methods, for example, metaheuristics, have been the major approaches to such kind of constrained COP. Although approximation algorithms do not guarantee achieving optimal solutions, they can attain near-optimal solutions with reasonable computation time.

In this chapter, hybrid EAs with integration of GA and EO are proposed to solve a typical sequence-dependent scheduling problem, which is illustrated with the production scheduling of a hot-strip mill (HSM) in the steel industry. In general, the manufacturing system under study has the following major features:

1. The manufacturing process could be either discrete, batch, or semibatch production.
2. Without loss of generality, here, we consider a manufacturing system with a single-production stage and a large number of products defined by multiple specification parameters, such as product grades, dimensions, etc.

3. The number of work orders to be scheduled is equal to or greater than the number of scheduled orders within the resulting schedule solution. It means that we may leave some work orders unscheduled under a given time horizon.
4. The objective is to construct an optimal schedule solution that minimizes the weighted sum of sequence-dependent transition costs, nonexecution penalties, earliness/tardiness (E/T) penalties, etc.
5. Multiple constraints, such as the feasible path between two consecutive orders, the capacity requirement of a single-scheduling solution, due delivery date, etc., are taken into consideration in practice.

8.1.1 An Overview of HSM Scheduling

An HSM produces hot-rolled products from steel slabs, and is one of the most important production lines in an integrated mill or minimill (Lee et al., 1996; Tang et al., 2001). Besides process control strategies such as automatic gauge control and rolling force prediction, production planning and scheduling also significantly affect the performance of an HSM in terms of product quality, throughput, and on-time delivery. As a consequence, HSM scheduling has become a critical task in a steel plant. Generally speaking, the aim of HSM scheduling is to construct an optimal production sequence consisting of a single-rolling round or multiple consecutive rounds (or the so-called campaign). A rolling round must optimize a set of given criteria and satisfy a series of constraints, such as the "coffin shape" of width profile, the smooth jumps in dimensions and hardness between adjacent coils, the minimal and maximal number of coils or footages, the maximal number of coils with the same width, the maximal number of short slabs, work-in-process (WIP) inventory, and on-time delivery.

Owing to the large number of manufacturing orders, multiple conflicting objectives, and various production constraints, the HSM-scheduling problem has been proven to be a typical NP-hard COP (Balas, 1989; Lopez et al., 1998; Chen et al., 2008). It is almost impossible to generate an optimal scheduling solution by human schedulers or traditional mathematical programming methods. Balas (1989) formulated the round-scheduling problem as a generalized TSP with multiple conflicting objectives and constraints. Kosiba et al. (1992) stated that the minimization of roller wears is equivalent to the composite objective of improving product quality, production rate, and profits, and the roller wears can be measured by a penalty function that is determined by the jump values in width, gauge, and hardness between adjacent coils. Assaf et al. (1997) incorporated four key submodels (i.e., rolling mill, reheat furnace, heat loss, and cost calculation), and developed an enumeration-based branching and pruning algorithm to generate scheduling solutions for the steel production sequence problem. Lopez et al. (1998) presented an aggressive heuristic to solve the HSM-scheduling problem. The heuristic repeatedly applies the tabu search method and the CROSS exchange operator in the cannibalization stage. Tang and Wang (2005) modeled the hot-rolling production-scheduling problem as a prize-collecting vehicle-routing problem (PCVRP), for which an iterated local search (ILS) algorithm was proposed on the basis of a very-large-scale neighborhood (VLSN) using cyclic transfer. Mattika et al. (2014) proposed two submodels to integrate a continuous caster and HSM planning in the steel industry in a hierarchical way.

Owing to the complexity of the scheduling models and the inefficiency of mathematical programming methods, various intelligent methods, including local search and greedy algorithm (Peterson et al., 1992; Verdejoa et al., 2009; Yadollahpour et al., 2009), GA (Fang and Tsai, 1998; Tang et al., 2000), tabu search (Stauffer and Liebling, 1997), and particle swarm optimization (Chen et al., 2008) have been widely applied in the past decades to solve the HSM-scheduling problem. Furthermore, scheduling systems with different features have also been developed. Cowling (2003) introduced a semiautomatic decision-support system featured with a multi-objective-scheduling model, and the model was solved by a variety of bespoke local search and tabu search methods. Knoop and Van Nerom (2003) proposed a scheduling architecture for an integrated steel plant. In the International Business Machine (IBM) research reports, the overall scheduling in the steel-manufacturing process was decomposed as primary production scheduling (Lee et al., 1996) and finishing line scheduling (Okano et al., 2004).

8.1.2 Production Process

Usually, a typical hot-rolling production line consists of reheating furnaces, a set of roughing mills and finishing mills, a watercooler, and a coiler as schematically shown in Figure 8.1.

In a hot-rolling mill, the steel slabs from continuous casters or slab yards are first charged to a working beam or push-type reheating furnace, and then the heated slabs discharged from the reheating furnace are processed through a set of roughing stands that make use of horizontal rollers to reduce the slab thickness (initially 20–30 cm) and vertical rollers to regulate the slab width, respectively. Subsequently, the intermediate slabs are loaded into a finishing mill, in which 6–7 stands can

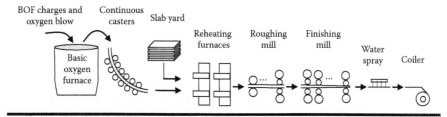

Figure 8.1   A typical hot-rolling production line. (Reprinted from *Computers and Operations Research*, 39 (2), Chen, Y. W. et al., Development of hybrid evolutionary algorithms for production scheduling of hot strip mill, 339–349, Copyright 2012, with permission from Elsevier.)

further reduce the gauge and width of slabs to desired specifications. After passing through the watercooler and the coiler, the raw slabs are finally converted into the finished hot-rolled products with a desired gauge (1.5–12 mm), width, mechanical and thermal properties, and surface quality.

8.1.3 Scheduling Objectives and Constraints

Production scheduling, in general terms, is the process of resource allocation and workload optimization for a period of time under a set of given criteria and constraints. In hot-rolling production, as discussed above, the primary aim of a scheduling solution is to construct an optimal production sequence that can take into consideration two subtasks simultaneously: (1) selecting a subset of manufacturing orders with the "make-to-inventory" or "make-to-order" principle; and (2) generating an optimal rolling sequence. Effective HSM scheduling is crucial to a steel plant's competitiveness, since the hot-rolling mill is one of the most important production lines in the steel industry (Lopez et al., 1998). In a hot-rolling mill, the finished products are produced by subjecting steel slabs to high pressures through a series of rollers. As a consequence, the working and backup rollers need to be replaced periodically due to abrasions. Usually, the set of slabs rolled between two consecutive replacements of working rollers is called a *round*, and the set of slabs rolled between two consecutive replacements of backup rollers is called a *campaign*. Figure 8.2 illustrates the empirical "coffin-shape" width profile of a rolling round and the multiple rolling rounds of a campaign. As shown in the figure, a coffin-shape rolling round mainly consists of two parts: the *warm-up* section, in which several slabs (5–15) are rolled from narrow to wide for warming up the rollers, and the *body* section, in which a number of slabs are rolled from wide to narrow for avoiding marks or grooves at the edge of rolled coilers.

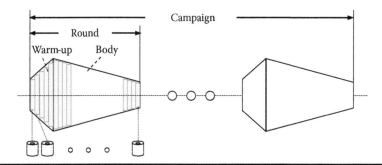

Figure 8.2 "Coffin-shape" width profile of a rolling round and the composition of a campaign. (Reprinted from *Computers and Operations Research*, 39 (2), Chen, Y. W. et al., Development of hybrid evolutionary algorithms for production scheduling of hot strip mill, 339–349, Copyright 2012, with permission from Elsevier.)

As the warm-up section consists of only a few slabs, it can be scheduled easily by various heuristic methods. In the following sections, we only consider the body section of the HSM scheduling without explicit declaration. In terms of their relative importance, the HSM-scheduling constraints can be classified into hard constraints and soft constraints. Hard constraints refer to those that cannot be violated in the final scheduling solution. In the HSM- scheduling problem, the main hard constraints are as follows:

1. A rolling round has a "coffin-shape" width profile, which starts with a warm-up section having the width pattern of narrow to wide, and follows a body section having the width pattern of wide to narrow.
2. The number of coils in a rolling round has lower and upper bounds due to the capacity of the rollers.
3. The changes in dimensions and hardness between adjacent coils should be smooth in both warm-up and body sections. This is because the rollers need to be adjusted for controlling the dimension and hardness jumps on the entry and outlet of mills.
4. The number of coils with the same width to be processed consecutively has an upper bound in a rolling round. Otherwise, the edges of slabs mark the rollers easily. In practice, this is also called groove constraint.
5. The number of short slabs in a rolling round has an upper bound, because reheating furnaces are generally designed for normal slabs and short ones are only used to fill in gaps.

Soft constraints, on the contrary, can be compromised and violated according to their priorities. In this chapter, the following soft constraints are considered for generating rolling rounds:

1. To provide a satisfactory delivery service and maintain a lower WIP inventory, the hot-rolled products should be processed near their due date, and not too early or too late for decreasing E/T penalties.
2. The hardness and desired surface quality of the coils processed in a warm-up section should not be too high.
3. The coils with high-quality requirements should not be processed in the rear part of a body section, because the performance of rollers degrades with the increase of rolling time.

In the HSM scheduling, there are a number of conflicting objectives to be considered. For example, product quality and production cost are two of the most common conflicting objectives. To improve product quality, the rollers need to be replaced more frequently, which incurs extra production cost and decreases the rate of throughput. Similarly, on-time delivery is in conflict with a low WIP inventory. Therefore, an effective scheduling system should be capable of making trade-offs among multiple conflicting objectives. In the HSM scheduling, the optimization objective is defined as the weighted combinations of the penalties for width, gauge, hardness jumps between adjacent coils, the penalties for the violation of short slab and groove constraints, and the due-date E/T penalties (Lopez et al., 1998; Tang et al., 2000). Generally, the HSM-scheduling system is situated in a dynamic environment. The manufacturing orders are only known for a limited period, and new ones transformed from customer orders are downloaded from upper-level enterprise resource planning (ERP) and production-planning systems periodically. This leads to the HSM scheduling being a periodic process (Stauffer and Liebling, 1997). Therefore, we can schedule rolling rounds one by one, and HSM scheduling can be simplified to generate an optimized rolling round by selecting and sequencing a subset of coils from existing manufacturing orders.

8.2 Problem Formulation

This section presents a mathematical HSM-scheduling model. Two scheduling subtasks as discussed above are considered simultaneously. Traditionally, the HSM-scheduling problem is solved by using a sequential strategy, which first selects a subset of manufacturing orders and then searches the optimal rolling round within the selected orders. Although this strategy is straightforward, it can only find a local optimum for the round scheduling. In this chapter, the proposed scheduling model formulates the two-coupling subtasks in an integrated way, and hence is beneficial to find the globally optimal scheduling solution. First of all, the scheduling solution is formulated as a directed graph $G(V, A)$, where the node set $V = (1, 2, ..., N)$ represents N-manufacturing orders (coils) with desired attributes and due date, and the arc set $A = \{(i, j): i \neq j\}$ denotes the transition trajectories between coil i and coil j. For each arc (i, j), let c_{ij} denote the sequence-dependent penalty incurred by the rolling coil j immediately after coil i. As shown in Table 8.1, the transition penalty can be measured according to the jump values in dimensions and hardness between adjacent coils.

To enhance the accuracy of the penalty structure, the jump values are scaled for different width or gauge rates. Undoubtedly, a gauge jump of 0.15 mm for 1.5-mm coils is more difficult to manipulate than the same jump for 7.5-mm coils.

To formulate a mathematical scheduling model, the parameters are defined as follows:

i, j number of coils
N total number of coils to be produced for manufacturing orders
w_i, g_i, l_i, h_i width, gauge, length, and hardness of coil i
u_i finished temperature of coil i in the outlet of the finishing mill
s_i logic symbol of short slabs, that is, if the raw slab of producing coil i is shorter than a predefined length, $s_i = 1$, otherwise, $s_i = 0$

Table 8.1   Transition Penalty for the Changes in Width, Gauge, and Hardness

Width Range (cm)	Inward Jump (cm)	Penalty	Gauge Range (mm)	Jump Value (mm)	Penalty	Hardness Jump	Penalty
85–100	0–3	1	1.2–3.0	0–0.03	3	1	5
85–100	3–6	3	1.2–3.0	0.03–0.06	7	2	15
85–100	6–9	10	1.2–3.0	0.06–0.09	15	3	35
–	–	–	–	–	–	4	50
85–100	50–	500	1.2–3.0	0.45	200	5	75
–	–	–	–	–	–	–	–
150–	50–	1000	12.0–	0.45–	1000	9	1000

Source: Reprinted from *Computers and Operations Research*, 39 (2), Chen, Y. W. et al., Development of hybrid evolutionary algorithms for production scheduling of hot strip mill, 339–349, Copyright 2012, with permission from Elsevier.

V width set of all coils to be produced, $V = \{v_1, v_2, ..., v_k\}$ ($k \leq n$)
N_v maximal number of coils with the same width to be processed consecutively in a rolling round
N_s maximal number of short slabs in a rolling round
L_l, L_u minimal and maximal lengths of coils in a rolling round
c_{ij} transition cost for the rolling coil j immediately after coil i
c_{ii} cost for not selecting coil i to the current- rolling round
d_i due date of coil i
p_i processing time of coil i
t_i starting time of processing coil i
T_s starting time of the current-rolling round

We also define the following decision variables:

$$x_{ij} = \begin{cases} 1, & \text{if coil } j \text{ immediately follows coil } i \\ 0, & \text{otherwise} \end{cases}$$

Since this chapter aims to solve the two subtasks of order selection and sequencing simultaneously, another decision variable x_{ii} is defined below by analogy with the prize-collecting TSP model (Lopez et al., 1998)

$$x_{ii} = \begin{cases} 1, & \text{if coil } i \text{ is not selected in the current rolling round} \\ 0, & \text{otherwise} \end{cases}$$

The sequence-dependent transition cost c_{ij} can be defined as $c_{ij} = p_{ij}^w + p_{ij}^g + p_{ij}^b + p_{ij}^t$, where p_{ij}^w, p_{ij}^g, p_{ij}^b and p_{ij}^t represent the width, gauge, hardness, and finished temperature transition cost, respectively. As discussed above, the transition cost can be measured by the jump values of the parameters of every two adjacent coils. For example, the finished temperature transition cost can be measured by a function $p_{ij}^t = p^t(u_j - u_i)$. In practice, p_{ij}^w, p_{ij}^g, and p_{ij}^b are usually defined by the penalty structure as shown in Table 8.1.

In the HSM scheduling, the optimization objective can be decomposed into two parts according to the information requirements of evaluating fitness (Han, 2005): (1) local evaluation function (LEF), which can be calculated by using only local information. The LEF is defined as $LEF(S) = \sum_{i=0}^{n} \sum_{j=0}^{n} c_{ij} x_{ij}$, which consists of the sequence-dependent transition costs and the nonexecution penalties (Refael and Mati, 2005); (2) global evaluation function (GEF), which needs to consider the overall configuration information of a solution. The E/T penalties can be included in this part, and therefore, the GEF is defined as $GEF(S) = \sum_{i=1}^{n}(e_i + r_i)$, where $e_i = \max\{0, d_i - t_i - p_i\}$ and $r_i = \max\{0, t_i + p_i - d_i\}$. To generate a rolling sequence, a virtual coil is added to the set of manufacturing orders, and it has no processing time and any sequence-dependent transition cost. A constraint is added to select the virtual coil as the starting node of a rolling round. As a result, a mathematical HSM-scheduling model can be formulated as follows:

Minimize

$$\lambda \sum_{i=0}^{n} \sum_{j=0}^{n} c_{ij} x_{ij} + (1 - \lambda) \sum_{i=1}^{n}(e_i + r_i) \tag{8.1}$$

subject to

$$\sum_{i=0}^{n} x_{ij} = 1, \quad j = 0, ..., n \tag{8.2}$$

$$\sum_{j=0}^{n} x_{ij} = 1, \quad i = 0, ..., n \tag{8.3}$$

$$L_l \le \sum_{i=0}^{n} l_i (1 - x_{ii}) \le L_u \tag{8.4}$$

$$\sum_{i=0}^{n}(1-x_{ii})*\mathrm{sgn}(w_i-\upsilon_k)\le N_\upsilon, \quad \forall \upsilon_k \in V \tag{8.5}$$

$$\sum_{i=0}^{n}s_i(1-x_{ii})\le N_s \tag{8.6}$$

$$x_{00}=0 \tag{8.7}$$

$$t_0=T_s \tag{8.8}$$

$$t_j=\sum_{i=0}^{n}x_{ij}(t_i+p_i)+x_{jj}(t_0+T), \quad j=1,\dots,n \tag{8.9}$$

$$x_{ij}\in\{0,1\}, \quad i,j=0,\dots,n \tag{8.10}$$

$$\mathrm{sgn}(x)=\begin{cases}1, & \text{if } x=0 \\ 0, & \text{otherwise}\end{cases} \tag{8.11}$$

Equation 8.1 is used to calculate the optimization objective $F(S)$ of any scheduling solution S, and the parameter $\lambda(0\le\lambda\le1)$ is the weight of measuring the relative importance of the two optimization parts. Equations 8.2 and 8.3 indicate that each coil can be processed only once or it is not selected in the current-rolling round. Equation 8.4 specifies the minimal and maximal capacities of a rolling round. Equations 8.5 and 8.6 represent the maximal number of same-width coils and that of short slabs, respectively. Equations 8.7 and 8.8 select the virtual coil as the starting node of a rolling round. Equation 8.9 establishes the relationship between variables t_j and x_{ij}, where constant T is greater than the total processing time of a rolling round. It means that if coil j is not selected in the current-rolling round, its processing time is not earlier than the starting time of the next rolling round.

Obviously, the scheduling model above is an NP-hard constrained COP. Considering a simple case, we suppose that M coils need to be scheduling to a rolling round, and there will be $N!/M!$ possible solutions without taking into account any constraints. If there are thousands of manufacturing orders, the complete enumeration of all possible solutions is computationally prohibitive, that is, no exact algorithm is capable of solving the optimization problem with reasonable computation time. Frequently, EAs as promising approximate techniques can be employed to solve this class of production-scheduling problem for finding desirable, although not necessary optimal solutions.

8.3 Hybrid Evolutionary Solutions with the Integration of GA and EO

The body section is the kernel part of a rolling round. The performance of an HSM-scheduling solution mainly depends on the optimization quality of the rolling sequence of the body section. Without specific explanation, the scheduling of the body section is usually regarded as the equivalent of the HSM scheduling as stated in many papers (Lopez et al., 1998; Tang et al., 2000). The mathematical model formulated in Section 8.2 describes the scheduling of the body section, and it is an NP-hard COP with a composite optimization objective. It is unrealistic to search for the optimal scheduling solution by using traditional mathematical programming methods. As one class of effective optimization techniques, intelligent algorithms have been widely employed to solve typical scheduling problems for effectively finding a desirable solution, although it may not necessarily be the optimal solution (Tang et al., 2001). In this chapter, effective hybrid evolutionary solutions are developed to solve the HSM-scheduling problem by combining the population-based search capacity of GA and the fine-grained local search efficacy of EO.

8.3.1 Global Search Algorithm: Modified GA for Order Selection and Sequencing

GA is inspired by the Darwinian evolution theory (Michalewicz, 1996). It is a class of population-based search method, and can explore the entire gene pools of solution configurations, in which the crossover operation performs global exchanges and the mutation operation enhances the diversity of the population. The iterative improvement is realized by a generate-test procedure (Han, 2005). In the past decades, the GA has been widely applied to the field of combinatorial optimization. In this section, a modified GA (MGA) is proposed to solve the HSM-scheduling problem.

8.3.1.1 Representation of Solutions

Generally, standard GA applications use binary strings or ordinal integers to represent a chromosome of a solution. In the HSM-scheduling problem, we define the chromosome with a chain of genes as a rolling round. Each gene represents a coil marked with coil ID. Suppose that the number of coils in a chromosome is equal to m and the total number of coils in the order book is n, the scheduling objective

Slot#:	1	2	3	4	5	6	7	8	9
Chromosome:	5	8	9	12	7	11	2	10	3

Figure 8.3 Illustration of chromosome representation. (With kind permission from Springer Science + Business Media: *International Journal of Advanced Manufacturing Technology*, Hybrid evolutionary algorithm with marriage of genetic algorithm and extremal optimization for production scheduling, 36, 2008, 959–968, Chen, Y. W. et al.)

is to select m coils from the n coils and generate an optimized rolling sequence. An example of a chromosome is shown in Figure 8.3.

The vector [5, 8, …, 3] denotes a possible rolling round, and each figure in the vector represents a particular coil ID. This chromosome can represent a possible scheduling solution, in which the rolling round consists of nine coils, but the total number of coils can be more than 12.

8.3.1.2 Population Initialization

Two issues are usually discussed for the population initialization: the population size and the method of initializing the population (Chang and Ramakrishna, 2002; Chen et al., 1998). To maintain the searching capability, the population size needs to increase exponentially with the dimension of the optimization problem, that is, the length of the chromosome. However, although a large population size can search the solution space effectively, it also incurs excessive computing time. On the contrary, a small quantity of individuals can't efficiently locate the optimal solution. As a result, an appropriate population size is crucial for finding the optimal solution efficiently and effectively.

In a large number of GA applications, heuristic and random methods are among the most popular approaches to generate the initial population. Well-designed heuristic methods can be used to produce some good individuals, which is beneficial to improve the convergence speed. However, if all initial individuals have good fitness values, the algorithm converges into locally optimal solutions easily and never explores the whole solution space due to lack of genetic diversity. On the contrary, if all initial individuals are generated randomly, it may take a large number of generations to improve the inferior individuals. A random initial population makes it difficult to obtain a good solution, particularly for a practical application with constraints. Therefore, in the proposed MGA, three different methods are employed simultaneously to generate the initial population.

First of all, a well-designed heuristic is proposed to generate a feasible solution. The constraints discussed in Section 8.1.3 are considered in this heuristic. The initial feasible individual can be used to improve the feasibility of all iterative-generated individuals and accelerate the convergence speed.

Algorithm 1: (Pseudocode for the heuristic to generate a feasible body section)

Step 1. Identify available slabs for the body section within this current- scheduling time horizon.

Step 2. Sequence the corresponding coils in descending width and ascending due date (i.e., the coils are first sequenced in descending width, and those with the same width are further sequenced in ascending due date).

Step 3. Group the coils with the same width, and calculate the number of groups N_w.

Step 4. Select N_v coils into a body set from each group sequentially, and calculate the number of selected coils N_{temp} in the body set.

Step 5. If N_{temp} is less than the required number N_{body} for a body section, it means no feasible body section can be generated with existing manufacturing orders, then go to *step 6*.

Else, generate a body section by sequentially selecting the coils in the body set with the step size $\lfloor N_{temp} / N_{body} \rfloor$.

Step 6. Stop the heuristic, and return the scheduling output. ■

Second, the nearest-neighbor search method, which first chooses a starting coil and then selects the next least-cost coil to the provisional sequence iteratively, is used to generate a proportion of individuals in the initial population.

Finally, the random insertion method is used to generate all other individuals. Starting from a randomly selected coil, a body section is generated by selecting the next unscheduled coil randomly and then inserting it in the least-cost position of the provisional sequence.

8.3.1.3 Fitness Function

The fitness function calculates how fit an individual is, and the "fittest" ones have more chances to be inherited into the next generation. In the HSM- scheduling problem, the fitness is defined as a composite optimization objective as discussed in Section 8.2.

8.3.1.4 Genetic Operators

8.3.1.4.1 Selection

The selection operator is used to improve the mean fitness values of the population by giving the better chromosomes higher probabilities to pass their genotypic information to the next generation. The selection schemes are usually characterized by a selection pressure, which is defined as the ratio of the selection probability of the best chromosome to that of an average chromosome. The rank-based selection scheme selects individuals based on the rank of their fitness values, and the selection probability is defined as $p_i = c(1 - c)^{i-1}$, where c denotes the selection pressure, and i is the rank number of a chromosome in the whole population (Michalewicz, 1996).

To generate new individuals, one parent is chosen from the feasible solution pool by the roulette-wheel method, and another parent is selected from the current population using a niche technique. The niche technique ensures a minimum difference between every two parent chromosomes and further maintains the genetic diversity of the new population. Thus, a similarity coefficient is defined as $c_{ij} = s_{ij}/n$, where n is the chromosome size, s_{ij} is the number of identical genes between chromosomes i and j. When a parent chromosome i is selected, only the chromosome j, whose similarity coefficient c_{ij} with the chromosome i is not higher than a predetermined value c_0, has the possibility to be selected as the second parent chromosome.

8.3.1.4.2 Crossover

The crossover operator transforms parent chromosomes for finding better child chromosomes. Partially matched crossover (PMX) and ordered crossover (OX) have been proved as effective crossover schemes for integer chromosomes (Larrañaga et al., 1999). In the HSM-scheduling problem, the integral range is equal to, or greater than the chromosome length. As a result, the genes in a child chromosome are not completely homologous to that of its parents, and the crossover operator in the MGA is also not completely identical to that of the standard PMX and OX. Here, we present a simple example to illustrate the PMX operator in the MGA. Given two parent chromosomes S_1 and S_2:

$$S_1: \quad 5-8-9-|12-7-11|-2-10-3$$

$$S_2: \quad 7-6-11-|1-9-10|-5-4-8$$

First, two cut points are chosen at random, and the genes bounded by the cut points are exchanged. As a result, one chromosome has some new partial genetic information from the other. The intermediate structures of the new solutions are

$$S_1': \quad 5-8-9-|1-9-10|-2-10-3$$

$$S_2': \quad 7-6-11-|12-7-11|-5-4-8$$

However, these two intermediate solutions are not necessarily feasible because some genes are repeated. The repeated genes can be replaced by mapping $|12 - 7 - 11|$ to $|1 - 9 - 10|$. And then, two new solutions are generated as follows:

$$S_1': \quad 5-8-7-|1-9-10|-2-11-3$$

$$S_2': \quad 9-6-10-|12-7-11|-5-4-8$$

One can see that the gene 1 is not included in the parent chromosome S_1, but it appears in its child chromosome S_1' after the PMX operation.

8.3.1.4.3 Local Search as Mutation

The mutation operator generates a new chromosome from a selected one. In the MGA, the *Or-opt* exchange is employed as the mutation operator. It is one of the chain-exchanging methods and attempts to improve the current solution by moving a chain of one or two consecutive nodes to a different location until no further improvement can be obtained. An example of the *Or-opt* exchange is given in Figure 8.4.

8.3.1.4.4 Repair Strategy

Sometimes, the crossover operation inevitably violates some constraints and generates nonfeasible solutions. Thus, a repair strategy is presented to maintain a feasible solution pool. The repair strategy is described as follows:

Algorithm 2: (Pseudocode for repair strategy)

Step 1. Sequence the selected coils in the current solution according to the descending width and ascending due date.

Step 2. If the groove constraint is satisfied, go to *step 3*. Else, delete a coil with a specific width violating the groove constraint, randomly select a new coil with a different width from manufacturing orders, and insert it to the current-rolling round with the least cost, and go to *step 2*.

Step 3. If the short-slab constraint is satisfied, go to *step 4*. Else, delete a short slab from the rolling round, select a nonshort slab from manufacturing orders randomly, and insert it to the current sequence with the least cost without violating the groove constraint, and then go to *step 3*.

Step 4. Stop this repair strategy, and return the scheduling solution.

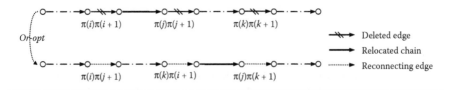

Figure 8.4 Example of the *Or-opt* exchange: moving chain $(\pi(i+1), \pi(j))$ to the position between $\pi(k)$ and $\pi(k+1)$. (Reprinted from *Computers and Operations Research*, 39 (2), Chen, Y. W. et al., Development of hybrid evolutionary algorithms for production scheduling of hot strip mill, 339–349, Copyright 2012, with permission from Elsevier.)

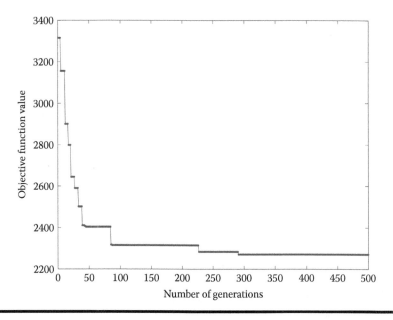

Figure 8.5 **Convergence curve of the modified GA. (Reprinted from *Computers and Operations Research*, 39 (2), Chen, Y. W. et al., Development of hybrid evolutionary algorithms for production scheduling of hot strip mill, 339–349, Copyright 2012, with permission from Elsevier.)**

After some criteria are satisfied, the MGA terminates the iterative process and reports the best schedule solution so far. The termination criteria can be a certain number of generations (*Gen*) or a given computation time. In the study, the algorithms are coded in C ++, and compile with *MS Visual Studio 6.0*. The initial parameters are set as follows: the population size $Pop = 200$, the selection pressure $c = 0.1$, the similarity coefficient threshold $c_0 = 0.5$, the crossover probability $p_c = 0.95$, and the mutation probability $p_m = 0.05$. By simulating a set of production data collected from a hot-rolling mill, the convergence curve of the proposed MGA is shown in Figure 8.5.

One can see that the proposed MGA can converge into a satisfactory solution within a few hundreds of generations.

8.3.2 Local Improving Algorithm: τ-EO

In the HSM-scheduling problem, the localized sequence-dependent transition costs play an important role in making the resulting schedule solution satisfactory. Thus, in this section, a novel local improving algorithm—τ-EO is presented to exploit the quality of a specific scheduling solution.

8.3.2.1 Introduction to EO

EO is inspired by the self-organized critical models in ecosystems (Boettcher and Percus, 1999). For a suboptimal solution, the EO algorithm eliminates the components with an extremely undesirable performance and then replaces them with randomly selected new components. Finally, a better solution may be generated by repeating such kinds of local search process. For a general minimization problem, the basic EO algorithm proceeds as follows:

Algorithm 3: (Pseudocode for the basic τ-EO algorithm)

Step 1. Initialize a solution S, and set $S_{best} = S$.
Step 2. For the current solution S
    a. Evaluate the local fitness λ_i for each variable x_i
    b. Find j with $\lambda_j \geq \lambda_i$ for all i, that is, x_j has the worst local fitness
    c. Choose at random $S' \in N(S)$ such that the "worst" x_j changes its state
    d. If the optimization objective $F(S') < F(S_{best})$, then set $S_{best} \leftarrow S'$
    e. Accept $S \leftarrow S'$ unconditionally, independently of $F(S') - F(S_{best})$.
Step 3. Repeat step 2 as long as desired.
Step 4. Return S_{best} and $F(S_{best})$. ■

It is obvious that the basic EO algorithm has no parameters, which can be adjusted for selecting better solutions. To improve its optimization performance and avoid the possible dead ends, a general modification of the EO algorithm called τ-EO is proposed by introducing an adjustable parameter (Boettcher and Percus, 2000). In the τ-EO algorithm, all variables x_i are ranked according to their fitness values λ_i, namely, find a permutation $\Pi: \lambda_{\Pi(1)} \geq \lambda_{\Pi(2)} \geq \cdots \geq \lambda_{\Pi(n)}$. Subsequently, each variable x_i to be updated is selected according to a probability distribution $P_k \propto k^{-\tau}, 1 \leq k \leq n$, where k is the rank of the variable x_i. The power-law distribution ensures that no ranks get excluded for further evolution while maintaining a bias against variables with bad fitness (Boettcher and Percus, 2000). In the past decade, the EO algorithm and its derivatives have been extensively applied to solve various COPs. Simulation results proved that the EO algorithm outperforms other state-of-the-art algorithms in many applications, such as graph bipartitioning, satisfiability (MAX-K-SAT), TSP problems, and some industrial applications (Boettcher and Percus, 1999; 2000; De Sousa et al., 2004; Chen et al., 2007).

8.3.2.2 EO for Improving the Body Section

Since the local transition costs between adjacent coils are the main parts of the scheduling optimization objective, the local search τ-EO algorithm is applied to improve the HSM-scheduling solution. The local fitness is defined as $\lambda_i = c_{p(i),i} + c_{i,s(i)}$, where $p(i)$ and $s(i)$ denotes the predecessor and the successor of coil i, respectively. It means that the local fitness of a scheduled coil i in a rolling round is the sum of two sequence-dependent penalties that can be calculated by using only local information.

Algorithm 4: (Pseudocode for local improving algorithm—τ-EO)

> *Step 1.* Initialize parameters and obtain an initial solution S, which can be inherited from other algorithms, and then calculate the optimization objective $F(S)$, set $S_{best} = S$.
>
> *Step 2.* For the current solution S, evaluate the local fitness λ_i for all scheduled coils and rank them according to their fitness values.
>
> *Step 3.* Select a coil $c(s)$ according to the power-law distribution $p_k(\tau_0)$, where τ_0 is a given parameter value.
>
> *Step 4.* Choose the best solution S' from a neighboring subspace $N(S)$ of the current solution S.
>
> *Step 5.* If $F(S') < F(S_{best})$, then set $S_{best} \leftarrow S'$.
>
> *Step 6.* Accept $S \leftarrow S'$ unconditionally.
>
> *Step 7.* If termination criteria are not satisfied, go to step 2; else go to the next step.
>
> *Step 8.* Return S_{best} and $F(S_{best})$.　　　　　　　　　　　　■

Note that in step 4, the subspace $N(S)$ can be generated through various strategies. In this chapter, the "route-improvement" algorithm that is similar to the perturbation moves (Cowling, 2003) is presented to generate the neighboring subspace. We take a scheduling solution S and improve it by slight perturbations. The perturbation is iterated until no further improvement is possible, and then the local optimum S' is obtained. The perturbation processes can be described as follows:

> *Delete* a selected coil $c(s)$ from the current- rolling sequence S
>
> *Select* an unscheduled coil $c(u)$, and insert it into the least-cost position
>
> *Accept* the solution S' as an element of $N(S)$ if the new rolling sequence is feasible and the local fitness $\lambda_{c(u)} < \lambda_{c(s)}$
>
> *Repeat* the above steps until all unscheduled orders have been processed

The above perturbation moves are usually employed by human schedulers. Therefore, the local optimum in the neighboring subspace is intuitively reasonable in real scheduling systems.

Because the τ-EO algorithm has only one parameter τ, its optimal choice is critical for improving the optimization performance. A number of experimental and theoretical research efforts have been devoted to analyzing the optimal parameter selection for different COPs (Boettcher and Percus, 2003; Chen et al., 2007). In the HSM-scheduling problem, a set of simulations indicate that the algorithm reaches the best solution with high probability at a prediction value $\tau_{opt} \approx 2.0$, and the objective function value seems to rise gradually around this τ_{opt} value for different scheduling instances as shown in Figure 8.6.

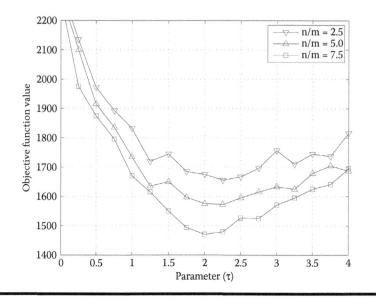

Figure 8.6 **Plot of the objective function value over the parameter τ under a number of scheduling instances.** (*Reprinted from Computers and Operations Research*, **39** (2), Chen, Y. W. et al., Development of hybrid evolutionary algorithms for production scheduling of hot strip mill, 339–349, Copyright 2012, with permission from Elsevier.)

Using the solution generated by the previous MGA as the initial solution, the τ-EO algorithm has the convergence curve as shown in Figure 8.7.

It is obvious that the algorithm can improve the initial solution significantly and converges into a solution within 2000 generations. Note that the optimization process takes less than 45 s.

8.3.2.3 Hybrid Evolutionary Algorithms

The hybrid EAs combine the MGA and the τ-EO algorithms in different ways. First of all, the best scheduling solution generated by the MGA is further optimized by the τ-EO algorithm. This scheme is quite straightforward but shows a weak integration. Second, multiple solutions of the MGA are improved by the τ-EO algorithm for increasing the diversity of initial solutions. Third, an integrated method is presented, in which the τ-EO algorithm acts as the mutation operator of the MGA. Simulation results show that the third combination scheme provides better results than other schemes. Below, we summarize and compare the simulation results of the above three hybrid evolutionary methods: GEO-1, GEO-2, and GEO-3.

1. GEO-1 (the best solution of the MGA is optimized by the τ-EO): Figure 8.8 compares the local fitness (i.e., local sequence-dependent transition costs) of the best solution of the MGA with that of the final solution of the GEO-1.

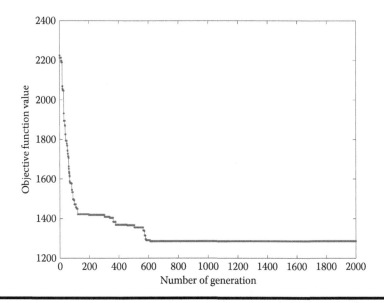

Figure 8.7 Convergence curve of the τ-EO algorithm. (Reprinted from *Computers and Operations Research*, 39 (2), Chen, Y. W. et al., Development of hybrid evolutionary algorithms for production scheduling of hot strip mill, 339–349, Copyright 2012, with permission from Elsevier.)

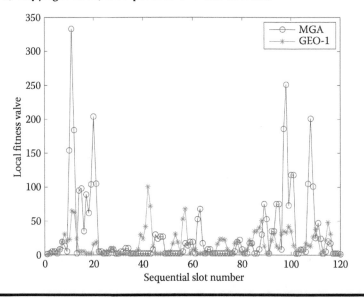

Figure 8.8 Local fitness comparison between the MGA and the GEO-1. (Reprinted from *Computers and Operations Research*, 39 (2), Chen, Y. W. et al., Development of hybrid evolutionary algorithms for production scheduling of hot strip mill, 339–349, Copyright 2012, with permission from Elsevier.)

It is obvious that the GEO-1 can generate a scheduling solution with much lower sequence-dependent transition costs by iteratively replacing the undesirable or underperformed coils in a rolling sequence.

2. GEO-2 (top 20 solutions of the MGA are further optimized by the τ-EO): In the GEO-2, the top 20 solutions in the final population of the MGA are improved by the τ-EO algorithm.

One can see from Figure 8.9 that the GEO-2 can considerably improve both local fitness (i.e., the LEF of evaluating the sequence-dependent transition costs and the nonexecution penalties) and global fitness (i.e., the GEF of evaluating the E/T penalties). It is worth noting that in the above figure, the multiple solutions of the MGA are ordered according to the value of the optimization objective $F(S)$. Obviously, an initial solution with a "worse" fitness value may be improved to a "better" scheduling solution by the GEO-2, such as the No. 6 solution of the top 20 solutions in Figure 8.9.

3. GEO-3 (using τ-EO as mutation operations in MGA): In the GEO-3, the τ-EO algorithm is used as the mutation operator of the MGA. So, a number of $Pop \times p_m$ solutions are optimized through the τ-EO algorithm in each generation of the MGA. Using the same initial solution, the evolutionary processes of the GEO-1, the GEO-2, and the GEO-3 are illustrated in Figure 8.10.

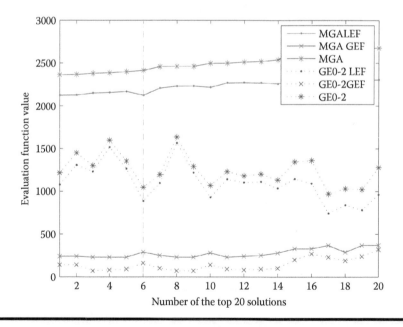

Figure 8.9 Fitness comparison of a multisolution generated by the MGA and the GEO-2. (Reprinted from *Computers and Operations Research*, 39 (2), Chen, Y. W. et al., Development of hybrid evolutionary algorithms for production scheduling of hot strip mill, 339–349, Copyright 2012, with permission from Elsevier.)

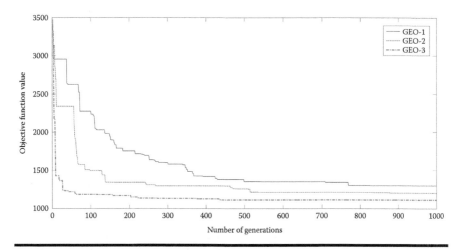

Figure 8.10 Comparison on the evolutionary processes of different combination schemes. (Reprinted from *Computers and Operations Research*, 39 (2), Chen, Y. W. et al., Development of hybrid evolutionary algorithms for production scheduling of hot strip mill, 339–349, Copyright 2012, with permission from Elsevier.)

In the simulation of a wide range of industrial data, the GEO-3 almost always provides the best scheduling solution. Therefore, the GEO-3 is selected as the optimization engine of our HSM-scheduling system.

8.3.3 Design of a HSM-Scheduling System

This scheduling system is defined as a configurable tool that solves the HSM-scheduling problem to maximize production throughput, improve product quality, and customer service (i.e., on-time delivery) levels. The schematic structure of the scheduling prototype system is presented in Figure 8.11.

The HSM-scheduling system first downloads manufacturing orders from the upper-level ERP systems, and then generates computer-aided scheduling scenarios for a single-rolling round or multiple rounds (i.e., a campaign) by the hybrid EA as discussed previously. The system assumes that human planners are allowed to guide the scheduling process and tune the optimization parameters for imposing human decision making. The semiautomatic scheduling process is described as follows:

Step 1. The human planner sets the scheduling time horizon to select available slabs according to the corresponding coils in manufacturing orders. The scheduling system allows planners to adjust the selection rules based on the attributes of manufacturing orders (e.g., due date, priority).

Step 2. Select the type of rolling rounds (i.e., coffin-type selection). In order to help the planner make a reasonable choice, the system provides detailed information of different round types.

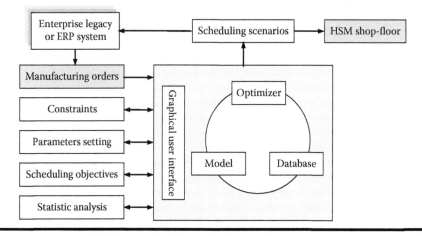

Figure 8.11   Schematic structure of the HSM-scheduling system. (Reprinted from *Computers and Operations Research*, 39 (2), Chen, Y. W. et al., Development of hybrid evolutionary algorithms for production scheduling of hot strip mill, 339–349, Copyright 2012, with permission from Elsevier.)

Step 3. Based on the site-specific production requirements, the planner can configure the constraint settings.

Step 4. Under the preconfigured conditions, the planner starts the optimization engine to generate scheduling solutions.

Step 5. Evaluate the generated scheduling solution. The planner can check the scheduling solutions by viewing the graphical width (or gauge, hardness, etc.) patterns and the statistical analysis results. The planner can also edit a specific rolling round by using graphical user interfaces, such as inserting, replacing, moving, or deleting slabs from a rolling round, or rescheduling some rolling rounds in the existing scheduling scenario.

Since the object-oriented software technology has gained wide recognition as a preferred approach for building and maintaining complex application programs, the scheduling system is developed under the object-oriented platform of *MS Visual Studio 6.0* and *MS SQL Server*. The main graphical user interface of the developed scheduling system is shown in Figure 8.12.

In the developed HSM-scheduling system, the human planner can view the detailed profiles of the scheduled rolling rounds in width, gauge, and hardness transitions and jumps through graphical user interfaces. He/she can also adjust the existing rolling rounds by inserting, replacing, moving, or deleting scheduled slabs. In addition, the scheduling solutions can be uploaded to the upper-level management systems and downloaded to the shop-floor manufacturing execution system.

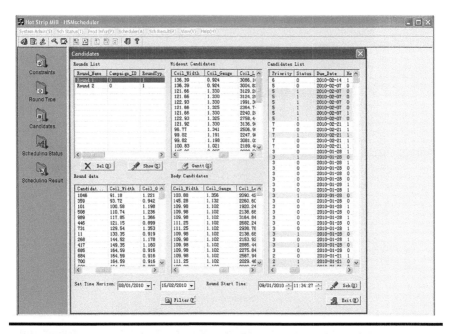

Figure 8.12 (See color insert.) **Main graphical user interface of the developed HSM-scheduling system. (Reprinted from** *Computers and Operations Research,* **39 (2), Chen, Y. W. et al., Development of hybrid evolutionary algorithms for production scheduling of hot strip mill, 339–349, Copyright 2012, with permission from Elsevier.)**

8.3.4 Computational Results

In this section, we first consider a set of production data, which consists of 1050 manufacturing orders with 336 short slabs and 75 warm-up slabs. The specifications of the corresponding coils mainly include width (range: 88.90–164.59 cm), gauge (range: 0.883–1.404 cm), hardness (range: 1–10), length, and finished temperature (range: 950–1450°C). The scheduling objective can be calculated on the basis of the penalty table and the mathematical model in Section 8.2. Figure 8.13 shows the width, gauge, and hardness transition patterns of a rolling round.

One can see in Figure 8.13 that the width profile of the rolling round follows a standard "coffin-shape" pattern, and the gauge and hardness transitions are smooth. It is worth noting that the scheduling of the rolling round only takes about 320 s on a Pentium *2.4-GHz CPU.* The campaign can also be constructed by generating rounds one by one. Figure 8.14 shows the width transition pattern of a rolling campaign. The optimization objective values of the five consecutive rolling rounds are 1112, 1733, 2639, 4107, and 3505, respectively.

During generating the rolling campaign, the optimization objectives for the MGA and the GEO-3 are reported in Table 8.2.

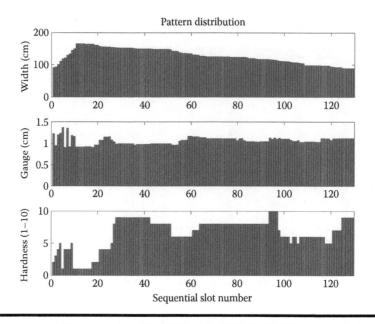

Figure 8.13 Width, gauge, and hardness transition patterns of a rolling round example. (Reprinted from *Computers and Operations Research*, 39 (2), Chen, Y. W. et al., Development of hybrid evolutionary algorithms for production scheduling of hot strip mill, 339–349, Copyright 2012, with permission from Elsevier.)

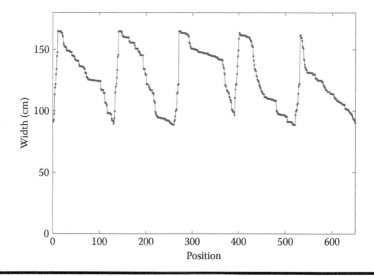

Figure 8.14 Width transition pattern of a rolling campaign. (Reprinted from *Computers and Operations Research*, 39 (2), Chen, Y. W. et al., Development of hybrid evolutionary algorithms for production scheduling of hot strip mill, 339–349, Copyright 2012, with permission from Elsevier.)

Table 8.2 Improvement Obtained by the Proposed GEO-3 without Updating Manufacturing Orders

Round No.	MGA			GEO-3			Improvement (%)
	LEF Value	GEF Value	Objective	LEF Value	GEF Value	Objective	
1	1916	90	2006	1032	80	1112	44.57
2	2443	230	2673	1583	150	1733	35.17
3	3973	260	4233	2539	100	2639	37.66
4	4947	370	5317	4007	100	4107	22.76
5	8162	220	8382	3385	120	3505	58.18

Source: Reprinted from *Computers and Operations Research*, 39 (2), Chen, Y. W. et al., Development of hybrid evolutionary algorithms for production scheduling of hot strip mill, 339–349, Copyright 2012, with permission from Elsevier.

Note that the rolling rounds usually become worse, without updating the manufacturing order pool. To simulate a dynamic environment, we import 200 new coils into the set of manufacturing orders after the scheduling of each rolling round. The optimization objectives for the heuristic algorithm 1 in Section 8.3.1, the MGA and the GEO-3 are reported in Table 8.3.

Table 8.3 Improvement Obtained by the Proposed GEO-3 with Updating Manufacturing Orders

Instance No.	Heuristic Algorithm 2 Objective	MGA Objective	GEO-3 Objective	Improvement (% over MGA)
1	4129	2404	1523	36.65
2	3824	1715	1341	21.81
3	3958	2253	1641	27.16
4	4007	1971	1374	30.29
5	4203	2685	2030	24.39

Source: Reprinted from *Computers and Operations Research*, 39 (2), Chen, Y. W. et al., Development of hybrid evolutionary algorithms for production scheduling of hot strip mill, 339–349, Copyright 2012, with permission from Elsevier.

In Tables 8.2 and 8.3, it is obvious that the proposed hybrid EA can considerably improve the optimization objective. The HSM-scheduling system equipped with the optimization engine can obtain an optimized rolling round within 600 s. Furthermore, extensive computational results on industrial data demonstrate that the developed HSM-scheduling system has superior performances in scheduling quality and computing efficiency.

8.4 Summary

In this chapter, the HSM-scheduling problem in the steel industry is studied. A mathematical model is formulated to describe two important scheduling subtasks: (1) selecting a subset of manufacturing orders, and (2) generating an optimal rolling sequence. In view of the complexity of the scheduling problem, hybrid EAs are proposed through the combination of GA and EO. With the help of the developed HSM-scheduling system, simulations are conducted to demonstrate that the hybrid EAs can generate optimized rolling rounds or campaigns efficiently. Although the hybrid EA in this chapter is developed to solve the HSM-scheduling problem, it has great potential in the areas of scheduling and optimization. Future work will emphasize the generalization of this hybrid evolutionary optimization method.

References

Achlioptas, D., Naor, A., and Peres, Y. 2005. Rigorous location of phase transitions in hard optimization problems. *Nature* 435: 759–764.

Ackley, D. H. 1987. *A Connectionist Machine for Genetic Hill Climbing*. Kluwer, Boston, MA.

Ahmed, E. and Elettreby, M. F. 2004. On multi-objective evolution model. *International Journal of Modern Physics C* 15 (9): 1189–1195.

Al Seyab, R. K. and Cao, Y. 2006. Nonlinear model predictive control for the ALSTOM gasifier. *Journal of Process Control* 16: 795–808.

Albert, R. and Barabási, A. L. 2000. Topology of evolving networks: Local events and universality. *Physical Review Letters* 85: 5234–5237.

Ali, S. and Smith, K. A. 2006. A meta-learning approach to automatic kernel selection for support vector machines. *Neurocomputing* 70: 173–186.

Altenberg, L. 1997. NK fitness landscapes. In: T. Bäck, D. B. Fogel and Z. Michalewicz (Eds.), *Handbook of Evolutionary Computation*. Oxford University Press, New York.

Ang, K. H., Chong, G., and Li, Y. 2005. PID control system analysis, design and technology. *IEEE Transactions on Control Systems Technology* 13 (4): 559–576.

Arifovic, J. and Gencay, R. 2001. Using genetic algorithms to select architecture of a feed-forward artificial neural network. *Physica A: Statistical Mechanics and Its Applications* 289: 574–594.

Armañanzas, R. and Lozano, J. A. 2005. A multi-objective approach to the portfolio optimization problem. *Proceedings of CEC'2005*, Edinburgh, UK, pp. 1388–1395.

Assaf, I., Chen, M., and Katzberg, J. 1997. Steel production schedule generation. *International Journal of Production Research* 35 (2): 467–477.

Azadehgan, V., Jafarian, N., and Jafarieh, F. 2011. A novel hybrid artificial bee colony with extremal optimization. *Proceedings of the 4th International Conference on Computer and Electrical Engineering (ICCEE 2011)*, Singapore, pp. 45–49.

Bak, P. 1996. *How Nature Works: The Science of Self-Organized Criticality*. Copernicus Press, New York.

Bak, P. and Sneppen, K. 1993. Punctuated equilibrium and criticality in a simple model of evolution. *Physical Review Letters* 71 (24): 4083–4086.

Bak, P., Tang, C., and Wiesenfeld, K. 1987. Self-organized criticality: An explanation of 1/f noise. *Physical Review Letters* 59: 381–384.

Balas, E. 1989. The prize collecting traveling salesman problem. *Networks* 19: 621–636.

Barabási, A. L. 2007. The architecture of complexity. *IEEE Control Systems Magazine* 27 (4): 33–42.

Barabási, A. L. and Oltvai, A. L. 2004. Network biology: Understanding the cell's functional organization. *Nature Reviews Genetics* 5: 101–113.

Bauke, H., Mertens, S., and Engel, A. 2003. Phase transition in multiprocessor scheduling. *Physical Review Letters* 90 (15): 158701-1–158701-4.

Baykasoglu, A. 2006. Applying multiple objective tabu search to continuous optimization problems with a simple neighborhood strategy. *International Journal for Numerical Methods in Engineering* 65: 406–424.

Beasley, J. E. 1990. OR-library: Distributing test problems by electronic mail. *Journal of the Operational Research Society* 41: 1069–1072.

Beausoleil, R. P. 2006. "MOSS" multi-objective scatter search applied to non-linear multiple criteria optimization. *European Journal of Operational Research* 169: 426–449.

Beck, J. C. and Jackson, K. 1997. Constrainedness and phase transition in job shop scheduling. *Technical Report*, School of Computer Sciences, Simon Fraster University, Burnaby, BC.

Biehl, M., Ahr, M., and Schlösser, E. 2000. Statistical physics of learning: Phase transitions in multilayered neural networks. *Advances in Solid State Physics* 40: 819–826.

Biroli, G., Cocco, S., and Monasson, R. 2002. Phase transitions and complexity in computer science: An overview of the statistical physics approach to the random satisfiability problem. *Physica A* 306: 381–394.

Blum, C. and Roli, A. 2003. Metaheuristics in combinatorial optimization: Overview and conceptual comparison. *ACM Computing Surveys* 35 (3): 268–308.

Boettcher, S. 2005a. Extremal optimization for the Sherrington–Kirkpatrick spin glass. *European Physics Journal B* 46: 501–505.

Boettcher, S. 2005b. Self-organizing dynamics for optimization. *Computational Science, Lecture Notes in Computer Science* 3515: 386–394.

Boettcher, S. and Frank, M. 2006. Optimizing at the ergodic edge. *Physica A* 367: 220–230.

Boettcher, S. and Percus, A. G. 1999. Extremal optimization: Methods derived from co-evolution. *Proceedings of the Genetic and Evolutionary Computation Conference*, Orlando, FL, pp. 825–832.

Boettcher, S. and Percus, A. G. 2000. Nature's way of optimizing. *Artificial Intelligence* 119: 275–286.

Boettcher, S. and Percus, A. G. 2001a. Optimization with extremal dynamics. *Physical Review Letters* 86 (23): 5211–5214.

Boettcher, S. and Percus, A. G. 2001b. Extremal optimization for graph partitioning. *Physical Review E* 64 (2): 1–13.

Boettcher, S. and Percus, A. G. 2003. Optimization with extremal dynamics. *Complexity* 8: 57–62.

Bonabeau, E., Dorigo, M., and Theraulaz, G. 2000. Inspiration for optimization from social insect behaviour. *Nature* 406: 39–42.

Bosman, P. A. N. and Thierens, D. 2003. The balance between proximity and diversity in multi-objective evolutionary algorithms. *IEEE Transactions on Evolutionary Computation* 7 (2): 174–188.

Burges, C. J. C. 1998. A tutorial on support vector machines for pattern recognition. *Knowledge Discovery and Data Mining* 2: 121–167.

Burke, E. K., Cowling, P. I., and Keuthen, R. 2001. Effective local and guided variable neighborhood search methods for the asymmetric traveling salesman problem. *Applications of Evolutionary Computing, Lecture Notes in Computer Science* 2037: 203–212.

Camacho, E. F. and Bordons, C. 1995. *Model Predictive Control in the Process Industry*. Springer-Verlag, Berlin.

Chaiyatham, T. and Ngamroo, I. 2012. A bee colony optimization based-fuzzy logic–PID control design of electrolyzer for microgrid stabilization. *International Journal of Innovative Computing, Information and Control* 8 (9): 6049–6066.

Chang, W. A. and Ramakrishna, R. S. 2002. A genetic algorithm for shortest path routing problem and the sizing of population. *IEEE Transactions on Evolutionary Computation* 6 (6): 566–579.

Chang, W. D. 2007. A multi-crossover genetic approach to multivariable PID controllers tuning. *Expert Systems with Applications* 33: 620–626.

Cheesman, P., Kanefsky, B., and Taylor, W. M. 1991. Where the really hard problems are. *Proceedings of the 12th International Joint Conference on Artificial Intelligence*, Sidney, Australia, pp. 163–169.

Chen, A. L., Yang, G. K., and Wu, Z. M. 2008. Production scheduling optimization algorithm for the hot rolling processes. *International Journal of Production Research* 46 (7): 1955–1973.

Chen, D. B. and Zhao, C. X. 2009. Particle swarm optimization with adaptive population size and its application. *Applied Soft Computing* 9: 39–48.

Chen, X., Wan, W., and Xu, X. 1998. Modeling rolling batch planning as vehicle routing problem with time windows. *Computers and Operations Research* 25 (12): 1127–1136.

Chen, Y. and Zhang, P. 2006. Optimized annealing of traveling salesman problem from the nth-nearest-neighbor distribution. *Physica A* 371: 627–631.

Chuang, C. C. and Lee, Z. J. 2009. Hybrid robust support vector machines for regression with outliers. *Applied Soft Computing* 11: 64–72.

Cirasella, J., Johnson, D. S., McGeoch, L. A., and Zhang, W. X. 2001. The asymmetric traveling salesman problem: Algorithms, instance generators, and tests. *Algorithm Engineering and Experimentation, Lecture Notes in Computer Science* 2153: 32–59.

Clauset, A., Shalizi, C. R., and Newman, M. E. J. 2009. Power-law distributions in empirical data. *SIAM Review* 51 (4): 661–703.

Coello, C. A. C. 1996. An empirical study of evolutionary techniques for multi-objective optimization in engineering design. PhD thesis, Department of Computer Science, Tulane University, New Orleans, LA.

Coello, C. A. C. 2005. Recent trend in evolutionary multi-objective optimization. In: A. Abraham, L. Jain., and R. Goldberg (Eds.), *Evolutionary Multi-Objective Optimization: Theoretical Advances and Applications*, Springer-Verlag, London, pp. 7–32.

Coello, C. A. C. 2006. Evolutionary multi-objective optimization: A historical view of the field. *IEEE Computational Intelligence Magazine* 1 (1): 28–36.

Coelho, L. S. and Pessôa, M. W. 2011. A tuning strategy for multivariable PI and PID controllers using differential evolution with chaotic Zaslavskii map. *Expert Systems with Applications* 38: 13694–13701.

Coello, C. A. C., Pulido, G. T., and Leehuga, M. S. 2004. Handling multiple objectives with particle swarm optimization. *IEEE Transactions on Evolutionary Computation* 8 (3): 256–279.

Cormen, T. H., Leiserson, C. E., Rivest, R. L., and Stein, C. 2001. *Introduction to Algorithms*. MIT Press, Cambridge, MA.

Cowling, P. 2003. A flexible decision support system for steel hot rolling mill scheduling. *Computers and Industrial Engineering* 45: 307–321.

Cox, I. J., Lewis, R. W., Ransing, R. S., Laszczewski, H., and Berni, G. 2002. Application of neural computing in basic oxygen steelmaking. *Journal of Material Processing Technology* 120: 310–315.

Daley, M. J. and Kari, L. 2002. DNA computing: Models and implementations. *Comments on Theoretical Biology* 7: 177–198.

Das, I. and Dennis, J. 1997. A close look at drawbacks of minimizing weighted sum of objectives for Pareto set generation in multicriteria optimization problems. *Structure Optimization* 14 (1): 63–69.

Davendra, D., Zelinka, I., and Senkerik, R. 2010. Chaos driven evolutionary algorithms for the task of PID control. *Computers and Mathematics with Applications* 60: 1088–1104.

Dawkins, R. 1976. *The Selfish Gene*. Oxford University Press, Oxford, UK.

Deb, K., Pratab, A., Agrawal, S., and Meyarivan, T. 2002. A fast and elitist multi-objective genetic algorithm: NSGA-II. *IEEE Transactions on Evolutionary Computation* 6 (2): 182–197.

Deb, K., Patrap, A., and Moitra, S. 2000. Mechanical component design for multi-objective using elitist non-dominated sorting GA. *KanGAL Report 200002*, Indian Institute of Technology, Kanpur, India.

De Castro, L. N. and Timmis, J. 2002. *Artificial Immune Systems: A New Computational Intelligence Approach*. Springer, London.

De Menezes, M. A. and Lima, A. R. 2003. Using entropy-based methods to study general constrained parameter optimization problems. *Physica A* 323: 428–434.

Dengiz, B., Alabas-Uslu, C., and Dengiz, O. 2009. A tabu search algorithm for the training of neural networks. *Journal of the Operational Research Society* 60: 282–291.

De Sousa, F. L. and Ramos, F. M. 2002. Function optimization using extremal dynamics. *Proceedings of the 4th International Conference on Inverse Problems in Engineering*, Rio de Janeiro, Brazil, pp. 1–5.

De Sousa, F. L., Vlassov, V., Ramos, F. M. 2004. Generalized extremal optimization: An application in heat pipe design. *Applied Mathematical Modeling* 28: 911–931.

Djoewahir, A., Tanaka, K., and Nakashima, S. 2013. Adaptive PSO-based self-tuning PID controller for ultrasonic motor. *International Journal of Innovative Computing, Information and Control* 9 (10): 3903–3914.

Dobson, C. M. 2003. Review article: Protein folding and misfolding. *Nature* 426: 884–890.

Dorigo, M. and Gambardella, L. M. 1997. Ant colony systems: A cooperative learning approach to the TSP. *IEEE Transactions on Evolutionary Computation* 1 (1): 53–66.

Dubolis, O. and Dequen, G. 2001. A backbone-search heuristic for efficient solving of hard 3-SAT formulae. *International Joint Conference on Artificial Intelligence*, Seattle, WA, pp. 248–253.

Duch, J. and Arenas, A. 2005. Community detection in complex networks using extremal optimization. *Physical Review E* 72: 27104.

Eberhart, R. C. and Kennedy, J. 1995. A new optimizer using particle swarm theory. *Proceedings of the 6th International Symposium on Micromachine and Human Science*, Nagoya, Japan, pp. 39–43.

Ehrlich, P. R. and Raven, P. H. 1964. Butterflies and plants: A study in coevolution. *Society for the Study of Evolution* 18: 586–608.

Elaoud, S., Loukil, T., and Teghem, J. 2007. The Pareto fitness genetic algorithm: Test function study. *European Journal of Operational Research* 177: 1703–1719.

Engel, A. 2001. Complexity of learning in artificial neural networks. *Theoretical Computer Science* 265: 285–306.

Engelbrecht, A. P. 2007. *Computational Intelligence, an Introduction*, 2nd ed. John Wiley & Sons, New York.

Engin, A. 2009. Selecting of the optimal feature subset and kernel parameters in digital modulation classification by using hybrid genetic algorithm–support vector machines: HGASVM. *Expert Systems with Application* 36: 1391–1402.

Fan, S. K. S. and Zahara, E. 2007. A hybrid simplex search and particle swarm optimization for unconstrained optimization. *European Journal of Operational Research* 181: 527–548.

Fang, H. L. and Tsai, C. H. 1998. A genetic algorithm approach to hot strip mill rolling scheduling problems. *Proceedings of the Tenth IEEE International Conference on Tools with Artificial Intelligence*, Taipei, Taiwan, pp. 264–271.

Fasih, A., Chedjou, C. J., and Kyamakya, K. 2009. Cellular neural networks-based genetic algorithm for optimizing the behavior of an unstructured robot. *International Journal of Computational Intelligence Systems* 2: 124–131.

Feng, M. X., Li, Q., and Zou, Z. S. 2008. An outlier identification and judgment method for an improved neural-network BOF forecasting model. *Steel Research International* 79: 323–332.

Fileti, A. M. F., Pacianotto, T. A., and Cunha, A. P. 2006. Neural modelling helps the BOS process to achieve aimed end-point conditions in liquid steel. *Engineering Applications of Artificial Intelligence* 19: 9–17.

Fonseca, C. M. and Fleming, P. J. 1993. Genetic algorithms for multi-objective optimization: Formulation, discussion and generalization. In: S. Forrest (Ed.), *Proceedings of the 5th International Conference on Genetic Algorithms*, Morgan Kaufmann, San Mateo, CA, pp. 416–423.

Fonseca, C. M. and Fleming, P. J. 1995. An overview of evolutionary algorithms in multi-objective optimization. *Evolutionary Computation* 1: 1–16.

Forrest, S. 1993. Genetic algorithms: Principles of natural selection applied to computation. *Science* 261 (5123): 872–878.

Franz, A. and Hoffmann, K. H. 2002. Optimal annealing schedules for a modified Tsallis statistics. *Journal of Computational Physics* 176: 196–204.

Fredman, M. L., Johnson, D. S., Mcgeoch, L. A., and Ostheimer, G. 1995. Data structures for traveling salesmen. *Journal of Algorithms* 18: 432–479.

Friedrichs, F. and Igel, C. 2005. Evolutionary tuning of multiple SVM parameters. *Neurocomputing* 64: 107–117.

Fukumizu, K. and Amari, S. 2000. Local minima and plateaus in hierarchical structures of multilayer perceptrons. *Neural Networks* 13: 317–327.

Fukuoka, Y., Matsuki, H., Minamitani, H., and Ishida, A. 1998. A modified back propagation method to avoid false local minima. *Neural Networks* 11: 1059–1072.

Gabrielli, A., Cafiero, R., Marsili, M., and Pietronero, L. 1997. Theory of self-organized criticality for problems with extremal dynamics. *Europhysics Letters* 38 (7): 491–496.

Galski, R. L., De Sousa, F. L., and Ramos, F. M. 2005. Application of a new multi-objective evolutionary algorithm to the optimum design of a remote sensing satellite constellation. *Proceedings of the 5th International Conference on Inverse Problems in Engineering: Theory and Practice*, Cambridge, Vol. II, G01.

Galski, R. L., De Sousa, F. L., Ramos, F. M., and Muraoka, I. 2004. Spacecraft thermal design with the generalized extremal optimization algorithm. *Proceedings of Inverse Problems, Design and Optimization Symposium*, Rio de Janeiro, Brazil.

Gao, W. F., Liu, S. Y., and Huang, L. L. 2012. Inspired artificial bee colony algorithm for global optimization problems. *Acta Electronica Sinica* 12: 2396–2403.

Garey, M. R. and Jonhson, D. S. 1979. *Computers and Intractability: A Guide to the Theory of NP-Completeness*. W. H. Freeman, New York.

Gent, I. P. and Walsh, T. 1996. The TSP phase transition. *Artificial Intelligence* 88: 349–358.

Goldberg, D. E. 1989. *Genetic Algorithms in Search, Optimization and Machine Learning*. Addison-Wesley, Reading, MA.

Goles, E., Latapy, M., Magnien, C., Morvan, M., and Phan, H. D. 2004. Sandpile models and lattices: A comprehensive survey. *Theoretical Computer Science* 322: 383–407.

Gutin, G. and Punnen, A. P. 2002. *The Traveling Salesman Problem and Its Variations*. Kluwer Academic Publishers, Boston.

Hagan, M. T. and Menhaj, M. B. 1994. Training feedforward networks with the Marquardt algorithm. *IEEE Transactions on Neural Networks* 5: 989–993.

Haken, H. 1977. *Synergetics*. Springer, Berlin, Germany.

Hamida, S. B. and Schoenauer, M. 2002. ASCHEA: New results using adaptive segregational constraint handling. *Proceedings of the Congress on Evolutionary Computation 2002 (CEC'2002)*, Hawaii, pp. 884–889.

Han, J. 2005. Local evaluation functions and global evaluation functions for computational evolution. *Complex Systems* 15 (4):307–347.

Han, J. and Cai, Q. S. 2003. Emergence from local evaluation function. *Journal of Systems Science and Complexity* 16 (3): 372–390.

Han, S. P. 1976. Superlinearly convergent variable metric algorithms for general nonlinear programming problems. *Mathematical Programming* 11: 263–282.

Hanne, T. 2007. A multi-objective evolutionary algorithm for approximating the efficient set. *European Journal of Operational Research* 176: 1723–1734.

Hansen, P. and Jaumard, B. 1990. Algorithms for the maximum satisfiability problem. *Computing* 44: 279–303.

Hart, W. E., Krasnogor, N., and Smith, J. E. 2004. *Recent Advances in Memetic Algorithms*. Springer, Berlin.

Hartmann, A. K. and Rieger, H. (Eds.) 2004. *New Optimization Algorithms in Physics*. Wiley-VCH, Weinheim.

Hartmann, A. K. and Weigt, M. 2005. *Phase Transitions in Combinatorial Optimization Problems: Basics, Algorithms and Statistical Mechanics*. Wiley-VCH, Weinheim, German.

Haykin, S. 1994. *A Comprehensive Foundation. Neural Networks*. Macmillan, New York.

He, L. and Mort, N. 2000. Hybrid genetic algorithms for telecommunications network back-up routing. *British Telecom Technology Journal* 18: 42–56.

Heilmann, F., Hoffmann, K. H., and Salamon, P. 2004. Best possible probability distribution over extremal optimization ranks. *Europhysics Letters* 66 (3): 305–310.

Helsgaun, K. 2000. An effective implementation of the Lin–Kernighan traveling salesman heuristic. *European Journal of Operational Research* 126: 106–130.

Henson, M. A. 1998. Nonlinear model predictive control: Current status and future directions. *Computers and Chemical Engineering* 23: 187–202.

Herroelen, W. and Reyck, B. D. 1999. Phase transition in project scheduling. *Journal of Operations Research Society* 50: 148–156.

Hoffmann, K. H., Heilmann, F., and Salamon, P. 2004. Fitness threshold accepting over extremal optimization ranks. *Physical Review E* 70 (4): 046704.

Hogg, T., Huberman, B. A., and Williams, C. P. 1996. Special issue on frontiers in problem solving: Phase transitions and complexity. *Artificial Intelligence* 81 (1–2).

Hohmann, W., Kraus, M., and Schneider, F. W. 2008. Learning and recognition in excitable chemical reactor networks. *Journal of Physical Chemistry A* 102: 3103–3111.

Holland, J. H. 1992. Genetic algorithms. *Scientific America* 267: 66–72.

Hordijk, W. 1997. A measure of landscapes. *Evolutionary Computation* 4 (4): 335–360.

Horn, J., Nafpliotis, N., and Goldberg, D. E. 1994. A niched Pareto genetic algorithm for multi-objective optimization. *Proceedings of the First IEEE Conference on Evolutionary Computation, IEEE World Congress on Computational Intelligence,* IEEE Service Center, Piscataway, NJ, pp. 82–87.

Hou, S. M. and Li, Y. R. 2009. Short-term fault prediction based on support vector machines with parameter optimization by evolution strategy. *Expert Systems with Applications* 36: 12383–12391.

Howley, T. and Madden, M. G. 2005. The genetic kernel support vector machine: Description and evaluation. *Artificial Intelligence Review* 24: 379–395.

Hsu, C. F., Chen, G. M., and Lee, T. T. 2007. Robust intelligent tracking control with PID-type learning algorithm. *Neurocomputing* 71 (1–3): 234–243.

Hsu, C. W., Chang, C. C., and Lin, C. J. 2004. A practical guide to support vector classification. *Technical Report,* Department of Computer Science and Information Engineering, National Taiwan University.

Huo, H. B., Zhu, X. J., and Cao, G. Y. 2007. Design for two-degree-of-freedom PID regulator based on improved generalized extremal optimization algorithm. *Journal of Shanghai Jiaotong University (Science)* E-12 (2): 148–153.

Hush, D. R. 1999. Training a sigmoidal node is hard. *Neural Computation* 11: 1249–1260.

Iruthayarajan, M. W. and Baskar, S. 2009. Evolutionary algorithms based design of multivariable PID controller. *Expert Systems with Applications* 36: 9159–9167.

Jacobs, R. A. 1988. Increased rates of convergence through learning rate adaptation. *Neural Networks* 1: 295–307.

Jain, A. K. and Mao, J. 1996. Artificial neural networks: A tutorial. *IEEE Computer* 29 (3): 31–44.

Jeng, J. T. 2006. Hybrid approach of selecting hyperparameters of support vector machine for regression. *IEEE Transactions on Systems, Man, and Cybernetics: Part B* 36: 699–709.

Jin, Y., Joshua, K., Lu, H., Liang, Y., and Douglas, B. K. 2008. The landscape adaptive particle swarm optimizer. *Applied Soft Computing* 8: 295–304.

Kall, P. and Wallace, S. W. 1994. *Stochastic Programming.* John Wiley & Sons, Chichester, UK.

Kanellakis, P. C. and Papadimitriou, C. H. 1980. Local search for the asymmetric traveling salesman problem. *Operations Research* 28 (5): 1086–1099.

Karaboga, D. 2005. *An Idea Based on Honey Bee Swarm for Numerical Optimization.* Erciyes University, Turkey.

Karaboga, D. and Akay, B. 2009. A comparative study of artificial bee colony algorithm. *Applied Mathematics and Computation* 214: 108–132.

Karaboga, D. and Akay, B. 2011. A modified artificial bee colony (ABC) algorithm for constrained optimization. *Applied Soft Computing* 11: 3021–3031.

Karaboga, D. and Basturk, B. 2008. On the performance of artificial bee colony (ABC) algorithm. *Applied Soft Computing* 8: 687–697.

Kauffman, S. A. 1993. *The Origins of Order: Self-Organization and Selection in Evolution.* Oxford University Press, Oxford, UK.

Kennedy, J. and Eberhart, R. C. 1995. Particle swarm optimization. *Proceedings of the 1995 IEEE International Conference on Neural Networks,* IEEE Service Center, Piscataway, pp. 1942–1948.

Kern, W. 1993. On the depth of combinatorial optimization problems. *Discrete Applied Mathematics* 43 (2): 115–129.

Kilby, P., Slaney, J., Thiébaux, S., and Walsh, T. 2005. Backbones and backdoors in satisfiability. *Proceedings of the 20th National Conference on Artificial Intelligence*, Pittsburgh, Pennsylvania, pp. 1368–1373.

Kim, T. H., Maruta, I., and Sugie, T. 2008. Robust PID controller tuning based on constrained particle swarm optimization. *Automatica* 44: 1104–1110.

Kinzel, W. 1998. Phase transitions of neural networks. *Philosophical Magazine B* 77: 1455–1477.

Kinzel, W. 1999. Statistical physics of neural networks. *Computer Physics Communications* 121: 86–93.

Kirkpatrick, S., Gelatt, C. D. Jr., and Vecchi, M. P. 1983. Optimization by simulated annealing. *Science* 220 (4598): 671–680.

Kirousis, L. M. and Kranakisdeg, E. 2005. Special issue on typical case complexity and phase transition. *Discrete Applied Mathematics* 153 (1–3): 1–182.

Knoop, P. and Van Nerom, L. 2003. Scheduling requirements for hot charge optimization in an integrated steel plant. *Proceedings of Industry Applications Conference, 38th IAS Annual Meeting*, Utah, USA, pp. 74–78.

Knowles, J. and Corne, D. 1999. The Pareto archived evolution strategy: A new baseline algorithm for multi-objective optimization. *Proceedings of the 1999 Congress on Evolutionary Computation*, IEEE Press, Piscataway, NJ, pp. 98–105.

Knowles, J. and Come, D. 2001. A comparative assessment of memetic, evolutionary, and constructive algorithms for the multiobjective D-MST problem. *2001 Genetic and Evolutionary Computation Workshop Proceeding*, San Francisco, pp. 162–167.

Korte, B. and Vygen, J. 2012. *Combinatorial Optimization: Theory and Algorithms*. Springer, Heidelberg, German.

Kosiba, E. D., Wright, J. R., and Cobbs, A. E. 1992. Discrete event sequencing as a traveling salesman problem. *Computers in Industry* 19 (3): 317–327.

Koza, J. R. 1998. *Genetic Programming*. MIT Press, Cambridge, MA.

Krasnogor, N. 2004. Self-generating metaheuristics in bioinformatics: The protein structure comparison case. *Genetic Programming and Evolvable Machines* 5: 181–201.

Krasnogor, N. and Gustafson, S. 2004. A study on the use of "self-generation" in memetic algorithms. *Natural Computing* 3: 53–76.

Krasnogor, N. and Smith, J. E. 2005. A tutorial for competent memetic algorithms: Model, taxonomy and design issues. *IEEE Transactions on Evolutionary Computation* 9: 474–488.

Ku, K. and Mak, M. 1998. Empirical analysis of the factors that affect the Baldwin effect. *Lecture Notes in Computer Science* 1498: 481–490.

Langton, C. G. 1998. *Artificial Life: An Overview*. MIT Press, Cambridge, MA.

Laporte, G. 2010. A concise guide to the traveling salesman problem. *Journal of the Operational Research Society* 61: 35–40.

Larrañaga, P., Kuijpers, C. M. H., Murga, R. H., Inza, I., and Dizdarevic, S. 1999. Genetic algorithms for the travelling salesman problem: A review of representations and operators. *Artificial Intelligence Review* 13: 129–170.

Lawler, E. J., Lenstra, J. K., Kan, A. H. G. R., and Shmoys, D. B. 1985. *The Traveling Salesman Problem: A Guided Tour of Combinatorial Optimization*. Wiley, New York.

Lee, C. Y. and Yao, X. 2001. Evolutionary algorithms with adaptive Lévy mutations. *Proceedings of the 2001 Congress on Evolutionary Computation*, Seoul, Korea, pp. 568–575.

Lee, H. S., Murthy, S. S., Haider, S. W., and Morse, D. V. 1996. Primary production scheduling at steelmaking industries. *IBM Journal of Research and Development* 40 (2): 231–252.

Li, N. J., Wang, W. J., Hsu, C. C. J., Chang, W., and Chou, H. G. 2014. Enhanced particle swarm optimizer incorporating a weighted particle. *Neurocomputing* 124: 218–227.

Li, S. J., Li, Y., Liu, Y., and Xu, Y. F. 2007. A GA-based NN approach for makespan estimation. *Applied Mathematics and Computation* 185: 1003–1014.

Lin, S. and Kernighan, B. W. 1973. An effective heuristic algorithm for the traveling salesman problem. *Operations Research* 21: 498–516.

Liu, H., Abraham, A., and Clerc, M. 2007. Chaotic dynamic characteristics in swarm intelligence. *Applied Soft Computing* 7: 1019–1026.

Liu, J., Han, J., and Tang, Y. Y. 2002. Multi-agent oriented constraint satisfaction. *Artificial Intelligence* 136: 101–144.

Liu, J., Jin, X., and Tsui, K. C. 2005. *Autonomy Oriented Computing: From Problem Solving to Complex Systems Modeling*. Kluwer Academic Publishers, Boston, MA.

Liu, J., Tang, Y. Y., and Cao, Y. C. 1997. An evolutionary autonomous agents approach to image feature extraction. *IEEE Transactions on Evolutionary Computation* 1 (2): 141–158.

Liu, J. and Tsui, K. C. 2006. Toward nature-inspired computing. *Communications of the ACM* 49 (10): 59–64.

Lopez, L., Carter, M. W., and Gendreau, M. 1998. The hot strip mill production scheduling problem: A tabu search approach. *European Journal of Operational Research* 106: 317–335.

Lorena, A. C. and De Carvalho, A. 2008. Evolutionary tuning of SVM parameter values in multiclass problems. *Neurocomputing* 71: 3326–3334.

Luenberger, D. G. 1984. *Linear and Nonlinear Programming*. Addison-Wesley, Reading, MA.

Mantegna, R. 1994. Fast, accurate algorithm for numerical simulation of Lévy stable stochastic process. *Physical Review E* 49: 4677–4683.

Mao, Y., Zhou, X., Pi, D., Sun, Y., and Wong, S. T. C. 2005. Parameters selection in gene selection using Gaussian kernel support vector machines by genetic algorithm. *Journal of Zhejiang University, Science B* 6: 961–973.

Martin, O. C., Monasson, R., and Zecchina, R. 2001. Statistical mechanics methods and phase transitions in optimization problems. *Theoretical Computer Science* 265: 3–67.

Martinez, M., Senent, J. S., and Blasco, X. 1998. Generalized predictive control using genetic algorithms (GAGPC). *Engineering Applications of Artificial Intelligence* 11: 355–367.

Martinsen, F., Biegler, L. T., and Foss, B. A. 2004. A new optimization algorithm with application to nonlinear MPC. *Journal of Process Control* 14: 853–865.

Mattika, I., Amorimb, P., and Günther, H. O. 2014. Hierarchical scheduling of continuous casters and hot strip mills in the steel industry: A block planning application. *International Journal of Production Research* 52 (9): 2576–2591.

Menaï, M. B. and Batouche, M. 2006. An effective heuristic algorithm for the maximum satisfiability problem. *Applied Intelligence* 24: 227–239.

Méndez, R. A., Valladares, A., Flores, J., Seligman, T. H., and Bohigas, O. 1996. Universal fluctuations of quasi-optimal paths of the traveling salesman problem. *Physica A* 232: 554–562.

Menhas, M. I., Wang, L., Fei, M., and Pan, H. 2012. Comparative performance analysis of various binary coded PSO algorithms in multivariable PID controller design. *Expert Systems with Applications* 39: 4390–4401.

Merz, P. 2000. Memetic algorithms for combinatorial optimization problems: Fitness landscapes and effective search strategies. PhD dissertation, Department of Electrical Engineering and Computer Science, University of Siegen, Germany.

Merz, P. 2004. Advanced fitness landscape analysis and the performance of memetic algorithms. *Evolutionary Computation* 12 (3): 303–325.

Merz, P. and Freisleben, B. 2000. Fitness landscape analysis and memetic algorithms for the quadratic assignment problem. *IEEE Transactions on Evolutionary Computation* 4 (4): 337–352.

Meza, G. R., Sanchis, J., Blasco, X., and Herrero, J. M. 2012. Multiobjective evolutionary algorithms for multivariable PI controller design. *Expert Systems with Applications* 39: 7895–7907.

Mézard, M., Parisi, G., and Zecchina, R. 2002. Analytic and algorithmic solution of random satisfiability problems. *Science* 297: 812–815.

Mezura-Montes, E. and Coello, C. A. C. 2005. A simple multimembered evolution strategy to solve constrained optimization problems. *IEEE Transactions on Evolutionary Computation* 9: 1–17.

Michalewicz, Z. 1996. *Genetic Algorithms + Data Structures = Evolution Programs.* Springer, Heidelberg.

Middleton, A. A. 2004. Improved extremal optimization for the using spin glass. *Physical Review E* 69: 055701R.

Miettinen, K. M. 1999. *Nonlinear Multi-Objective Optimization.* Kluwer Academic Publishers, Boston, MA.

Miller, D. L. and Pekny, J. F. 1989. Results from a parallel branch and bound algorithm for the asymmetric traveling salesman problem. *Operations Research Letters* 8 (3): 129–135.

Min, J. H. and Lee, Y. C. 2005. Bankruptcy prediction using support vector machine with optimal choice of kernel function parameters. *Expert Systems with Applications* 28: 603–614.

Molga, M. and Smutnicki, C. 2005. Test functions for optimization needs. Available at http://www.zsd.ict.pwr.wroc.pl/files/docs/functions.pdf.

Monasson, R., Zecchina, R., Kirkpatrick, S., Selman, B., and Troyansky, L. 1999. Determining computational complexity from characteristic "phase transitions". *Nature* 400: 133–137.

Moscato, P. 1989. On evolution, search, optimization, genetic algorithms, and martial arts: Towards memetic algorithms. *Technical Report Caltech Concurrent Computation Program*, Report 826, California Institute of Technology, Pasadena, CA.

Moscato, P. and Cotta, C. 2003. *A Gentle Introduction to Memetic Algorithms, Handbook of Metaheuristics.* Kluwer Academic Publishers, Boston.

Moscato, P., Mendes, A., and Berretta, R. 2007. Benchmarking a memetic algorithm for ordering microarray data. *Biosystems* 88: 56–75.

Mosetti, G., Jug, G., and Scalas, E. 2007. Power laws from randomly sampled continuous-time random walks. *Physica A* 375: 233–238.

Murty, K. G. and Kabadi, S. N. 1987. Some NP-complete problems in quadratic and nonlinear programming. *Mathematical Programming* 39: 117–129.

Nocedal, J. and Stephen, J. W. 2006. *Numerical Optimization.* Springer, New York.

Okano, H., Davenport, A. J., Trumbo, M., Reddy, C., Yoda, K., and Amano, M. 2004. Finishing line scheduling in the steel industry. *IBM Journal of Research and Development* 48 (5/6): 811–830.

Ong, Y. S. and Keane, A. 2004. Meta-Lamarckian learning in memetic algorithms. *IEEE Transactions on Evolutionary Computation* 8: 99–110.

Onnen, C., Babuška, R., Kaymak, U., Sousa, J. M., Verbruggen, H. B., and Isermann, R. 1997. Genetic algorithms for optimization in predictive control. *Control Engineering Practice* 5: 1363–1372.

Paczuski, M., Maslov, S., and Bak, P. 1996. Avalanche dynamics in evolution, growth, and depinning models. *Physical Review E* 53 (1): 414–443.

Pai, P. F. and Hong, W. C. 2005. Support vector machines with simulated annealing algorithms in electricity load forecasting. *Energy Conversion and Management* 46: 2669–2688.

Papadimitriou, C. H. 1994. *Computational Complexity*, 1st ed. Addison-Wesley, USA.

Papadimitriou, C. H. and Steiglitz, K. 1998. *Combinatorial Optimization: Algorithms and Complexity*. Courier Dover Publications, USA.

Patrick, S. and Michalewicz, Z. 2008. *Advances in Meta-Heuristics for Hard Optimization*. Springer, Heidelberg.

Pearson, R. K. 2006. Nonlinear empirical modeling techniques. *Computers and Chemical Engineering* 30: 1514–1528.

Peterson, C. M., Sorensen, K. L., and Vidal, R. V. V. 1992. Inter-process synchronization in steel production. *International Journal of Production Research* 30: 1415–1425.

Potocnik, P. and Grabec, I. 2002. Nonlinear model predictive control of a cutting process. *Neurocomputing* 43: 107–126.

Potschka, H. 2010. Targeting regulation of ABC efflux transporters in brain diseases: A novel therapeutic approach. *Pharmacology and Therapeutics* 125: 118–127.

Potvin, J. Y. 1996. Genetic algorithms for the traveling salesman problem. *Annals of Operations Research* 63: 339–370.

Powell, M. J. D. 1978. A fast algorithm for nonlinearly constrained optimization calculations. *Lecture Notes in Mathematics* 630: 144–157.

Qiao, J. F. and Wang, H. D. 2008. A self-organizing fuzzy neural network and its applications to function approximation and forecast modeling. *Neurocomputing* 71: 564–569.

Qin, S. J. and Badgwell, T. A. 2003. A survey of industrial model predictive control technology. *Control Engineering Practice* 11: 733–764.

Ramos, V., Fernandes, C., and Rosa, A. C. 2005. Societal memory and his speed on tracking extrema over dynamic environments using self-regulatory swarms. *Proceedings of the 1st European Symposium on Nature Inspired Smart Information Systems*, Albufeira, Portugal.

Rangel, L. P. 2012. Putative role of an ABC transporter in Fonsecaea pedrosoi multidrug resistance. *International Journal of Antimicrobial Agents* 40: 409–415.

Refael, H. and Mati, S. 2005. Machine scheduling with earliness, tardiness and non-execution penalties. *Computers and Operations Research* 32: 683–705.

Reidys, C. M. and Stadler, P. F. 2002. Combinatorial landscapes. *SIAM Review* 44 (1): 3–54.

Reinelt, G. 1991. TSPLIB—A traveling salesman problem library. *ORSA Journal on Computing* 3 (4): 376–384.

Reyaz-Ahmed, A., Zhang, Y. Q., and Harrison, R. W. 2009. Granular decision tree and evolutionary neural SVM for protein secondary structure prediction. *International Journal of Computational Intelligence Systems* 2: 343–352.

Rigler, A. K., Irvine, J. M., and Vogl, T. P. 1991. Rescaling of variables in back propagation learning. *Neural Networks* 4: 225–229.

Rogers, A., Prügel-Bennett, A., and Jennings, N. R. 2006. Phase transitions and symmetry breaking in genetic algorithms with crossover. *Theoretical Computer Science* 358: 121–141.

Rosenberg, R. S. 1967. Simulation of genetic populations with biochemical properties. PhD thesis, University of Michigan, Ann Arbor, MI.

Rumelhart, D. E., Hinton, G. E., and Williams, R. J. 1986a. Learning internal representations by error propagation. In: *Parallel Distributed Processing: Exploration in the Microstructure of Cognition*. J. L. McClelland and D. E. Rumelhart, Eds, MIT Press, Cambridge, MA, pp. 318–362.

Rumelhart, D. E., Hinton, G. E., and Williams, R. J. 1986b. Learning representations by back propagating errors. *Nature* 323: 533–536.

Runarsson, T. P. and Yao, X. 2000. Stochastic ranking for constrained evolutionary optimization. *IEEE Transactions on Evolutionary Computation* 4: 284–294.

Salomon, R. 1998. Evolutionary algorithms and gradient search: Similarities and differences. *IEEE Transactions on Evolutionary Computation* 2: 45–55.

Sarker, R., Liang, K. H., and Newton, C. 2002. A new multi-objective evolutionary algorithm. *European Journal of Operational Research* 140: 12–23.

Schaffer, J. D. 1985. Multiple objective optimization with vector evaluated genetic algorithms. *Proceedings of the First International Conference on Genetic Algorithms*, Lawrence Erlbaum, Hillsdale, NJ, pp. 93–100.

Schneider, J. 2003. Searching for backbones—A high-performance parallel algorithm for solving combinatorial optimization problem. *Future Generation Computer Systems* 19: 121–131.

Schneider, J., Froschhammer, C., Morgenstern, I., Husslein, T., and Singer, J. M. 1996. Searching for backbones-efficient parallel algorithm for the traveling salesman problem. *Computer Physics Communications* 96: 173–188.

Sedkia, A., Ouazar, D., and Mazoudi, E. El. 2009. Evolving neural network using real coded genetic algorithm for daily rainfall-runoff forecasting. *Expert Systems with Applications* 36: 4523–4527.

Seitz, S., Alava, M., and Orponen, P. 2005. Focused local search for random 3-satisfiability. *Journal of Statistical Mechanics: Theory and Experiment* 23(6): 524–536.

Selman, B. 2008. Computational science: A hard statistical view. *Nature* 451: 639–640.

Selman, B. and Kautz, H. A. 1993. An empirical study of greedy local search for satisfiability testing. *Proceedings of the 11th National Conference on Artificial Intelligence*, Washington, D.C., pp. 46–51.

Selman, B., Kautz, H. A., and Cohen, B. 1994. Noise strategies for improving local search. *Proceedings of the 12th National Conference on Artificial Intelligence*, Seattle, WA, pp. 337–343.

Senthil Arumugam, M., Rao, M. V. C., and Tan, A. W. C. 2009. A novel and effective particle swarm optimization like algorithm with extrapolation technique. *Applied Soft Computing* 9: 308–320.

Seung, H. S., Sompolinsky, H., and Tishby, N. 1992. Statistical mechanics of learning from examples. *Physical Review A* 45: 6056–6091.

Shelokar, P. S., Siarry, P. V., Jayaraman, K., and Kulkarni, B. D. 2007. Particle swarm and ant colony algorithms hybridized for improved continuous optimization. *Applied Mathematics and Computation* 188: 129–142.

Shi, X. H., Liang, Y. C., Lee, H. P., Lu, C., and Wang, L. M. 2005. An improved GA and a novel PSO–GA-based hybrid algorithm. *Information Processing Letters* 93: 255–261.

Singer, J., Gent, I. P., and Smaill, A. 2000. Backbone fragility and the local search cost peak. *Journal of Artificial Intelligence Research* 12: 235–270.

Slaney, J. and Walsh, T. 2001. Backbones in optimization and approximation. *Proceedings of the International Joint Conference on Artificial Intelligence, Seattle, WA*, Morgan Kaufmann, San Mateo, CA, pp. 254–259.

Smith, J. E. 2007. Coevolving memetic algorithms: A review and progress report. *IEEE Transactions on Systems, Man, and Cybernetics, Part B: Cybernetics* 37: 6–17.

Sneppen, K. 1995. Extremal dynamics and punctuated co-evolution. *Physica A* 221: 168–179.

Song, Y., Chen, Z. Q., and Yuan, Z. Z. 2007. New chaotic PSO-based neural network predictive control for nonlinear process. *IEEE Transactions on Neural Networks* 18: 595–600.

Srinivas, N. and Deb, K. 1994. Multi-objective optimization using nondominated sorting in genetic algorithms. *Evolutionary Computation* 2 (3): 221–248.

Srinivasan, D. and Seow, T. H. 2006. Particle swarm inspired evolutionary algorithm (PS-EA) for multi-criteria optimization problems. *Proceedings of the Evolutionary Multiobjective Optimization*, Springer, Berlin, pp. 147–165.

Stauffer, L. and Liebling, T. M. 1997. Rolling horizon scheduling in a rolling-mill. *Annals of Operations Research* 69: 323–349.

Steve, G. 1998. Support vector machines classification and regression. *ISIS Technical Report*, Image, Speech, and Intelligent Systems Group, University of Southampton.

Strogatz, S. H. 2001. Exploring complex networks. *Nature* 410: 268–276.

Suganthan, P. N., Hansen, N., Liang, J. J., Deb, K., Chen, Y. P., Auger, A., and Tiwari, S. 2005. Problem definitions and evaluation criteria for the CEC 2005 special session on real-parameter optimization. *Technical Report*, Nanyang Technological University, Singapore.

Szedmak, S. 2001. How to find more efficient initial solution for searching. *RUTCOR Research Report 49-2001*, Rutgers Center for Operations Research, Rutgers University, Piscataway, NJ.

Taherdangkoo, M., Paziresh, M., Yazdi, M., and Bagheri, M. H. 2012. An efficient algorithm for function optimization: Modified stem cells algorithm. *Central European Journal of Engineering* 3: 36–50.

Tahk, M. J., Woo, H. W., and Park, M. S. 2007. A hybrid optimization method of evolutionary and gradient search. *Engineering Optimization* 39: 87–104.

Tang, L. X., Liu, J. Y., Rong, A. Y., and Yang, Z. H. 2000. A multiple traveling salesman problem model for hot rolling scheduling in Shanghai Baoshan Iron & Steel Complex. *European Journal of Operational Research* 124: 267–282.

Tang, L. X., Liu, J. Y., Rong, A. Y., and Yang, Z. H. 2001. A review of planning and scheduling systems and methods for integrated steel production. *European Journal of Operational Research* 133: 1–20.

Tang, L. X. and Wang, X. P. 2005. Iterated local search algorithm based on very large-scale neighborhood for prize-collecting vehicle routing problem. *The International Journal of Advanced Manufacturing Technology* 12: 1433–3015.

Tang, X., Zhuang, L., and Jiang, C. 2009. Prediction of silicon content in hot metal using support vector regression based on chaos particle swarm optimization. *Expert Systems with Applications* 36: 11853–11857.

Tao, J., Wang, X., and Chai, T. Y. 2002. Intelligent control method and application for BOF steelmaking. *Proceedings of the 15th IFAC World Congress*, Barcelona, Spain, pp. 1071–1076.

Telelis, O. and Stamatopoulos, P. 2002. Heuristic backbone sampling for maximum satisfiability. *Proceedings of the 2nd Hellenic Conference on Artificial Intelligence*, Thessaloniki, Greece, pp. 129–139.

Thadani, K., Ashutosh, Jayaraman, V. K., and Sundararajan, V. 2006. Evolutionary selection of kernels in support vector machines. *Proceedings of the International Conference on Advanced Computing and Communications*, Surathkal, India, pp. 19–24.

Tsallis, C. and Stariolo, D. A. 1996. Generalized simulated annealing. *Physica A* 233 (1–2): 395–406.

TSPLIB. Available at http://www.iwr.uni-heidelberg.de/groups/comopt/software/TSPLIB95/.

Vapnik, V. 1995. *The Nature of Statistical Learning Theory*. Springer-Verlag, New York.

Venkateswarlu, C. and Reddy, A. D. 2008. Nonlinear model predictive control of reactive distillation based on stochastic optimization. *Industrial and Engineering Chemistry Research* 47: 6949–6960.

Verdejoa, V. V., Alarcób, M. A. P., and Sorlí, M. P. L. 2009. Scheduling in a continuous galvanizing line. *Computers and Operations Research* 36: 280–296.

Wang, Q. G., Zou, B., Lee, T. H., and Qiang, B. 1997. Auto-tuning of multivariable PID controllers from decentralized relay feedback. *Automatica* 33 (3): 319–330.

Wang, X., Wang, Z. J., and Tao, J. 2006. Multiple neural network modeling method for carbon and temperature estimation in basic oxygen furnace. *Lecture Notes in Computer Science* 3973: 870–875.

Werbos, P. 1974. Beyond regression: New tools for prediction and analysis in the behavioral sciences. PhD dissertation, Harvard University.

Wilson, R. B. 1963. A simplicial algorithm for concave programming. PhD dissertation, Graduate School of Business Administration, Harvard University.

Wolpert, D. H. and Macready, W. G. 1997. No free lunch theorems for optimization. *IEEE Transactions on Evolutionary Computation* 1: 67–82.

Wu, C. H., Tzeng, G. H., and Lin, R. H. 2009. A novel hybrid genetic algorithm for kernel function and parameter optimization in support vector regression. *Expert Systems with Applications* 36: 4725–4735.

Wu, Q. 2010. A hybrid-forecasting model based on Gaussian support vector machine and chaotic particle swarm optimization. *Expert Systems with Applications* 37: 2388–2394.

Xu, C. W. and Lu, Y. Z. 1987. Fuzzy model identification and self-learning for dynamic systems. *IEEE Transactions on Systems, Man and Cybernetics SMC* 17 (4): 683–689.

Xu, K. and Li, W. 2006. Many hard examples in exact phase transitions. *Theoretical Computer Science* 355: 191–302.

Yadollahpour, M. R., Bijari, M., Kavosh, S., and Mahnam, M. 2009. Guided local search algorithm for hot strip mill scheduling problem with considering hot charge rolling. *International Journal of Advanced Manufacturing Technology* 45: 1215–1231.

Yan, D., Ahmad, S. Z., and Yang, D. 2013. Matthew effect, ABC analysis and project management of scale-free information systems. *The Journal of Systems and Software* 86: 247–254.

Yan, X. H., Zhu, Y. L., and Zou, W. P. 2011. A hybrid artificial bee colony algorithm for numerical function optimization. *Proceedings of the 11th International Conference on Hybrid Intelligent Systems*, Malacca, Malaysia, pp. 127–132.

Yang, B. and Liu, J. 2007. An autonomy oriented computing (AOC) approach to distributed network community mining. *Proceedings of the 1st International Conference on Self-Adaptive and Self-Organizing Systems*, Boston, MA, pp. 151–160.

Yao, X. and Islam, Md. M. 2008. Evolving artificial neural network ensembles. *IEEE Computational Intelligence Magazine* 3: 31–42.

Yao, X., Liu, Y., and Lin, G. 1999. Evolutionary programming made faster. *IEEE Transactions on Evolutionary Computation* 3 (2): 82–102.

Ye, J. 2008. Adaptive control of nonlinear PID-based analog neural networks for a nonholonomic mobile robot. *Neurocomputing* 71 (7–9): 1561–1565.

Zadel, L. A. 1965. Fuzzy sets. *Information and Control* 8: 338–353.

Zhang, J., Chung, S. H. H., and Lo, W. L. 2007. Clustering-based adaptive crossover and mutation probabilities for genetic algorithms. *IEEE Transactions on Evolutionary Computation* 11: 326–335.

Zhang, J. H., Zhuang, J., Du, H. F., and Wang, S. A. 2009. Self-organizing genetic algorithm based tuning of PID controllers. *Information Sciences* 179: 1007–1018.

Zhang, L., Zhou, W., and Jiao, L. 2004. Wavelet support vector machine. *IEEE Transactions on Systems, Man, and Cybernetics, Part B: Cybernetics* 34: 34–39.

Zhang, M., Luo, W., and Wang, X. 2008. Differential evolution with dynamic stochastic selection for constrained optimization. *Information Sciences* 178: 3043–3074.

Zhang, N. G. and Zeng, C. 2008. Reference energy extremal optimization: A stochastic search algorithm applied to computational protein design. *Journal of Computational Chemistry* 29: 1762–1771.

Zhang, W. X. 2001. Phase transitions and backbones of 3-SAT and maximum 3-SAT. *Proceedings of the 7th International Conference on Principles and Practice of Constraint Programming*, Paphos, Cyprus, pp. 153–167.

Zhang, W. X. 2002. Phase transitions, backbones, measurement accuracy, and phase-aware approximation: The ATSP as a case study. *Proceedings of CP-AI-OR*, Le Croisic, France, pp. 345–357.

Zhang, W. X. 2004. Phase transitions and backbones of the asymmetric traveling salesman problem. *Journal of Artificial Intelligence Research* 21: 471–497.

Zhang, W. X. and Looks, M. 2005. A novel local search algorithm for the traveling salesman problem that exploit backbones. *Proceedings of the 19th International Joint Conference on Artificial Intelligence*, Morgan Kaufmann Publishers, San Francisco, pp. 343–351.

Zhao, S. Z., Iruthayarajan, M. W., Baskar, S., and Suganthan, P. N. 2011. Multi-objective robust PID controller tuning using two lbests multi-objective particle swarm optimization. *Information Sciences* 181: 3323–3335.

Zhu, W. and Ali, M. M. 2009. Solving nonlinearly constrained global optimization problem via an auxiliary function method. *Journal of Computational and Applied Mathematics* 230: 491–503.

Zitzler, E., Deb, K., and Thiele, L. 2000. Comparison of multi-objective evolutionary algorithms: Empirical results. *Evolutionary Computation* 8 (2): 173–195.

Zitzler, E., Laumanns, M., and Thiele, L. 2001. SPEA2: Improving the performance of the strength Pareto evolutionary algorithm. *Technical Report* 103, Computer Engineering and Communication Networks Lab (TIK), Swiss Federal Institute of Technology (ETH), Zurich, Gloriastrasse 35, CH-8092, Zurich.

Zitzler, E. and Thiele, L. 1998. Multi-objective optimization using evolutionary algorithms—A comparative case study. *Proceedings of the 7th International Conference on Parallel Problem Solving from Nature, PPSN-V [M]*, Springer, Berlin.

Zitzler, E. and Thiele, L. 1999. Multi-objective evolutionary algorithms: A comparative case study and the strength Pareto approach. *IEEE Transactions on Evolutionary Computation* 3 (4): 257–271.

List of Authors-Related Publications

Chen, M. R. 2008. Studies on optimization methods with extremal dynamics and applications. PhD thesis, Shanghai Jiao Tong University, Shanghai, China.

Chen, M. R. and Lu, Y. Z. 2008. A novel elitist multi-objective optimization algorithm: Multi-objective extremal optimization. *European Journal of Operational Research* 3 (188): 637–651.

Chen, M. R., Lu, Y. Z., and Yang, G. 2006. Population-based extremal optimization with adaptive Lévy mutation for constrained optimization. *Proceedings of the 2006 International Conference on Computational Intelligence and Security (CIS'2006)*, Guangzhou, China, pp. 258–261.

Chen, M. R., Lu, Y. Z., and Yang, G. K. 2007. Multi-objective extremal optimization with applications to engineering design. *Journal of Zhejiang University, Science A* 8 (12): 1905–1911. DOI:10.1631/jzus.2007.A1905.

Chen, M. R., Lu, Y. Z., and Yang, G. K. 2008. Multiobjective optimization using population-based extremal optimization. *Neural Computing and Applications* 17 (2): 101–109.

Chen, M. R., Weng, J., and Li, X. 2009. Multi-objective extremal optimization for portfolio optimization problem. *Proceedings of the 2009 IEEE International Conference on Intelligent Computing and Intelligent Systems (ICIS 2009)*, Shanghai, China, pp. 552–556.

Chen, M. R., Weng, J., Li, X., and Zhang, X. 2014a. Handling multiple objectives with integration of particle swarm optimization and extremal optimization. *Proceedings of the Eighth International Conference on Intelligent Systems and Knowledge Engineering (ISKE 2013)*, Shenzhen, China, Vol. 277, pp. 287–297.

Chen, M. R., Zeng, W., Zeng, G. Q., Li, X., and Luo, J. P. 2014b. A novel artificial bee colony algorithm with integration of extremal optimization for numerical optimization problems. *Proceedings of the 2014 IEEE Congress on Evolutionary Computation (CEC 2014)*, Beijing, China, pp. 242–249.

Chen, P. 2011. Extremal dynamics based memetic algorithm and its applications in non-linear predictive control. PhD thesis, Shanghai Jiao Tong University, Shanghai, China.

Chen, P. and Lu, Y. Z. 2010a. Memetic algorithm-based neural network learning for basic oxygen furnace endpoint prediction. *Journal of Zhejiang University, Science A* 11: 841–848. DOI:10.1631/jzus.A0900664.

Chen, P. and Lu, Y. Z. 2010b. Nonlinear model predictive control with the integration of support vector machine and extremal optimization. *Proceedings of the 8th world Congress on Intelligent Control and Automation*, Jinan, China, pp. 3167–3172.

Chen, P. and Lu, Y. Z. 2011a. Memetic algorithms based real-time optimization for nonlinear model predictive control. *International Conference on System Science and Engineering*, Macau, China, pp. 119–124.

Chen, P. and Lu, Y. Z. 2011b. Extremal optimization for optimizing kernel function and its parameters in support vector regression. *Journal of Zhejiang University, Science C* 12: 297–306. DOI:10.1631/jzus.C1000110.

Chen, P., Lu, Y. Z., and Chen, Y. W. 2010. Extremal optimization combined with LM gradient search for MLP network learning. *International Journal of Computational Intelligence Systems* 3: 622–631.

Chen, Y. W. 2008. Self-organizing optimization with extremal dynamics: Theory, algorithms, and applications. PhD thesis, Shanghai Jiao Tong University, Shanghai, China.

Chen, Y. W. and Lu, Y. Z. 2007. Gene optimization: Computationalintelligence from the natures and micro-mechanisms of hard computational systems. *Proceedings of the International Conference on Life System Modeling and Simulation, LNCS 4688*, Shanghai, China, pp. 182–190.

Chen, Y. W., Lu, Y. Z., and Chen, P. 2007. Optimization with extremal dynamics for the traveling salesman problem. *Physica A* 385: 115–123.

Chen, Y. W., Lu, Y. Z., Ge, M., Yang, G. K., and Pan, C. C. 2012. Development of hybrid evolutionary algorithms for production scheduling of hot strip mill. *Computers and Operations Research* 39 (2): 339–349.

Chen, Y. W., Zhu, Y. J., Yang, G. K., and Lu, Y. Z. 2011. Improved extremal optimization for the asymmetric traveling salesman problem. *Physica A* 390 (2011): 4459–4465.

Chen, Y. W., Lu, Y. Z., and Yang, G. 2008. Hybrid evolutionary algorithm with marriage of genetic algorithm and extremal optimization for production scheduling. *International Journal of Advanced Manufacturing Technology* 36: 959–968.

Li, X., Luo, J., Chen, M. R., and Wang, N. 2012. An improved shuffled frog-leaping algorithm with extremal optimisation for continuous optimisation. *Information Sciences* 192: 143–151.

Liu, J., Chen, Y. W. 2012. Toward understanding the optimization of complex systems. *Artificial Intelligence Review* 38: 313–324.

Liu, J., Chen, Y. W., Yang, G. K., and Lu, Y. Z. 2011. Self-organized combinatorial optimization. *Expert Systems with Applications* 38: 10532–10540.

Lu, Y. Z. 1996. *Industrial Intelligent Control: Fundamentals and Applications.* John Wiley & Sons, New York.

Lu, Y. Z., Chen, M. R., and Chen, Y. W. 2007. Studies on extremal optimization and its applications in solving real world optimization problems. *Proceedings of the 2007 IEEE Symposium on Foundations of Computational Intelligence (FOCI 2007)*, Hawaii, pp. 162–168.

Luo, J. and Chen, M. R. 2014. Multi-phase modified shuffled frog leaping algorithm with extremal optimization for the MDVRP and the MDVRPTW. *Computers and Industrial Engineering* 72: 84–97.

Zeng, G. Q. 2011. Research on modified extremal optimization algorithms and their applications in combinatorial optimization problems. PhD thesis, Zhejiang University, Hangzhou, China.

Zeng, G. Q., Chen, J., Chen, M. R., Dai, Y. X., Li, L. M., Lu, K. D., and Zheng, C. W. 2015. Design of multivariable PID controllers using real-coded population-based extremal optimization. *Neurocomputing* 151: 1343–1353.

Zeng, G. Q., Chen, J., Dai, Y. X., Li, L. M., Zheng, C. W., and Chen, M. R. 2015. Design of fractional order PID controller for automatic regulator voltage system based on multi-objective extremal optimization. *Neurocomputing* 160: 173–184.

Zeng, G. Q., Lu, K. D., Dai, Y. X., Zhang, Z. J., Chen, M. R., Zheng, C. W., Peng, W. W., and Wu, D. 2014. Binary-coded extremal optimization for the design of PID controllers. *Neurocomputing* 138: 180–188.

Zeng, G. Q., Lu, Y. Z., Dai, Y.-X., Wu, Z. G., Mao, W. J., Zhang, Z. J., and Zheng, C. W. 2012. Backbone guided extremal optimization for the hard maximum satisfiability problem. *International Journal of Innovative Computing, Information and Control* 8 (12): 8355–8366.

Zeng, G. Q., Lu, Y. Z., Mao., and W. J. 2010a. A novel stochastic method with modified extremal optimization and nearest neighbor search for hard combinatorial problems. *Proceedings of the 8th World Congress on Intelligent Control and Automation*, Jinan, China, pp. 2903–2908.

Zeng, G. Q., Lu, Y. Z., and Mao, W. J. 2010b. Multistage extremal optimization for hard travelling salesman problem. *Physica A* 389 (21): 5037–5044.

Zeng, G. Q., Lu, Y. Z., and Mao, W. J. 2011. Modified extremal optimization for the hard maximum satisfiability problem. *Journal of Zhejiang University, Science C* 12 (7): 589–596.

Zeng, G. Q., Lu, Y. Z., Mao, W. J., and Chu, J. 2010c. Study on probability distributions for evolution in modified extremal optimization. *Physica A* 389 (9): 1922–1930.

Author Index

Subject Index